Solutions Manual
for Econometrics

Springer
*Berlin
Heidelberg
New York
Barcelona
Budapest
Hong Kong
London
Milan
Paris
Santa Clara
Singapore
Tokyo*

Badi H. Baltagi

Solutions Manual for Econometrics

With 14 Figures
and 3 Tables

 Springer

Professor Badi H. Baltagi
Texas A&M University
Department of Economics
College Station
Texas 77343-4228
USA

ISBN 3-540-63896-2 Springer-Verlag Berlin Heidelberg New York

Cataloging-in-Publication Data applied for
Die Deutsche Bibliothek – CIP-Einheitsaufnahme
Baltagi, Badi H.: Solutions manual for econometrics / Badi H. Baltagi. – Berlin; Heidelberg; New York; Barcelona; Budapest; Hongkong; London; Mailand; Paris; Santa Clara; Singapur; Tokio: Springer, 1998
ISBN 3-540-63896-2

This work is subject to copyright. All rights are reserved, whether the whole or part of the material is concerned, specifically the rights of translation, reprinting, reuse of illustrations, recitation, broadcasting, reproduction on microfilm or in any other way, and storage in data banks. Duplication of this publication or parts thereof is permitted only under the provisions of the German Copyright Law of September 9, 1965, in its current version, and permission for use must always be obtained from Springer-Verlag. Violations are liable for prosecution under the German Copyright Law.

© Springer-Verlag Berlin · Heidelberg 1998
Printed in Germany

The use of general descriptive names, registered names, trademarks, etc. in this publication does not imply, even in the absence of a specific statement, that such names are exempt from the relevant protective laws and regulations and therefore free for general use.

Product liability: The publishers cannot guarantee the accuracy of any information about the application of operative techniques and medications contained in this book. In every individual case the user must check such information by consulting the relevant literature.

Cover-Design: Erich Kirchner, Heidelberg

SPIN 10661751   42/2202-5 4 3 2 1 0 – Printed on acid-free paper

PREFACE

This manual provides solutions to selected exercises from each chapter of *Econometrics* by Badi H. Baltagi starting with Chapter 2. For the empirical exercises some SAS® programs are provided to replicate the results. Most graphs are plotted using EViews. Some of the problems and solutions are obtained from *Econometric Theory* (ET) and these are reprinted with the permission of Cambridge University Press. I would like to thank Peter C.B. Phillips, and the editors of the Problems and Solutions section, Alberto Holly and Juan Dolado for this useful service to the econometrics profession. I would also like to thank my colleague James M. Griffin for providing many empirical problems and data sets. I have also used three empirical data sets from Lott and Ray (1992). The reader is encouraged to apply these econometric techniques to their own data sets and to replicate the results of published articles. Some journals/authors provide data sets upon request or are readily available on the web. Other empirical examples are given in Lott and Ray (1992) and Berndt (1991). Finally I would like to thank my students Wei-Wen Xiong, Ming-Jang Weng and Kiseok Nam who solved several of these exercises.

Please report any errors, typos or suggestions to: Badi H. Baltagi, Department of Economics, Texas A&M University, College Station, Texas 77843-4228. Telephone (409) 845-7380, Fax (409) 847-8757, or send EMAIL to Badi@econ.tamu.edu.

Table of Contents

Preface V

Chapter 2
A Review of Some Basic Statistical Concepts 1

Chapter 3
Simple Linear Regression 22

Chapter 4
Multiple Regression Analysis 40

Chapter 5
Violations of the Classical Assumptions 69

Chapter 6
Distributed Lags and Dynamic Models 108

Chapter 7
The General Linear Model: The Basics 137

Chapter 8
Regression Diagnostics and Specification Tests 163

Chapter 9
Generalized Least Squares 189

Chapter 10
Seemingly Unrelated Regressions 203

Chapter 11
Simultaneous Equations Model 227

Chapter 12
Pooled Time-Series of Cross-Section Data 264

Chapter 13
Limited Dependent Variables 283

Chapter 14
Time Series Models 298

CHAPTER 2
A Review of Some Basic Statistical Concepts

2.1 *Variance and Covariance of Linear Combinations of Random Variables.*

a. Let $Y = a+bX$, then $E(Y) = E(a+bX) = a+bE(X)$. Hence,

$$\text{var}(Y) = E[Y-E(Y)]^2 = E[a+bX-a-bE(X)]^2 = E[b(X-E(X))]^2$$
$$= b^2 E[X-E(X)]^2 = b^2 \text{var}(X).$$

Only the multiplicative constant b matters for the variance, not the additive constant a.

b. Let $Z = a+bX+cY$, then $E(Z) = a+bE(X) + cE(Y)$ and

$$\text{var}(Z) = E[Z-E(Z)]^2 = E[a+bX+cY-a-bE(X)-cE(Y)]^2$$
$$= E[b(X-E(X))+c(Y-E(Y))]^2$$
$$= b^2 E[X-E(X)]^2 + c^2 E[Y-E(Y)]^2 + 2bc\, E[X-E(X)][Y-E(Y)]$$
$$= b^2 \text{var}(X) + c^2 \text{var}(Y) + 2bc\, \text{cov}(X,Y).$$

c. Let $Z = a+bX+cY$, and $W = d+eX+fY$, then $E(Z) = a+bE(X)+cE(Y)$

$E(W) = d+eE(X)+fE(Y)$

and

$$\text{cov}(Z,W) = E[Z-E(Z)][W-E(W)]$$
$$= E[b(X-E(X))+c(Y-E(Y))][e(X-E(X))+f(Y-E(Y))]$$
$$= be\, \text{var}(X) + cf\, \text{var}(Y) + (bf+ce)\, \text{cov}(X,Y).$$

2.2 *Independence and Simple Correlation.*

a. Assume that X and Y are continuous random variables. The proof is similar if X and Y are discrete random variables only the integrals are replaced by summation signs. If X and Y are independent, then $f(x,y) = f_1(x)f_2(y)$ where $f_1(x)$ is the marginal probability density function (p.d.f.) of X and $f_2(y)$ is the marginal p.d.f. of Y. In this case,

$$E(XY) = \iint xy f(x,y) dx dy = \iint xy f_1(x) f_2(y) dx dy = \left(\int x f_1(x) dx\right)\left(\int y f_2(y) dy\right)$$
$$= E(X)E(Y)$$

Hence, $cov(X,Y) = E[X-E(X)][Y-E(Y)] = E(XY) - E(X)E(Y)$
$= E(X)E(Y) - E(X)E(Y) = 0$.

b. If $Y = a+bX$, then $E(Y) = a+bE(X)$ and

$cov(X,Y) = E[X-E(X)][Y-E(Y)] = E[X-E(X)][a+bX-a-bE(X)] = b\ var(X)$

which takes the sign of b since var(X) is always positive. Hence,

$$\text{correl } (X,Y) = \rho_{XY} = \frac{cov(X,Y)}{\sqrt{var(X)var(Y)}} = \frac{b\ var(X)}{\sqrt{var(X)var(Y)}}$$

but $var(Y) = b^2\ var(X)$ from problem 1 part (a). Hence,

$$\rho_{XY} = \frac{b\ var(X)}{\sqrt{b^2(var(X))^2}} = \pm 1 \text{ depending on the sign of b.}$$

2.3 *Zero Covariance Does Not Necessarily Imply Independence.*

X	P(X)
-2	1/5
-1	1/5
0	1/5
1	1/5
2	1/5

$$E(X) = \sum_{X=-2}^{2} X(P(X)) = \frac{1}{5}[(-2)+(-1)+0+1+2] = 0$$

$$E(X^2) = \sum_{X=-2}^{2} X^2 P(X) = \frac{1}{5}[4+1+0+1+4] = 2$$

and var(X) = 2. For $Y = X^2$, $E(Y) = E(X^2) = 2$ and

$$E(X^3) = \sum_{X=-2}^{2} X^3 P(X) = \frac{1}{5}[(-2)^3+(-1)^3+0+1^3+2^3] = 0.$$

In fact, any odd moment of X is zero. Therefore, $E(YX) = E(X^2 \cdot X) = E(X^3) = 0$
and $cov(Y,X) = E(X-E(X))(Y-E(Y)) = E(X-0)(Y-2) = E(XY)-2E(X)$
$= E(XY) = E(X^3) = 0$.

Hence, $\rho_{XY} = \dfrac{cov(X,Y)}{\sqrt{var(X)var(Y)}} = 0$.

2.4 The *Binomial Distribution*.

a. Pr[X=5 or 6] = Pr[X=5] + Pr[X=6]
$$= b(n=20, X=5, \theta=0.1) + b(n=20, X=6, \theta=0.1)$$
$$= \binom{20}{5}(0.1)^5(0.9)^{15} + \binom{20}{6}(0.1)^6(0.9)^{14}$$
$$= 0.0319 + 0.0089 = 0.0408.$$

This can be easily done with a calculator, on the computer or using the Binomial tables, see Freund (1992).

b. $\binom{n}{n-X} = \dfrac{n!}{(n-X)!\,(n-n+X)!} = \dfrac{n!}{(n-X)!\,X!} = \binom{n}{X}$

Hence,

$$b(n, n-X, 1-\theta) = \binom{n}{n-X}(1-\theta)^{n-X}(1-1+\theta)^{n-n+X} = \binom{n}{X}(1-\theta)^{n-X}\theta^X = b(n, X, \theta).$$

c. Using the MGF for the Binomial distribution given in problem 14 part (a), we get

$$M_X(t) = [(1-\theta) + \theta e^t]^n.$$

Differentiating with respect to t yields $M'_X(t) = n[(1-\theta) + \theta e^t]^{n-1}\theta e^t$.

Therefore, $M'_X(0) = n\theta = E(X)$.

Differentiating $M'_X(t)$ again with respect to t yields

$$M''_X(t) = n(n-1)[(1-\theta)+\theta e^t]^{n-2}(\theta e^t)^2 + n[(1-\theta)+\theta e^t]^{n-1}\theta e^t.$$

Therefore $M''_X(0) = n(n-1)\theta^2 + n\theta = E(X^2)$.

Hence $\text{var}(X) = E(X^2) - (E(X))^2 = n\theta + n^2\theta^2 - n\theta^2 - n^2\theta^2 = n\theta(1-\theta).$

An alternative proof for $E(X) = \sum_{X=0}^{n} X b(n, X, \theta)$. This entails factorial moments and the reader is referred to Freund (1992).

d. The likelihood function is given by

$$L(\theta) = f(X_1, ..., X_n; \theta) = \theta^{\sum_{i=1}^{n} X_i}(1-\theta)^{n - \sum_{i=1}^{n} X_i}$$

so that $\log L(\theta) = \left(\sum_{i=1}^{n} X_i\right)\log\theta + \left(n - \sum_{i=1}^{n} X_i\right)\log(1-\theta)$

$$\frac{\partial \log L(\theta)}{\partial \theta} = \frac{\sum_{i=1}^{n} X_i}{\theta} - \frac{(n - \sum_{i=1}^{n} X_i)}{(1-\theta)} = 0.$$

Solving for θ one gets

$$\sum_{i=1}^{n} X_i - \theta \sum_{i=1}^{n} X_i - \theta n + \theta \sum_{i=1}^{n} X_i = 0$$

so that $\hat{\theta}_{mle} = \sum_{i=1}^{n} X_i/n = \bar{X}$.

e. $E(\bar{X}) = \sum_{i=1}^{n} E(X_i)/n = n\theta/n = \theta$.

Hence, $\bar{X}$ is unbiased for θ. $\text{var}(\bar{X}) = \text{var}(X_i)/n = \theta(1-\theta)/n$ which goes to zero as $n \to \infty$. Hence, the sufficient condition for $\bar{X}$ to be consistent for θ is satisfied.

f. The joint probability function in part (d) can be written as
$$f(X_1,..,X_n;\theta) = \theta^{n\bar{X}}(1-\theta)^{n-n\bar{X}} = h(\bar{X},\theta)\, g(X_1,..,X_n)$$
where $h(\bar{X},\theta) = \theta^{n\bar{X}}(1-\theta)^{n-n\bar{X}}$ and $g(X_1,..,X_n) = 1$ for all X_i's. The latter function is independent of θ in form and domain. Hence, by the factorization theorem, $\bar{X}$ is sufficient for θ.

g. $\bar{X}$ was shown to be MVU for θ for the Bernoulli case in Example 2 in the text.

h. From part (d), $L(0.2) = (0.2)^{\sum_{i=1}^{n} X_i}(0.8)^{n-\sum_{i=1}^{n} X_i}$ while $L(0.6) = (0.6)^{\sum_{i=1}^{n} X_i}(0.4)^{n-\sum_{i=1}^{n} X_i}$

with the likelihood ratio

$$\frac{L(0.2)}{L(0.6)} = \left(\frac{1}{3}\right)^{\sum_{i=1}^{n} X_i} 2^{n-\sum_{i=1}^{n} X_i}$$

The uniformly most powerful critical region C of size $\alpha \leq 0.05$ is given by

$\left(\frac{1}{3}\right)^{\sum_{i=1}^{n} X_i} 2^{n-\sum_{i=1}^{n} X_i} \leq k$ inside C. Taking logarithms of both sides

$-\sum_{i=1}^{n} X_i(\log 3) + (n - \sum_{i=1}^{n} X_i)\log 2 \leq \log k$

solving $-(\sum_{i=1}^{n} X_i)\log 6 \leq K'$ or $\sum_{i=1}^{n} X_i \geq K$

where K is determined by making the size of $C = \alpha \leq 0.05$. In this case,

$\sum_{i=1}^{n} X_i \sim b(n,\theta)$ and under $H_o = \theta = 0.2$. Therefore, $\sum_{i=1}^{n} X_i \sim b(n=20, \theta=0.2)$.

Hence, $\alpha = \Pr[b(n=20, \theta=0.2) \geq K] \leq 0.05$

From the Binomial tables for $n = 20$ and $\theta = 0.2$, $K = 7$ gives

$\Pr[b(n=20, \theta=0.2) \geq 7] = 0.0322$.

Hence, $\sum_{i=1}^{n} X_i \geq 7$ is our required critical region.

i. The likelihood ratio test is

$$\frac{L(0.2)}{L(\hat{\theta}_{mle})} = \frac{(0.2)^{\sum_{i=1}^{n} X_i} (0.8)^{n - \sum_{i=1}^{n} X_i}}{(\bar{X})^{\sum_{i=1}^{n} X_i} (1-\bar{X})^{n - \sum_{i=1}^{n} X_i}}$$

so that $LR = -2\log L(0.2) + 2\log L(\hat{\theta}_{mle})$
$= -2[\sum_{i=1}^{n} X_i (\log 0.2 - \log \bar{X}) - 2(n - \sum_{i=1}^{n} X_i)(\log 0.8 - \log(1-\bar{X}))]$.

This is given in Example 5 in the text for a general θ_o. The Wald statistic is given by $\dfrac{(\bar{X}-0.2)^2}{\bar{X}(1-\bar{X})/n}$

and the LM statistic is given by $LM = \dfrac{(\bar{X}-0.2)^2}{(0.2)(0.8)/n}$

Although, the three statistics, LR, LM and W look different, they are all based on $|\bar{X}-2| \geq k$ and for a finite n, the same exact critical value could be obtained from the binomial distribution.

2.5 d. *The Wald, LR, and LM Inequality.* This is based on Baltagi (1995). The likelihood is given by equation (2.1) in the text.

$$L(\mu, \sigma^2) = (1/2\pi\sigma^2)^{n/2} e^{-(1/2\sigma^2) \sum_{i=1}^{n} (X_i - \mu)^2} \tag{1}$$

It is easy to show that the score is given by

$$S(\mu, \sigma^2) = \begin{pmatrix} \dfrac{n(\bar{X}-\mu)}{\sigma^2} \\ \dfrac{\sum_{i=1}^{n}(X_i-\mu)^2 - n\sigma^2}{2\sigma^4} \end{pmatrix}, \tag{2}$$

and setting $S(\mu,\sigma^2) = 0$ yields $\hat{\mu} = \overline{X}$ and $\hat{\sigma}^2 = \sum_{i=1}^{n}(X_i - \overline{X})^2/n$. Under H_o, $\tilde{\mu} = \mu_o$ and $\tilde{\sigma}^2 = \sum_{i=1}^{n}(X_i - \mu_o)^2/n$. Therefore,

$$\log L(\tilde{\mu}, \tilde{\sigma}^2) = -\frac{n}{2}\log \tilde{\sigma}^2 - \frac{n}{2}\log 2\pi - \frac{n}{2} \tag{3}$$

and

$$\log L(\hat{\mu}, \hat{\sigma}^2) = -\frac{n}{2}\log \hat{\sigma}^2 - \frac{n}{2}\log 2\pi - \frac{n}{2}. \tag{4}$$

Hence,

$$LR = n \log \left[\frac{\sum_{i=1}^{n}(X_i - \mu_o)^2}{\sum_{i=1}^{n}(X_i - \overline{X})^2} \right]. \tag{5}$$

It is also known that the information matrix is given by

$$I\begin{pmatrix} \mu \\ \sigma^2 \end{pmatrix} = \begin{bmatrix} \frac{n}{\sigma^2} & 0 \\ 0 & \frac{n}{2\sigma^4} \end{bmatrix} \tag{6}$$

Therefore,

$$W = (\hat{\mu} - \mu_o)^2 \hat{I}_{11} = \frac{n^2(\overline{X} - \mu_o)^2}{\sum_{i=1}^{n}(X_i - \overline{X})^2}, \tag{7}$$

where $\hat{I}_{11}$ denotes the (1,1) element of the information matrix evaluated at the unrestricted maximum likelihood estimates. It is easy to show from (1) that

$$\log L(\tilde{\mu}, \hat{\sigma}^2) = -\frac{n}{2}\log \hat{\sigma}^2 - \frac{n}{2}\log 2\pi - \frac{\sum_{i=1}^{n}(X_i - \mu_o)^2}{2\hat{\sigma}^2}. \tag{8}$$

Hence, using (4) and (8), one gets

$$-2\log[L(\tilde{\mu},\hat{\sigma}^2)/L(\hat{\mu},\hat{\sigma}^2)] = \frac{\sum_{i=1}^{n}(X_i - \mu_o)^2 - n\hat{\sigma}^2}{\hat{\sigma}^2} = W, \tag{9}$$

and the last equality follows from (7). Similarly,

$$S^2(\tilde{\mu}, \tilde{\sigma}^2)\tilde{I}^{11} = \frac{n^2(\bar{X}-\mu_o)^2}{\sum_{i=1}^{n}(X_i - \mu_o)^2}, \tag{10}$$

where $\tilde{I}^{11}$ denotes the (1,1) element of the inverse of the information matrix evaluated at the restricted maximum likelihood estimates. From (1), we also get

$$\log L(\hat{\mu}, \tilde{\sigma}^2) = -\frac{n}{2}\log \tilde{\sigma}^2 - \frac{n}{2}\log 2\pi - \frac{\sum_{i=1}^{n}(X_i - \bar{X})^2}{2\tilde{\sigma}^2}. \tag{11}$$

Hence, using (3) and (11), one gets

$$-2 \log[L(\tilde{\mu}, \tilde{\sigma}^2)/L(\hat{\mu}, \hat{\sigma}^2)] = n - \frac{\sum_{i=1}^{n}(X_i - \bar{X})^2}{\hat{\sigma}^2} = LM, \tag{12}$$

where the last equality follows from (10). $L(\tilde{\mu}, \tilde{\sigma}^2)$ is the restricted maximum; therefore, $\log L(\tilde{\mu}, \hat{\sigma}^2) \leq \log L(\tilde{\mu}, \tilde{\sigma}^2)$, from which we deduce that $W \geq LR$. Also, $L(\hat{\mu}, \hat{\sigma}^2)$ is the unrestricted maximum; therefore $\log L(\hat{\mu}, \hat{\sigma}^2) \geq \log L(\hat{\mu}, \tilde{\sigma}^2)$, from which we deduce that $LR \geq LM$.

An alternative derivation of this inequality shows first that

$$\frac{LM}{n} = \frac{W/n}{1+(W/n)} \quad \text{and} \quad \frac{LR}{n} = \log\left(1+\frac{W}{n}\right).$$

Then one uses the fact that $y \geq \log(1+y) \geq y/(1+y)$ for $y = W/n$.

2.6 *Poisson Distribution.*

a. Using the MGF for the Poisson derived in problem 14 part (c) one gets

$$M_X(t) = e^{\lambda(e^t - 1)}.$$

Differentiating with respect to t yields $M'_X(t) = e^{\lambda(e^t - 1)}\lambda e^t$.

Evaluating $M'_X(t)$ at $t = 0$, we get $M'_X(0) = E(X) = \lambda$.

Similarly, differentiating $M'_X(t)$ once more with respect to t, we get

$$M''_X(t) = e^{\lambda(e^t - 1)}(\lambda e^t)^2 + e^{\lambda(e^t - 1)}\lambda e^t$$

evaluating it at t = 0 gives $M_X''(0) = \lambda^2 + \lambda = E(X^2)$
so that $\text{var}(X) = E(X^2) - (E(X))^2 = \lambda^2 + \lambda - \lambda^2 = \lambda$.

Hence, the mean and variance of the Poisson are both equal to λ.

b. The likelihood function is

$$L(\lambda) = \frac{e^{-n\lambda} \lambda^{\sum_{i=1}^{n} X_i}}{X_1! X_2! .. X_n!}$$

so that $\log L(\lambda) = -n\lambda + (\sum_{i=1}^{n} X_i) \log \lambda - \sum_{i=1}^{n} \log X_i!$

$$\frac{\partial \log L(\lambda)}{\partial \lambda} = -n + \frac{\sum_{i=1}^{n} X_i}{\lambda} = 0.$$

Solving for λ, yields $\hat{\lambda}_{mle} = \bar{X}$.

c. The method of moments equates $E(X)$ to $\bar{X}$ and since $E(X) = \lambda$ the solution is $\hat{\lambda} = \bar{X}$, same as the ML method.

d. $E(\bar{X}) = \sum_{i=1}^{n} E(X_i)/n = n\lambda/n = \lambda$. Therefore $\bar{X}$ is unbiased for λ. Also, $\text{var}(\bar{X}) = \frac{\text{var}(X_i)}{n} = \frac{\lambda}{n}$ which tends to zero as $n \to \infty$. Therefore, the sufficient condition for $\bar{X}$ to be consistent for λ is satisfied.

e. The joint probability function can be written as

$$f(X_1,..,X_n;\lambda) = e^{-n\lambda} \lambda^{n\bar{X}} \frac{1}{X_1!..X_n!} = h(\bar{X},\lambda) g(X_1,..,X_n)$$

where $h(\bar{X},\lambda) = e^{-n\lambda} \lambda^{n\bar{X}}$ and $g(X_1,..,X_n) = \frac{1}{X_1!..X_n!}$. The latter is independent of λ in form and domain. Therefore, $\bar{X}$ is a sufficient statistic for λ.

f. $\log f(X;\lambda) = -\lambda + X\log\lambda - \log X!$

and $\frac{\partial \log f(X;\lambda)}{\partial \lambda} = -1 + \frac{X}{\lambda}$

$$\frac{\partial^2 \log f(X;\lambda)}{\partial \lambda^2} = \frac{-X}{\lambda^2}.$$

The Cramér-Rao lower bound for any unbiased estimator $\hat{\lambda}$ of λ is given by

$$\text{var}(\hat{\lambda}) \geq -\frac{1}{nE\left(\frac{\partial^2 \log f(X;\lambda)}{\partial \lambda^2}\right)} = \frac{\lambda^2}{nE(X)} = \frac{\lambda}{n}.$$

But $\text{var}(\bar{X}) = \lambda/n$, see part (d). Hence, $\bar{X}$ attains the Cramér-Rao lower bound.

g. The likelihood ratio is given by

$$\frac{L(2)}{L(4)} = \frac{e^{-2n} 2^{\sum_{i=1}^{n} X_i}}{e^{-4n} 4^{\sum_{i=1}^{n} X_i}}$$

The uniformly most powerful critical region C of size $\alpha \leq 0.05$ is given by

$$e^{2n} \left(\frac{1}{2}\right)^{\sum_{i=1}^{n} X_i} \leq k \text{ inside C}$$

Taking logarithms of both sides and rearranging terms, we get

$$-\left(\sum_{i=1}^{n} X_i\right) \log 2 \leq K'$$

or $\sum_{i=1}^{n} X_i \geq K$

where K is determined by making the size of C = $\alpha \leq 0.05$. In this case, $\sum_{i=1}^{n} X_i \sim \text{Poisson}(n\lambda)$ and under H_o; $\lambda = 2$. Therefore, $\sum_{i=1}^{n} X_i \sim \text{Poisson}(\lambda=18)$. Hence $\alpha = \Pr[\text{Poisson}(18) \geq K] \leq 0.05$. From the Poisson tables, for $\lambda = 18$, K = 26 gives $\Pr[\text{Poisson}(18) \geq 26] = 0.0446$. Hence, $\sum_{i=1}^{n} X_i \geq 26$ is our required critical region.

h. The likelihood ratio test is

$$\frac{L(2)}{L(\bar{X})} = \frac{e^{-2n} 2^{\sum_{i=1}^{n} X_i}}{e^{-n\bar{X}} (\bar{X})^{\sum_{i=1}^{n} X_i}} = e^{n(\bar{X}-2)} \left(\frac{2}{\bar{X}}\right)^{\sum_{i=1}^{n} X_i}$$

so that

$$LR = -2\log L(2) + 2\log L(\bar{X}) = -2n(\bar{X}-2) - 2\sum_{i=1}^{n} X_i[\log 2 - \log \bar{X}].$$

In this case, $C(\lambda) = \left|\frac{\partial^2 \log L(\lambda)}{\partial \lambda^2}\right| = -\frac{\sum_{i=1}^{n} X_i}{\lambda^2}$

and

$$I(\lambda) = -E\left[\frac{\partial^2 \log L(\lambda)}{\partial \lambda^2}\right] = \frac{n\lambda}{\lambda^2} = \frac{n}{\lambda}.$$

The Wald statistic is based upon

$$W = (\bar{X}-2)^2 \, I(\hat{\lambda}_{mle}) = (\bar{X}-2)^2 \, n/\bar{X}$$

using the fact that $\hat{\lambda}_{mle} = \bar{X}$. The LM statistic is based upon

$$LM = S^2(2) \, I^{-1}(2) = \frac{n^2(\bar{X}-2)^2}{4} \cdot \frac{2}{n} = \frac{n(\bar{X}-2)^2}{2}.$$

Note that all three test statistics are based upon $|\bar{X}-2| \geq K$ and for finite n the same exact critical value could be obtained using the fact that $\sum_{i=1}^{n} X_i$ has a Poisson distribution, see part (g).

2.7 *The Geometric Distribution.*

a. Using the MGF for the Geometric distribution derived in problem 14 part (d), one gets $M_X(t) = \dfrac{\theta e^t}{[1-(1-\theta)e^t]}$.

Differentiating it with respect to t yields

$$M'_X(t) = \frac{\theta e^t[1-(1-\theta)e^t] + (1-\theta)e^t \theta e^t}{[1-(1-\theta)e^t]^2} = \frac{\theta e^t}{[1-(1-\theta)e^t]^2}$$

evaluating $M'_X(t)$ at $t = 0$, we get $M'_X(0) = E(X) = \dfrac{\theta}{\theta^2} = \dfrac{1}{\theta}$.

Similarly, differentiating $M'_X(t)$ once more with respect to t, we get

$$M''_X(t) = \frac{\theta e^t[1-(1-\theta)e^t]^2 + 2[1-(1-\theta)e^t](1-\theta)e^t \theta e^t}{[1-(1-\theta)e^t]^4}$$

evaluating $M''_X(t)$ at $t = 0$, we get

Chapter 2: A Review of Some Basic Statistical Concepts

$$M''_X(0) = E(X^2) = \frac{\theta^3 + 2\theta^2(1-\theta)}{\theta^4} = \frac{2\theta^2 - \theta^3}{\theta^4} = \frac{2-\theta}{\theta^2}$$

so that $var(X) = E(X^2) - (E(X))^2 = \dfrac{2-\theta}{\theta^2} - \dfrac{1}{\theta^2} = \dfrac{1-\theta}{\theta^2}$.

b. The likelihood function is given by

$$L(\theta) = \theta^n (1-\theta)^{\sum_{i=1}^{n} X_i - n}$$

so that $\log L(\theta) = n\log\theta + (\sum_{i=1}^{n} X_i - n)\log(1-\theta)$

$$\frac{\partial \log L(\theta)}{\partial \theta} = \frac{n}{\theta} - \frac{\sum_{i=1}^{n} X_i - n}{(1-\theta)} = 0$$

solving for θ one gets $n(1-\theta) - \theta \sum_{i=1}^{n} X_i + n\theta = 0$

or $n = \theta \sum_{i=1}^{n} X_i$

which yields $\hat{\theta}_{mle} = n / \sum_{i=1}^{n} X_i = 1/\overline{X}$.

The method of moments estimator equates $E(X) = \overline{X}$

so that $\dfrac{1}{\hat{\theta}} = \overline{X}$ or $\hat{\theta} = 1/\overline{X}$

which is the same as the MLE.

2.8 The *Uniform Density*.

a. $E(X) = \int_0^1 x\,dx = \dfrac{1}{2}[x^2]_0^1 = \dfrac{1}{2}$

$E(X^2) = \int_0^1 x^2\,dx = \dfrac{1}{3}[x^3]_0^1 = \dfrac{1}{3}$

so that $var(X) = E(X^2) - (E(X))^2 = \dfrac{1}{3} - \dfrac{1}{4} = \dfrac{4-3}{12} = \dfrac{1}{12}$.

b. $Pr[0.1 < X < 0.3] = \int_{0.1}^{0.3} dx = 0.3 - 0.1 = 0.2$. It does not matter if we include the equality signs $Pr[0.1 \leq X \leq 0.3]$ since this is a continuous random variable. Note that this integral is the area of the rectangle, for X between 0.1 and 0.3 and height equal to 1. This it is just the length of this rectangle, i.e., $0.3 - 0.1 = 0.2$.

2.9 The *Exponential Distribution*.

a. Using the MGF for the exponential distribution derived in problem 14 part (e), we get $M_X(t) = \dfrac{1}{(1-\theta t)}$.

Differentiating with respect to t yields $M'_X(t) = \dfrac{\theta}{(1-\theta t)^2}$.

Therefore $M'_X(0) = \theta = E(X)$.

Differentiating $M'_X(t)$ with respect to t yields $M''_X(t) = \dfrac{2\theta^2(1-\theta t)}{(1-\theta t)^4} = \dfrac{2\theta^2}{(1-\theta t)^3}$.

Therefore $M''_X(0) = 2\theta^2 = E(X^2)$.

Hence $\text{var}(X) = E(X^2) - (E(X))^2 = 2\theta^2 - \theta^2 = \theta^2$.

b. The likelihood function is given by

$$L(\theta) = \left(\dfrac{1}{\theta}\right)^n e^{-\sum_{i=1}^{n} X_i/\theta}$$

so that

$$\log L(\theta) = -n\log\theta - \sum_{i=1}^{n} X_i/\theta$$

$$\dfrac{\partial \log L(\theta)}{\partial \theta} = \dfrac{-n}{\theta} + \dfrac{\sum_{i=1}^{n} X_i}{\theta^2} = 0$$

solving for θ one gets $\sum_{i=1}^{n} X_i - n\theta = 0$

so that $\hat{\theta}_{mle} = \bar{X}$.

c. The method of moments equates $E(X) = \bar{X}$. In this case, $E(X) = \theta$, hence $\hat{\theta} = \bar{X}$ is the same as MLE.

d. $E(\bar{X}) = \sum_{i=1}^{n} E(X_i)/n = n\theta/n = \theta$. Hence, $\bar{X}$ is unbiased for θ. Also, $\text{var}(\bar{X}) = \text{var}(X_i)/n = \theta^2/n$ which goes to zero as $n \to \infty$. Hence, the sufficient condition for $\bar{X}$ to be consistent for θ is satisfied.

e. The joint p.d.f. is given by

$$f(X_1,...,X_n;\theta) = \left(\dfrac{1}{\theta}\right)^n e^{-\sum_{i=1}^{n} X_i/\theta} = e^{-n\bar{X}/\theta}\left(\dfrac{1}{\theta}\right)^n = h(\bar{X};\theta)\, g(X_1,...,X_n)$$

where $h(\bar{X};\theta) = e^{-n\bar{X}/\theta}\left(\dfrac{1}{\theta}\right)^n$ and $g(X_1,..,X_n) = 1$ independent of θ in form and domain. Hence, by the factorization theorem, $\bar{X}$ is a sufficient for θ.

f. $\log f(X;\theta) = -\log\theta - \dfrac{X}{\theta}$

and $\dfrac{\partial \log f(X;\theta)}{\partial \theta} = \dfrac{-1}{\theta} + \dfrac{X}{\theta^2} = \dfrac{X-\theta}{\theta^2}$

$\dfrac{\partial^2 \log f(X;\theta)}{\partial \theta^2} = \dfrac{1}{\theta^2} - \dfrac{2X\theta}{\theta^4} = \dfrac{\theta - 2X}{\theta^3}$

The Cramér-Rao lower bound for any unbiased estimator $\hat{\theta}$ of θ is given by

$$\mathrm{var}(\hat{\theta}) \geq \dfrac{-1}{nE\left(\dfrac{\partial^2 \log f(X;\theta)}{\partial \theta^2}\right)} = \dfrac{-\theta^3}{nE(\theta - 2X)} = \dfrac{\theta^2}{n}.$$

But $\mathrm{var}(\bar{X}) = \theta^2/n$, see part (d). Hence, $\bar{X}$ attains the Cramér-Rao lower bound.

g. The likelihood ratio is given by $\dfrac{L(1)}{L(2)} = \dfrac{e^{-\sum_{i=1}^{n} X_i}}{2^n e^{-\sum_{i=1}^{n} X_i/2}} = \left(\dfrac{1}{2}\right)^n e^{-\sum_{i=1}^{n} X_i/2}$.

The uniformly most powerful critical region C of size $\alpha \leq 0.05$ is given by

$\left(\dfrac{1}{2}\right)^n e^{-\sum_{i=1}^{n} X_i/2} \leq k$ inside C.

Taking logarithms of both sides and rearranging terms, we get

$-n\log 2 - (\sum_{i=1}^{n} X_i/2) \leq K'$

or $\sum_{i=1}^{n} X_i \geq K$

where K is determined by making the size of $C = \alpha \leq 0.05$. In this case, $\sum_{i=1}^{n} X_i$ is distributed as a Gamma p.d.f. with $\beta = \theta$ and $\alpha = n$. Under H_o; $\theta = 1$. Therefore, $\sum_{i=1}^{n} X_i \sim \mathrm{Gamma}(\alpha = n, \beta = 1)$. Hence

$\Pr[\mathrm{Gamma}(\alpha = n, \beta = 1) \geq K] \leq 0.05$

K should be determined from the integral $\int_{K}^{\infty} \dfrac{1}{\Gamma(n)} x^{n-1} e^{-x} dx = 0.05$

for $n = 20$.

h. The likelihood ratio test is

$$\frac{L(1)}{L(\bar{X})} = \frac{e^{-\sum_{i=1}^{n} X_i}}{\left(\frac{1}{\bar{X}}\right)^n e^{-n}}$$

so that $LR = -2\log L(1) + 2\log L(\bar{X}) = 2\sum_{i=1}^{n} X_i - 2n\log \bar{X} - 2n$.

In this case,

$$C(\theta) = \left|\frac{\partial^2 \log L(\theta)}{\partial \theta^2}\right| = \left|\frac{n}{\theta^2} - \frac{2\sum_{i=1}^{n} X_i}{\theta^3}\right| = \left|\frac{n\theta - 2\sum_{i=1}^{n} X_i}{\theta^3}\right|$$

and

$$I(\theta) = -E\left[\frac{\partial^2 \log L(\theta)}{\partial \theta^2}\right] = \frac{n}{\theta^2}.$$

The Wald statistic is based upon

$$W = (\bar{X} - 1)^2 I(\hat{\theta}_{mle}) = (\bar{X} - 1)^2 n / \bar{X}^2$$

using the fact that $\hat{\theta}_{mle} = \bar{X}$. The LM statistic is based upon

$$LM = S^2(1) I^{-1}(1) = (\sum_{i=1}^{n} X_i - n)^2 / n = n(\bar{X} - 1)^2.$$

All three test statistics are based upon $|\bar{X} - 1| \geq k$ and, for finite n, the same exact critical value could be obtained using the fact that $\sum_{i=1}^{n} X_i$ is Gamma ($\alpha = 1$, and $\beta = n$) under H_o, see part (g).

2.10 The *Gamma Distribution*.

a. Using the MGF for the Gamma distribution derived in problem 14 part (f), we get $M_X(t) = (1-\beta t)^{-\alpha}$.

Differentiating with respect to t yields $M'_X(t) = -\alpha(1-\beta t)^{-\alpha-1}(-\beta) = \alpha\beta(1-\beta t)^{-\alpha-1}$.

Therefore $M'_X(0) = \alpha\beta = E(X)$.

Differentiating $M'_X(t)$ with respect to t yields

$M''_X(t) = -\alpha\beta(\alpha+1)(1-\beta t)^{-\alpha-2}(-\beta) = \alpha\beta^2(\alpha+1)(1-\beta t)^{-\alpha-2}$.

Therefore $M''_X(0) = \alpha^2\beta^2 + \alpha\beta^2 = E(X^2)$.

Hence $var(X) = E(X^2) - (E(X))^2 = \alpha\beta^2$.

Chapter 2: A Review of Some Basic Statistical Concepts   15

b. The method of moments equates $E(X) = \bar{X} = \alpha\beta$

and $E(X^2) = \sum_{i=1}^{n} X_i^2/n = \alpha^2\beta^2 + \alpha\beta^2$.

These are two non-linear equations in two unknowns. Substitute $\alpha = \bar{X}/\beta$ into the second equation, one gets $\sum_{i=1}^{n} X_i^2/n = \bar{X}^2 + \bar{X}\beta$

Hence, $\hat{\beta} = \sum_{i=1}^{n}(X_i - \bar{X})^2 / n\bar{X}$

and $\hat{\alpha} = \dfrac{n\bar{X}^2}{\sum_{i=1}^{n}(X_i - \bar{X})^2}$.

c. For $\alpha = 1$ and $\beta = \theta$, we get

$$f(X; \alpha=1, \beta=\theta) = \frac{1}{\Gamma(1)\theta} X^{1-1} e^{-X/\theta} \quad \text{for } X > 0 \quad \text{and } \theta > 0,$$

$$= \frac{1}{\theta} e^{-X/\theta} \quad \text{for } X > 0$$

which is the exponential p.d.f.

d. For $\alpha = r/2$ and $\beta = 2$, the Gamma $(\alpha=r/2, \beta=2)$ is a χ_r^2. Hence, from part (a), we get $E(X) = \alpha\beta = (r/2)(2) = r$

and $\text{var}(X) = \alpha\beta^2 = (r/2)(4) = 2r$.

The expected value of a χ_r^2 is r and its variance is 2r.

e. The joint p.d.f. for $\alpha = r/2$ and $\beta = 2$ is given by

$$f(X_1,..,X_n; \alpha=r/2, \beta=2) = \left(\frac{1}{\Gamma(r/2)2^{r/2}}\right)^n (X_1,..,X_n)^{\frac{r}{2}-1} e^{-\sum_{i=1}^{n} X_i/2}$$

$$= h(X_1,..,X_n; r)\, g(X_1,..,X_n)$$

where $h(X_1,..,X_n; r) = \left(\dfrac{1}{\Gamma(r/2)2^{r/2}}\right)^n (X_1,..,X_n)^{\frac{r}{2}-1}$

and $g(X_1,..,X_n) = e^{-\sum_{i=1}^{n} X_i/2}$ independent of r in form and domain. Hence, by the factorization theorem $(X_1,..,X_n)$ is a sufficient statistic for r.

f. Let $X_1,..,X_m$ denote independent $N(0,1)$ random variables. Then, $X_1^2,..,X_m^2$ will be independent χ_1^2 random variables and $Y = \sum_{i=1}^{m} X_i^2$ will be χ_m^2. The sum of m independent χ_1^2 random variables is a χ_m^2 random variable.

2.12 The *t-distribution with r Degrees of Freedom.*

a. If $X_1,...,X_n$ are IIN$(0,\sigma^2)$, then $\bar{X} \sim N(\mu,\sigma^2/n)$ and $z = \dfrac{(\bar{X}-\mu)}{\sigma/\sqrt{n}}$ is $N(0,1)$.

b. $(n-1)s^2/\sigma^2 \sim \chi^2_{n-1}$. Dividing our $N(0,1)$ random variable z in part (a), by the square-root of our χ^2_{n-1} random variable in part (b), divided by its degrees of freedom, we get

$$t = \dfrac{(\bar{X}-\mu)/\sigma/\sqrt{n}}{\sqrt{\dfrac{(n-1)s^2}{\sigma^2}/(n-1)}} = \dfrac{(\bar{X}-\mu)}{s/\sqrt{n}}.$$

Using the fact that $\bar{X}$ is independent of s^2, this has a t-distribution with $(n-1)$ degrees of freedom.

c. The 95% confidence interval for μ would be based on the t-distribution derived in part (b) with $n - 1 = 15$ degrees of freedom.

$$t = \dfrac{\bar{X}-\mu}{s/\sqrt{n}} = \dfrac{20-\mu}{2/\sqrt{16}} = \dfrac{20-\mu}{1/2} = 40 - 2\mu$$

$\Pr[-t_{\alpha/2} < t < t_{\alpha/2}] = 1-\alpha = 0.95$

From the t-tables with 15 degrees of freedom, $t_{0.025} = 2.131$. Hence

$\Pr[-2.131 < 40-2\mu < 2.131] = 0.95$.

rearranging terms, one gets $\Pr[37.869/2 < \mu < 42.131/2] = 0.95$ or

$\Pr[18.9345 < \mu < 21.0655] = 0.95$.

2.13 The *F-distribution.*

$(n_1-1)s_1^2/\sigma_1^2 = 24(15.6)/\sigma_1^2 \sim \chi^2_{24}$

also $(n_2-1)\,s_2^2/\sigma_2^2 = 30(18.9)/\sigma_2^2 \sim \chi^2_{30}$. Therefore, under H_o: $\sigma_1^2 = \sigma_2^2$,

$F = s_2^2/s_1^2 = \dfrac{18.9}{15.6} = 1.2115 \sim F_{30,24}$.

Using the F-tables with 30 and 24 degrees of freedom, we find $F_{.05,30,24} = 1.94$. Since the observed F-statistic 1.2115 is less than 1.94, we do not reject H_o that the variance of the two shifts is the same.

2.14 *Moment Generating Function (MGF).*

a. For the *Binomial Distribution.*

$$M_X(t) = E(e^{Xt}) = \sum_{X=0}^{n} \binom{n}{X} e^{Xt}\theta^X(1-\theta)^{n-X}$$

$$= \sum_{X=0}^{n} \binom{n}{X} (\theta e^t)^X(1-\theta)^{n-X} = [(1-\theta)+\theta e^t]^n$$

where the last equality uses the binomial expansion

$$(a+b)^n = \sum_{X=0}^{n} \binom{n}{X} a^X b^{n-X}$$

with $a = \theta e^t$ and $b = (1-\theta)$. This is the fundamental relationship underlying the binomial probability function and what makes it a proper probability function.

b. For the *Normal Distribution*.

$$M_X(t) = E(e^{Xt}) = \int_{-\infty}^{+\infty} e^{xt} \frac{1}{\sigma\sqrt{2\pi}} e^{-\frac{1}{2\sigma^2}(x-\mu)^2} dx$$

$$= \frac{1}{\sigma\sqrt{2\pi}} \int_{-\infty}^{+\infty} e^{-\frac{1}{2\sigma^2}\{x^2-2\mu x+\mu^2 - xt 2\sigma^2\}} dx$$

$$= \frac{1}{\sigma\sqrt{2\pi}} \int_{-\infty}^{+\infty} e^{-\frac{1}{2\sigma^2}\{x^2-2(\mu+t\sigma^2)x+\mu^2\}} dx$$

completing the square

$$M_X(t) = \frac{1}{\sigma\sqrt{2\pi}} \int_{-\infty}^{+\infty} e^{-\frac{1}{2\sigma^2}\{[x-(\mu+t\sigma^2)]^2 - (\mu+t\sigma^2)^2+\mu^2\}} dx = e^{-\frac{1}{2\sigma^2}[\mu^2-\mu^2-2\mu t\sigma^2 - t^2\sigma^4]}$$

The remaining integral integrates to 1 using the fact that the Normal density is proper and integrates to one. Hence $M_X(t) = e^{\mu t + \frac{1}{2}\sigma^2 t^2}$ after some cancellations.

c. For the *Poisson Distribution*.

$$M_X(t) = E(e^{Xt}) = \sum_{X=0}^{\infty} e^{Xt} \frac{e^{-\lambda}\lambda^X}{X!} = \sum_{X=0}^{\infty} \frac{e^{-\lambda}(\lambda e^t)^X}{X!} = e^{-\lambda} \sum_{X=0}^{\infty} \frac{(\lambda e^t)^X}{X!}$$

$$= e^{\lambda e^t - \lambda} = e^{\lambda(e^t - 1)}$$

where the fifth equality follows from the fact that $\sum_{X=0}^{\infty} \frac{a^X}{X!} = e^a$ and in this case $a = \lambda e^t$. This is the fundamental relationship underlying the Poisson distribution and what makes it a proper probability function.

d. For the *Geometric Distribution*.

$$M_X(t) = E(e^{Xt}) = \sum_{X=1}^{\infty} \theta(1-\theta)^{X-1} e^{Xt} = \theta \sum_{X=1}^{\infty} (1-\theta)^{X-1} e^{(X-1)t} e^t$$

$$= \theta e^t \sum_{X=1}^{\infty} [(1-\theta)e^t]^{X-1} = \frac{\theta e^t}{1-(1-\theta)e^t}$$

where the last equality uses the fact that $\sum_{X=1}^{\infty} a^{X-1} = \frac{1}{1-a}$ and in this case $a = (1-\theta)e^t$. This is the fundamental relationship underlying the Geometric distribution and what makes it a proper probability function.

e. For the *Exponential Distribution*.

$$M_X(t) = E(e^{Xt}) = \int_0^{\infty} \frac{1}{\theta} e^{-X/\theta} e^{Xt} dx = \frac{1}{\theta} \int_0^{\infty} e^{-X[\frac{1}{\theta} - t]} dx$$

$$= \frac{1}{\theta} \int_0^{\infty} e^{-X[\frac{1-\theta t}{\theta}]} dx = \frac{1}{\theta} \cdot \frac{-\theta}{(1-\theta t)} \left[e^{-X(\frac{1-\theta t}{\theta})} \right]_0^{\infty} = (1-\theta t)^{-1}$$

f. For the *Gamma Distribution*.

$$M_X(t) = E(e^{Xt}) = \int_0^{\infty} \frac{1}{\Gamma(\alpha)\beta^{\alpha}} X^{\alpha-1} e^{-X/\beta} e^{Xt} dx$$

$$= \frac{1}{\Gamma(\alpha)\beta^{\alpha}} \int_0^{\infty} X^{\alpha-1} e^{-X(\frac{1}{\beta} - t)} dx = \frac{1}{\Gamma(\alpha)\beta^{\alpha}} \int_0^{\infty} X^{\alpha-1} e^{-X(\frac{1-\beta t}{\beta})} dx$$

The Gamma density is proper and integrates to one using the fact that $\int_0^{\infty} X^{\alpha-1} e^{-X/\beta} dx = \Gamma(\alpha)\beta^{\alpha}$. Using this fundamental relationship for the last integral, we get $M_X(t) = \frac{1}{\Gamma(\alpha)\beta^{\alpha}} \cdot \Gamma(\alpha) \left(\frac{\beta}{1-\beta t} \right)^{\alpha} = (1-\beta t)^{-\alpha}$

where we substituted $\beta/(1-\beta t)$ for the usual β. The χ_r^2 distribution is Gamma with $\alpha = \frac{r}{2}$ and $\beta = 2$. Hence, its MGF is $(1-2t)^{-r/2}$.

g. This was already done in the solutions to problems 5, 6, 7, 9 and 10.

2.15 *Moment Generating Function Method.*

a. If $X_1, ..., X_n$ are independent Poisson distributed with parameters (λ_i) respectively, then from problem 14 part (c), we have

Chapter 2: A Review of Some Basic Statistical Concepts

$$M_{X_i}(t) = e^{\lambda_i(e^t-1)} \quad \text{for} \quad i = 1,2,..,n$$

$Y = \sum_{i=1}^{n} X_i$ has $M_Y(t) = \prod_{i=1}^{n} M_{X_i}(t)$ since the X_i's are independent. Hence,

$$M_Y(t) = e^{\sum_{i=1}^{n} \lambda_i(e^t-1)}$$ which we recognize as a Poisson with parameter $\sum_{i=1}^{n} \lambda_i$.

b. If $X_1,..,X_n$ are independent $N(\mu_i,\sigma_i^2)$, then from problem 14 part (b), we have

$$M_{X_i}(t) = e^{\mu_i t + \frac{1}{2}\sigma_i^2 t^2} \quad \text{for} \quad i = 1,2,..,n$$

$Y = \sum_{i=1}^{n} X_i$ has $M_Y(t) = \prod_{i=1}^{n} M_{X_i}(t)$ since the X_i's are independent. Hence,

$$M_Y(t) = e^{(\sum_{i=1}^{n}\mu_i)t + \frac{1}{2}(\sum_{i=1}^{n}\sigma_i^2)t^2}$$ which we recognize as Normal with mean $\sum_{i=1}^{n}\mu_i$ and variance $\sum_{i=1}^{n}\sigma_i^2$.

c. If $X_1,..,X_n$ are $IIN(\mu,\sigma^2)$, then $Y = \sum_{i=1}^{n} X_i$ is $N(n\mu, n\sigma^2)$ from part (b) using the equality of means and variances. Therefore, $\bar{X} = Y/n$ is $N(\mu,\sigma^2/n)$.

d. If $X_1,..,X_n$ are independent χ^2 distributed with parameters (r_i) respectively, then from problem 14 part (f), we get

$$M_{X_i}(t) = (1-2t)^{-r_i/2} \quad \text{for} \quad i = 1,2,..,n$$

$Y = \sum_{i=1}^{n} X_i$ has $M_Y(t) = \prod_{i=1}^{n} M_{X_i}(t)$ since the X_i's are independent. Hence,

$$M_Y(t) = (1-2t)^{-\sum_{i=1}^{n} r_i/2}$$ which we recognize as χ^2 with degrees of freedom $\sum_{i=1}^{n} r_i$.

2.16 *Best Linear Prediction.* This is based on Amemiya (1994).

a. The mean squared error predictor is given by

$$MSE = E(Y-\alpha-\beta X)^2 = E(Y^2) + \alpha^2 + \beta^2 E(X^2) - 2\alpha E(Y) - 2\beta E(XY) + 2\alpha\beta E(X)$$

minimizing this MSE with respect to α and β yields the following first-order conditions:

$$\partial MSE/\partial\alpha = 2\alpha - 2E(Y) + 2\beta E(X) = 0$$

$$\partial MSE/\partial\beta = 2\beta E(X^2) - 2E(XY) + 2\alpha E(X) = 0.$$

Solving these two equations for α and β yields $\hat{\alpha} = \mu_Y - \hat{\beta}\mu_X$ from the first equation, where $\mu_Y = E(Y)$ and $\mu_X = E(X)$. Substituting this in the second equation one gets

$$\hat{\beta}E(X^2) - E(XY) + \mu_Y\mu_X - \hat{\beta}\mu_X^2 = 0$$

$$\hat{\beta}\text{var}(X) = E(XY) - \mu_X\mu_Y = \text{cov}(X,Y).$$

Hence, $\hat{\beta} = \text{cov}(X,Y)/\text{var}(X) = \sigma_{XY}/\sigma_X^2 = \rho\sigma_Y/\sigma_X$, since $\rho = \sigma_{XY}/\sigma_X\sigma_Y$. The *best* predictor is given by $\hat{Y} = \hat{\alpha} + \hat{\beta}X$.

b. Substituting $\hat{\alpha}$ into the *best* predictor one gets

$$\hat{Y} = \hat{\alpha} + \hat{\beta}X = \mu_Y + \hat{\beta}(X-\mu_X) = \mu_Y + \rho\frac{\sigma_Y}{\sigma_X}(X-\mu_X)$$

one clearly deduces that $E(\hat{Y}) = \mu_Y$ and $\text{var}(\hat{Y}) = \rho^2\frac{\sigma_Y^2}{\sigma_X^2}\text{var}(X) = \rho^2\sigma_Y^2$. The

prediction error $\hat{u} = Y - \hat{Y} = (Y-\mu_Y) - \rho\frac{\sigma_Y}{\sigma_X}(X-\mu_X)$ with $E(\hat{u}) = 0$ and

$$\text{var}(\hat{u}) = E(\hat{u}^2) = \text{var}(Y) + \rho^2\frac{\sigma_Y^2}{\sigma_X^2}\text{var}(X) - 2\rho\frac{\sigma_Y}{\sigma_X}\sigma_{XY}$$

$$= \sigma_Y^2 + \rho^2\sigma_Y^2 - 2\rho^2\sigma_Y^2 = \sigma_Y^2(1-\rho^2).$$

This is the proportion of the var(Y) that is not explained by the *best linear predictor* $\hat{Y}$.

c. $\text{cov}(\hat{Y},\hat{u}) = \text{cov}(\hat{Y}, Y-\hat{Y}) = \text{cov}(\hat{Y},Y) - \text{var}(\hat{Y})$

But $\text{cov}(\hat{Y},Y) = E(\hat{Y}-\mu_Y)(Y-\mu_Y) = E\left(\rho\frac{\sigma_Y}{\sigma_X}(X-\mu_X)(Y-\mu_Y)\right)$

$$= \rho\frac{\sigma_Y}{\sigma_X}\text{cov}(X,Y) = \rho^2\frac{\sigma_X\sigma_Y^2}{\sigma_X} = \rho^2\sigma_Y^2$$

Hence, $\text{cov}(\hat{Y},\hat{u}) = \rho^2\sigma_Y^2 - \rho^2\sigma_Y^2 = 0$.

2.17 The *Best Predictor*.

a. The problem is to minimize $E[Y-h(X)]^2$ with respect to $h(X)$. Add and subtract $E(Y/X)$ to get

$$E\{[Y-E(Y/X)] + [E(Y/X)-h(X)]\}^2 = E[Y-E(Y/X)]^2 + E[E(Y/X)-h(X)]^2$$

and the cross-product term $E[Y-E(Y/X)][E(Y/X)-h(X)]$ is zero because of the law of iterated expectations, see the Appendix to this chapter or Amemiya (1994). In fact, this says that expectations can be written as

$$E = E_X E_{Y/X}$$

and for the cross-product term given above $E_{Y/X}[Y-E(Y/X)][E(Y/X)-h(X)]$ is clearly zero. Hence, $E[Y-h(X)]^2$ is expressed as the sum of two positive terms. The first term is not affected by our choice of $h(X)$. The second term however is zero for $h(X) = E(Y/X)$. Clearly, this is the *best predictor* of Y based on X.

b. In the Appendix to this chapter, we considered the bivariate Normal distribution and showed that $E(Y/X) = \mu_Y + \rho \dfrac{\sigma_Y}{\sigma_X}(X-\mu_X)$. In part (a), we showed that this is the *best predictor* of Y based on X. But, in this case, this is exactly the form for the *best linear predictor* of Y based on X derived in problem 16. Hence, for the bivariate Normal density, the *best predictor* is identical to the *best linear predictor* of Y based on X.

CHAPTER 3
Simple Linear Regression

3.1 For least squares, the first-order conditions of minimization, given by equations (3.2) and (3.3), yield immediately the first two numerical properties of OLS estimates, i.e., $\sum_{i=1}^{n} e_i = 0$ and $\sum_{i=1}^{n} e_i X_i = 0$. Now consider $\sum_{i=1}^{n} e_i \hat{Y}_i = \hat{\alpha} \sum_{i=1}^{n} e_i + \hat{\beta} \sum_{i=1}^{n} e_i X_i = 0$ where the first equality uses $\hat{Y}_i = \hat{\alpha} + \hat{\beta} X_i$ and the second equality uses the first two numerical properties of OLS. Using the fact that $e_i = Y_i - \hat{Y}_i$, we can sum both sides to get $\sum_{i=1}^{n} e_i = \sum_{i=1}^{n} Y_i - \sum_{i=1}^{n} \hat{Y}_i$, but $\sum_{i=1}^{n} e_i = 0$, therefore we get $\sum_{i=1}^{n} Y_i = \sum_{i=1}^{n} \hat{Y}_i$. Dividing both sides by n, we get $\overline{Y} = \overline{\hat{Y}}$.

3.2 Minimizing $\sum_{i=1}^{n} (Y_i - \alpha)^2$ with respect to α yields $-2 \sum_{i=1}^{n} (Y_i - \alpha) = 0$. Solving for α yields $\hat{\alpha}_{ols} = \overline{Y}$. Averaging $Y_i = \alpha + u_i$ we get $\overline{Y} = \alpha + \overline{u}$. Hence $\hat{\alpha}_{ols} = \alpha + \overline{u}$ with $E(\hat{\alpha}_{ols}) = \alpha$ since $E(\overline{u}) = \sum_{i=1}^{n} E(u_i)/n = 0$ and $var(\hat{\alpha}_{ols}) = E(\hat{\alpha}_{ols} - \alpha)^2 = E(\overline{u})^2 = var(\overline{u}) = \sigma^2/n$. The residual sum of squares is $\sum_{i=1}^{n} (Y_i - \hat{\alpha}_{ols})^2 = \sum_{i=1}^{n} (Y_i - \overline{Y})^2 = \sum_{i=1}^{n} y_i^2$.

3.3 a. Minimizing $\sum_{i=1}^{n} (Y_i - \beta X_i)^2$ with respect to β yields $-2 \sum_{i=1}^{n} (Y_i - \beta X_i) X_i = 0$. Solving for β yields $\hat{\beta}_{ols} = \sum_{i=1}^{n} Y_i X_i / \sum_{i=1}^{n} X_i^2$. Substituting $Y_i = \beta X_i + u_i$ yields $\hat{\beta}_{ols} = \beta + \sum_{i=1}^{n} X_i u_i / \sum_{i=1}^{n} X_i^2$ with $E(\hat{\beta}_{ols}) = \beta$ since X_i is nonstochastic and $E(u_i) = 0$. Also,
$$var(\hat{\beta}_{ols}) = E(\hat{\beta}_{ols} - \beta)^2 = E(\sum_{i=1}^{n} X_i u_i / \sum_{i=1}^{n} X_i^2)^2 = \sigma^2 \sum_{i=1}^{n} X_i^2 / (\sum_{i=1}^{n} X_i^2)^2 = \sigma^2 / \sum_{i=1}^{n} X_i^2.$$

b. From the first-order condition in part (a), we get $\sum_{i=1}^{n} e_i X_i = 0$, where $e_i = Y_i - \hat{\beta}_{ols} X_i$. However, $\sum_{i=1}^{n} e_i$ is not necessarily zero. Therefore $\sum_{i=1}^{n} Y_i$ is not necessarily equal to $\sum_{i=1}^{n} \hat{Y}_i$, see problem 1. $\sum_{i=1}^{n} e_i \hat{Y}_i = \hat{\beta}_{ols} \sum_{i=1}^{n} e_i X_i = 0$, using $\hat{Y}_i = \hat{\beta}_{ols} X_i$ and $\sum_{i=1}^{n} e_i X_i = 0$. Therefore, only two of the numerical properties

Chapter 3: Simple Linear Regression 23

considered in problem 1 hold for this regression without a constant.

3.4 $E(\sum_{i=1}^{n} x_i u_i)^2 = \sum_{i=1}^{n}\sum_{j=1}^{n} x_i x_j E(u_i u_j) = \sum_{i=1}^{n} x_i^2 \text{var}(u_i) + \sum\sum_{i\neq j} x_i x_j \text{cov}(u_i, u_j) = \sigma^2 \sum_{i=1}^{n} x_i^2$ since $\text{cov}(u_i, u_j) = 0$ for $i \neq j$ and $\text{var}(u_i) = \sigma^2$ using assumptions 2 and 3.

3.5 a. From equation (3.4), we know that $\hat{\alpha}_{ols} = \bar{Y} - \hat{\beta}_{ols}\bar{X}$, substituting $\bar{Y} = \alpha + \beta\bar{X} + \bar{u}$ we get $\hat{\alpha}_{ols} = \alpha + (\beta - \hat{\beta}_{ols})\bar{X} + \bar{u}$. Therefore, $E(\hat{\alpha}_{ols}) = \alpha$ since $E(\hat{\beta}_{ols}) = \beta$, see equation (3.5), and $E(\bar{u}) = 0$.

b. $\text{var}(\hat{\alpha}_{ols}) = E(\hat{\alpha}_{ols} - \alpha)^2 = E[(\beta - \hat{\beta}_{ols})\bar{X} + \bar{u}]^2$.

$= \text{var}(\hat{\beta}_{ols})\bar{X}^2 + \text{var}(\bar{u}) + 2\bar{X}\text{cov}(\hat{\beta}_{ols}, \bar{u})$

But from (3.6), $\text{var}(\hat{\beta}_{ols}) = \sigma^2 / \sum_{i=1}^{n} x_i^2$. Also, $\text{var}(\bar{u}) = \sigma^2/n$ and

$\text{cov}(\hat{\beta}_{ols}, \bar{u}) = E\left(\dfrac{\sum_{i=1}^{n} x_i u_i}{\sum_{i=1}^{n} x_i^2} \cdot \dfrac{\sum_{i=1}^{n} u_i}{n}\right) = \sigma^2 \sum_{i=1}^{n} x_i / n \sum_{i=1}^{n} x_i^2 = 0$

where the first equality follows from equation (3.5). The second equality uses the fact that $\text{cov}(u_i, u_j) = 0$ for $i \neq j$ and $\text{var}(u_i) = \sigma^2$ from assumptions 2 and 3. The last equality follows from the fact that $\sum_{i=1}^{n} x_i = 0$. Therefore,

$\text{var}(\hat{\alpha}_{ols}) = \sigma^2[(1/n) + (\bar{X}^2 / \sum_{i=1}^{n} x_i^2)] = \sigma^2 \sum_{i=1}^{n} X_i^2 / (n \sum_{i=1}^{n} x_i^2)$.

c. $\hat{\alpha}_{ols}$ is unbiased for α and $\text{var}(\hat{\alpha}_{ols}) \to 0$ as $n \to \infty$, since $\sum_{i=1}^{n} x_i^2/n$ has a finite limit as $n \to \infty$ by assumption 4. Therefore, $\hat{\alpha}_{ols}$ is consistent for α.

d. Using (3.5) and part (a), we get

$\text{cov}(\hat{\alpha}_{ols}, \hat{\beta}_{ols}) = E(\hat{\alpha}_{ols} - \alpha)(\hat{\beta}_{ols} - \beta) = -\bar{X}\text{var}(\hat{\beta}_{ols}) + \text{cov}(\bar{u}, \hat{\beta}_{ols}) = -\sigma^2\bar{X}/\sum_{i=1}^{n} x_i^2$

The last equality uses the fact that $\text{cov}(\bar{u}, \hat{\beta}_{ols}) = 0$ from part (b).

3.6 a. $\hat{\alpha}_{ols} = \bar{Y} - \hat{\beta}_{ols}\bar{X} = \sum_{i=1}^{n} Y_i/n - \bar{X}\sum_{i=1}^{n} w_i Y_i = \sum_{i=1}^{n} \lambda_i Y_i$ where $\lambda_i = (1/n) - \bar{X}w_i$,

$\hat{\beta}_{ols} = \sum_{i=1}^{n} w_i Y_i$ and w_i is defined above (3.7).

b. $\sum_{i=1}^{n}\lambda_i = 1 - \bar{X}\sum_{i=1}^{n}w_i = 1$ where $\sum_{i=1}^{n}w_i = 0$ from (3.7), and

$$\sum_{i=1}^{n}\lambda_i X_i = \sum_{i=1}^{n}X_i/n - \bar{X}\sum_{i=1}^{n}X_i w_i = \bar{X} - \bar{X} = 0$$

using $\sum_{i=1}^{n}w_i X_i = 1$ from (3.7).

c. Any other linear estimator of α, can be written as

$$\tilde{\alpha} = \sum_{i=1}^{n}b_i Y_i = \alpha\sum_{i=1}^{n}b_i + \beta\sum_{i=1}^{n}b_i X_i + \sum_{i=1}^{n}b_i u_i$$

where the last equality is obtained by substituting equation (3.1) for Y_i. For $E(\tilde{\alpha})$ to equal α, we must have $\sum_{i=1}^{n}b_i = 1$, $\sum_{i=1}^{n}b_i X_i = 0$. Hence, $\tilde{\alpha} = \alpha + \sum_{i=1}^{n}b_i u_i$.

d. Let $b_i = \lambda_i + f_i$, then $1 = \sum_{i=1}^{n}b_i = \sum_{i=1}^{n}\lambda_i + \sum_{i=1}^{n}f_i$. From part (b), $\sum_{i=1}^{n}\lambda_i = 1$, therefore $\sum_{i=1}^{n}f_i = 0$. Similarly, $0 = \sum_{i=1}^{n}b_i X_i = \sum_{i=1}^{n}\lambda_i X_i + \sum_{i=1}^{n}f_i X_i$. From part (b), $\sum_{i=1}^{n}\lambda_i X_i = 0$, therefore $\sum_{i=1}^{n}f_i X_i = 0$.

e. $\text{var}(\tilde{\alpha}) = E(\tilde{\alpha} - \alpha)^2 = E(\sum_{i=1}^{n}b_i u_i)^2 = \sigma^2 \sum_{i=1}^{n}b_i^2$ using the expression of $\tilde{\alpha}$ from part (c) and assumptions 2 and 3 with $\text{var}(u_i) = \sigma^2$ and $\text{cov}(u_i, u_j) = 0$ for $i \neq j$. Substituting, $b_i = \lambda_i + f_i$ from part (d), we get

$$\text{var}(\tilde{\alpha}) = \sigma^2(\sum_{i=1}^{n}\lambda_i^2 + \sum_{i=1}^{n}f_i^2 + 2\sum_{i=1}^{n}\lambda_i f_i).$$

For $\lambda_i = (1/n) - \bar{X}w_i$ from part (a), we get $\sum_{i=1}^{n}\lambda_i f_i = \sum_{i=1}^{n}f_i/n - \bar{X}\sum_{i=1}^{n}w_i f_i = 0$.

This uses $\sum_{i=1}^{n}f_i = 0$ and $\sum_{i=1}^{n}f_i X_i = 0$ from part (d). Recall, $w_i = x_i/\sum_{i=1}^{n}x_i^2$ and so $\sum_{i=1}^{n}w_i f_i = 0$. Hence, $\text{var}(\tilde{\alpha}) = \sigma^2(\sum_{i=1}^{n}\lambda_i^2 + \sum_{i=1}^{n}f_i^2) = \text{var}(\hat{\alpha}_{ols}) + \sigma^2\sum_{i=1}^{n}f_i^2$. Any linear unbiased estimator $\tilde{\alpha}$ of α, will have a larger variance than $\hat{\alpha}_{ols}$ as long as $\sum_{i=1}^{n}f_i^2 \neq 0$. $\sum_{i=1}^{n}f_i^2 = 0$ when $f_i = 0$ for all i, which means that $b_i = \lambda_i$ for all i and $\tilde{\alpha} = \sum_{i=1}^{n}b_i Y_i = \sum_{i=1}^{n}\lambda_i Y_i = \hat{\alpha}_{ols}$. This proves that $\hat{\alpha}_{ols}$ is BLUE for α.

3.7 a. $\frac{\partial \log L}{\partial \alpha} = 0$ from (3.9) leads to $-\frac{1}{2\sigma^2}\frac{\partial RSS}{\partial \alpha} = 0$ where

$$RSS = \sum_{i=1}^{n}(Y_i - \alpha - \beta X_i)^2.$$

Solving this first-order condition yields $\hat{\alpha}_{ols}$. Therefore, $\hat{\alpha}_{mle} = \hat{\alpha}_{ols}$.

Similarly, $\frac{\partial \log L}{\partial \beta} = 0$ from (3.9) leads to $-\frac{1}{2\sigma^2}\frac{\partial RSS}{\partial \beta} = 0$. Solving for β yields $\hat{\beta}_{ols}$. Hence, $\hat{\beta}_{mle} = \hat{\beta}_{ols}$.

b. $\frac{\partial \log L}{\partial \sigma^2} = -\frac{n}{2}\cdot\frac{1}{\sigma^2} + \frac{\sum_{i=1}^{n}(Y_i - \alpha - \beta X_i)^2}{2\sigma^4}$

Setting $\frac{\partial \log L}{\partial \sigma^2} = 0$, yields $\hat{\sigma}^2_{mle} = \sum_{i=1}^{n}(Y_i - \hat{\alpha} - \hat{\beta}X_i)^2/n = \sum_{i=1}^{n}e_i^2/n$ where $\hat{\alpha}$ and $\hat{\beta}$ are the MLE estimates, which in this case are identical to the OLS estimates.

3.9 a. $R^2 = \dfrac{\sum_{i=1}^{n}\hat{y}_i^2}{\sum_{i=1}^{n}y_i^2} = \dfrac{\hat{\beta}_{ols}^2\sum_{i=1}^{n}x_i^2}{\sum_{i=1}^{n}y_i^2} = \dfrac{\left(\sum_{i=1}^{n}x_i y_i\right)^2 \left(\sum_{i=1}^{n}x_i^2\right)}{\left(\sum_{i=1}^{n}x_i^2\right)^2\left(\sum_{i=1}^{n}y_i^2\right)} = \dfrac{(\sum_{i=1}^{n}x_i y_i)^2}{\sum_{i=1}^{n}x_i^2 \sum_{i=1}^{n}y_i^2} = r_{xy}^2$

where the second equality substitutes $\hat{y}_i = \hat{\beta}_{ols} x_i$ and the third equality substitutes $\hat{\beta}_{ols} = \sum_{i=1}^{n}x_i y_i / \sum_{i=1}^{n}x_i^2$.

b. Multiply both sides of $y_i = \hat{y}_i + e_i$ by $\hat{y}_i$ and sum, one gets $\sum_{i=1}^{n}y_i \hat{y}_i = \sum_{i=1}^{n}\hat{y}_i^2 + \sum_{i=1}^{n}\hat{y}_i e_i$. The last term is zero by the numerical properties of least squares. Hence,

$\sum_{i=1}^{n}y_i \hat{y}_i = \sum_{i=1}^{n}\hat{y}_i^2$. Therefore

$$r_{y\hat{y}}^2 = \dfrac{\left(\sum_{i=1}^{n}y_i \hat{y}_i\right)^2}{\sum_{i=1}^{n}y_i^2 \sum_{i=1}^{n}\hat{y}_i^2} = \dfrac{\left(\sum_{i=1}^{n}\hat{y}_i^2\right)^2}{\sum_{i=1}^{n}y_i^2 \sum_{i=1}^{n}\hat{y}_i^2} = \dfrac{\sum_{i=1}^{n}\hat{y}_i^2}{\sum_{i=1}^{n}y_i^2} = R^2.$$

3.11 *Optimal Weighting of Unbiased Estimators.* This is based on Baltagi (1995).

All three estimators of β can be put in the linear form $\hat{\beta} = \sum_{i=1}^{n} w_i Y_i$ where

$$w_i = X_i / \sum_{i=1}^{n} X_i^2 \qquad \text{for } \hat{\beta}_1,$$

$$= (1/\sum_{i=1}^{n} X_i) \qquad \text{for } \hat{\beta}_2,$$

$$= (X_i - \bar{X}) / \sum_{i=1}^{n} (X_i - \bar{X})^2 \qquad \text{for } \hat{\beta}_3.$$

All satisfy $\sum_{i=1}^{n} w_i X_i = 1$, which is necessary for $\hat{\beta}_i$ to be unbiased for $i = 1,2,3$.

Therefore

$$\hat{\beta}_i = \beta + \sum_{i=1}^{n} w_i u_i \qquad \text{for } i = 1,2,3$$

with $E(\hat{\beta}_i) = \beta$ and $\text{var}(\hat{\beta}_i) = \sigma^2 \sum_{i=1}^{n} w_i^2$ since the u_i's are IID$(0,\sigma^2)$. Hence, $\text{var}(\hat{\beta}_1) = \sigma^2 / \sum_{i=1}^{n} X_i^2$, $\text{var}(\hat{\beta}_2) = \sigma^2 / n\bar{X}^2$ and $\text{var}(\hat{\beta}_3) = \sigma^2 / \sum_{i=1}^{n} (X_i - \bar{X})^2$.

a.

$$\text{cov}(\hat{\beta}_1, \hat{\beta}_2) = E(\sum_{i=1}^{n} u_i / \sum_{i=1}^{n} X_i)(\sum_{i=1}^{n} X_i u_i / \sum_{i=1}^{n} X_i^2)$$

$$= \sigma^2 \sum_{i=1}^{n} (X_i / \sum_{i=1}^{n} X_i^2)(1/\sum_{i=1}^{n} X_i) = \sigma^2 / \sum_{i=1}^{n} X_i^2 = \text{var}(\hat{\beta}_1) > 0.$$

Also,

$$\rho_{12} = \text{cov}(\hat{\beta}_1, \hat{\beta}_2)/[\text{var}(\hat{\beta}_1)\text{var}(\hat{\beta}_2)]^{1/2} = [\text{var}(\hat{\beta}_1)/\text{var}(\hat{\beta}_2)]^{1/2} = (n\bar{X}^2 / \sum_{i=1}^{n} X_i^2)^{1/2}$$

with $0 < \rho_{12} \le 1$. Samuel-Cahn (1994) showed that whenever the correlation coefficient between two unbiased estimators is equal to the square root of the ratio of their variances, the optimal combination of these two unbiased estimators is the one that weights the estimator with the smaller variance by 1. In this example, $\hat{\beta}_1$ is the best linear unbiased estimator (BLUE). Therefore, as expected, when we combine $\hat{\beta}_1$ with any other unbiased estimator of β, the optimal weight α^* turns out to be 1 for $\hat{\beta}_1$ and zero for the other linear unbiased estimator.

b. Similarly, $\text{cov}(\hat{\beta}_1, \hat{\beta}_3) = E(\sum_{i=1}^{n} X_i u_i / \sum_{i=1}^{n} X_i^2)[\sum_{i=1}^{n}(X_i - \bar{X})u_i / \sum_{i=1}^{n}(X_i - \bar{X})^2]$

$= \sigma^2 \sum_{i=1}^{n}(X_i / \sum_{i=1}^{n} X_i^2)[(X_i - \bar{X}) / \sum_{i=1}^{n}(X_i - \bar{X})^2]$

$= \sigma^2 / \sum_{i=1}^{n} X_i^2 = \text{var}(\hat{\beta}_1) > 0$. Also

$\rho_{13} = \text{cov}(\hat{\beta}_1, \hat{\beta}_3) / [\text{var}(\hat{\beta}_1)\text{var}(\hat{\beta}_3)]^{1/2} = [\text{var}(\hat{\beta}_1) / \text{var}(\hat{\beta}_3)]^{1/2}$

$= [\sum_{i=1}^{n}(X_i - \bar{X})^2 / \sum_{i=1}^{n} X_i^2]^{1/2} = (1 - \rho_{12}^2)^{1/2}$

with $0 < \rho_{13} \leq 1$. For the same reasons given in part (a), the optimal combination of $\hat{\beta}_1$ and $\hat{\beta}_3$ is that for which $\alpha^* = 1$ for $\hat{\beta}_1$ and $1 - \alpha^* = 0$ for $\hat{\beta}_3$.

c. Finally, $\text{cov}(\hat{\beta}_2, \hat{\beta}_3) = E(\sum_{i=1}^{n} u_i / \sum_{i=1}^{n} X_i)[\sum_{i=1}^{n}(X_i - \bar{X})u_i / \sum_{i=1}^{n}(X_i - \bar{X})^2]$

$= \sigma^2 \sum_{i=1}^{n}(X_i - \bar{X})/(\sum_{i=1}^{n} X_i)[\sum_{i=1}^{n}(X_i - \bar{X})^2] = 0$.

Hence $\rho_{23} = 0$. In this case, it is a standard result, see Samuel-Cahn (1994), that the optimal combination of $\hat{\beta}_2$ and $\hat{\beta}_3$ weights these unbiased estimators in proportion to $\text{var}(\hat{\beta}_3)$ and $\text{var}(\hat{\beta}_2)$, respectively. In other words

$\hat{\beta} = \{\text{var}(\hat{\beta}_2) / [\text{var}(\hat{\beta}_2) + \text{var}(\hat{\beta}_3)]\} \hat{\beta}_3 + \{\text{var}(\hat{\beta}_3) / [\text{var}(\hat{\beta}_2) + \text{var}(\hat{\beta}_3)]\} \hat{\beta}_2$

$= (1 - \rho_{12}^2)\hat{\beta}_3 + \rho_{12}^2 \hat{\beta}_2$ since $\text{var}(\hat{\beta}_3) / \text{var}(\hat{\beta}_2) = \rho_{12}^2 / (1 - \rho_{12}^2)$ and

$\rho_{12}^2 = n\bar{X}^2 / \sum_{i=1}^{n} X_i^2$ while $1 - \rho_{12}^2 = \sum_{i=1}^{n}(X_i - \bar{X})^2 / \sum_{i=1}^{n} X_i^2$. Hence,

$\hat{\beta} = [\sum_{i=1}^{n}(X_i - \bar{X})^2 / \sum_{i=1}^{n} X_i^2]\hat{\beta}_3 + [n\bar{X}^2 / \sum_{i=1}^{n} X_i^2]\hat{\beta}_2$

$= (1/\sum_{i=1}^{n} X_i^2)[\sum_{i=1}^{n}(X_i - \bar{X})Y_i + n\bar{X}\bar{Y}] = \sum_{i=1}^{n} X_i Y_i / \sum_{i=1}^{n} X_i^2 = \hat{\beta}_1$.

See also the solution by Trenkler (1996).

3.12 *Efficiency as Correlation.* This is based on Zheng (1994). Since $\hat{\beta}$ and $\tilde{\beta}$ are linear unbiased estimators of β, it follows that $\hat{\beta} + \lambda(\hat{\beta} - \tilde{\beta})$ for any λ is a linear unbiased estimator of β. Since $\hat{\beta}$ is the BLU estimator of β,

$$\text{var}[\hat{\beta} + \lambda(\hat{\beta} - \tilde{\beta})] = E[\hat{\beta} + \lambda(\hat{\beta} - \tilde{\beta})]^2 - \beta^2$$

is minimized at $\lambda = 0$. Setting the derivative of $\text{var}[\hat{\beta} + \lambda(\hat{\beta} - \tilde{\beta})]$ with respect to λ at $\lambda = 0$, we have $2E[\hat{\beta}(\hat{\beta} - \tilde{\beta})] = 0$, or $E(\hat{\beta}^2) = E(\hat{\beta}\tilde{\beta})$. Thus, the squared correlation between $\hat{\beta}$ and $\tilde{\beta}$ is

$$\frac{[\text{cov}(\hat{\beta},\tilde{\beta})]^2}{\text{var}(\hat{\beta})\text{var}(\tilde{\beta})} = \frac{[E[(\hat{\beta}-\beta)(\tilde{\beta}-\beta)]]^2}{\text{var}(\hat{\beta})\text{var}(\tilde{\beta})} = \frac{[E(\hat{\beta}\tilde{\beta}) - \beta^2]^2}{\text{var}(\hat{\beta})\text{var}(\tilde{\beta})}$$

$$= \frac{[E(\hat{\beta}^2) - \beta^2]^2}{\text{var}(\hat{\beta})\text{var}(\tilde{\beta})} = \frac{[\text{var}(\hat{\beta})]^2}{\text{var}(\hat{\beta})\text{var}(\tilde{\beta})} = \frac{\text{var}(\hat{\beta})}{\text{var}(\tilde{\beta})},$$

where the third equality uses the result that $E(\hat{\beta}^2) = E(\hat{\beta}\tilde{\beta})$. The final equality gives $\text{var}(\hat{\beta})/\text{var}(\tilde{\beta})$ which is the relative efficiency of $\hat{\beta}$ and $\tilde{\beta}$.

3.13 a. Adding 5 to each observation of X_i, adds 5 to the sample average $\bar{X}$ and it is now 12.5. This means that $x_i = X_i - \bar{X}$ is unaffected. Hence $\sum_{i=1}^{n} x_i^2$ is the same and since Y_i is unchanged, we conclude that $\hat{\beta}_{ols}$ is still the same at 0.8095. However, $\hat{\alpha}_{ols} = \bar{Y} - \hat{\beta}_{ols}\bar{X}$ is changed because $\bar{X}$ is changed. This is now $\hat{\alpha}_{ols} = 6.5 - (0.8095)(12.5) = -3.6188$. It has decreased by $5\hat{\beta}_{ols}$ since $\bar{X}$ increased by 5 while $\hat{\beta}_{ols}$ and $\bar{Y}$ remained unchanged. It is easy to see that $\hat{Y}_i = \hat{\alpha}_{ols} + \hat{\beta}_{ols}X_i$ remains the same. When X_i increased by 5, with $\hat{\beta}_{ols}$ the same, this increases $\hat{Y}_i$ by $5\hat{\beta}_{ols}$. But $\hat{\alpha}_{ols}$ decreases $\hat{Y}_i$ by $-5\hat{\beta}_{ols}$. The net effect on $\hat{Y}_i$ is zero. Since Y_i is unchanged, this means $e_i = Y_i - \hat{Y}_i$ is unchanged. Hence $s^2 = \sum_{i=1}^{n} e_i^2/n-2$ is unchanged. Since x_i is unchanged $s^2/\sum_{i=1}^{n} x_i^2$ is unchanged and $\text{se}(\hat{\beta}_{ols})$ and the t-statistic for $H_o^a; \beta = 0$ are unchanged. The

$$\text{var}(\hat{\alpha}_{ols}) = s^2 \left[\frac{1}{n} + \frac{\overline{X}^2}{\sum_{i=1}^{n} x_i^2} \right] = 0.311905 \left[\frac{1}{10} + \frac{(12.5)^2}{52.5} \right] = 0.95697$$

with its square root given by $\hat{se}(\hat{\alpha}_{ols}) = 0.97825$. Hence the t-statistic for H_0^b ; $\alpha = 0$ is $-3.6189/0.97825 = -3.699$. This is significantly different from zero. $R^2 = 1 - (\sum_{i=1}^{n} e_i^2 / \sum_{i=1}^{n} y_i^2)$ is unchanged. Hence, only $\hat{\alpha}_{ols}$ and its standard error and t-statistic are affected by an additive constant of 5 on X_i.

b. Adding 2 to each observation of Y_i, adds 2 to the sample average $\overline{Y}$ and it is now 8.5. This means that $y_i = Y_i - \overline{Y}$ is unaffected. Hence $\hat{\beta}_{ols}$ is still the same at 0.8095. However, $\hat{\alpha}_{ols} = \overline{Y} - \hat{\beta}_{ols} \overline{X}$ is changed because $\overline{Y}$ is changed. This is now $\hat{\alpha}_{ols} = 8.5 - (0.8095)(7.5) = 2.4286$ two more than the old $\hat{\alpha}_{ols}$ given in the numerical example. It is easy to see that $\hat{Y}_i = \hat{\alpha}_{ols} + \hat{\beta}_{ols} X_i$ has increased by 2 for each i = 1,2,..,10 since $\hat{\alpha}_{ols}$ increased by 2 while $\hat{\beta}_{ols}$ and X_i are the same. Both Y_i and $\hat{Y}_i$ increased by 2 for each observation. Hence, $e_i = Y_i - \hat{Y}_i$ is unchanged. Also, $s^2 = \sum_{i=1}^{n} e_i^2 / n-2$ and $s^2 / \sum_{i=1}^{n} x_i^2$ are unchanged. This means that $se(\hat{\beta}_{ols})$ and the t-statistic for H_0^a; $\beta = 0$ are unchanged. The

$$\text{var}(\hat{\alpha}_{ols}) = s^2 \left[\frac{1}{n} + \frac{\overline{X}^2}{\sum_{i=1}^{n} x_i^2} \right]$$

is unchanged and therefore $se(\hat{\alpha}_{ols})$ is unchanged. However, the t-statistic for H_0^b; $\alpha = 0$ is now $2.4286/0.60446 = 4.018$ which is now statistically significant. $R^2 = 1 - (\sum_{i=1}^{n} e_i^2 / \sum_{i=1}^{n} y_i^2)$ is unchanged. Again, only $\hat{\alpha}_{ols}$ is affected by an additive constant of 2 on Y_i.

c. If each X_i is multiplied by 2, then the old $\bar{X}$ is multiplied by 2 and it is now 15. This means that each $x_i = X_i - \bar{X}$ is now double what it was in Table 3.1. Since y_i is the same, this means that $\sum_{i=1}^{n} x_i y_i$ is double what it was in Table 3.1 and $\sum_{i=1}^{n} x_i^2$ is four times what it was in Table 3.1. Hence,

$$\hat{\beta}_{ols} = \sum_{i=1}^{n} x_i y_i / \sum_{i=1}^{n} x_i^2 = 2(42.5)/4(52.5) = 0.8095/2 = 0.40475.$$

In this case, $\hat{\beta}_{ols}$ is half what it was in the numerical example. $\hat{\alpha}_{ols} = \bar{Y} - \hat{\beta}_{ols}\bar{X}$ is the same since $\hat{\beta}_{ols}$ is half what it used to be while $\bar{X}$ is double what it used to be. Also, $\hat{Y}_i = \hat{\alpha}_{ols} + \hat{\beta}_{ols} X_i$ is the same, since X_i is doubled while $\hat{\beta}_{ols}$ is half what it used to be and $\hat{\alpha}_{ols}$ is unchanged. Therefore, $e_i = Y_i - \hat{Y}_i$ is unchanged and $s^2 = \sum_{i=1}^{n} e_i^2 / n-2$ is also unchanged. Now, $s^2 / \sum_{i=1}^{n} x_i^2$ is one fourth what it used to be and $se(\hat{\beta}_{ols})$ is half what it used to be. Since, $\hat{\beta}_{ols}$ and $se(\hat{\beta}_{ols})$ have been both reduced by half, the t-statistic for $H_o^a; \beta = 0$ remains unchanged. The

$$var(\hat{\alpha}) = s^2 \left[\frac{1}{n} + \frac{\bar{X}^2}{\sum_{i=1}^{n} x_i^2} \right]$$

is unchanged since $\bar{X}^2$ and $\sum_{i=1}^{n} x_i^2$ are now both multiplied by 4. Hence, $se(\hat{\alpha}_{ols})$ and the t-statistic for $H_o^b; \alpha = 0$ are unchanged. $R^2 = 1 - \left(\sum_{i=1}^{n} e_i^2 / \sum_{i=1}^{n} y_i^2 \right)$ is also unchanged. Hence, only $\hat{\beta}$ is affected by a multiplicative constant of 2 on X_i.

3.14 a. Dependent Variable: LNC

Analysis of Variance

Source	DF	Sum of Squares	Mean Square	F Value	Prob > F
Model	1	0.04693	0.04693	1.288	0.2625
Error	44	1.60260	0.03642		
C Total	45	1.64953			

Root MSE	0.19085	R-square	0.0285	
Dep Mean	4.84784	Adj R-sq	0.0064	
C.V.	3.93675			

Parameter Estimates

Variable	DF	Parameter Estimate	Standard Error	T for H0: Parameter=0	Prob > |T|
INTERCEP	1	5.931889	0.95542530	6.209	0.0001
LNY	1	-0.227003	0.19998321	-1.135	0.2625

The income elasticity is -0.227 which is negative! Its standard error is (0.1999) and the t-statistic for testing this income elasticity is zero is -1.135 which is insignificant with a p-value of 0.26. Hence, we cannot reject the null hypothesis. $R^2 = 0.0285$ and $s = 0.19085$. This regression is not very useful. The income variable is not significant and the R^2 indicates that the regression explains only 2.8% of the variation in consumption.

b. Plot of Residuals, and the 95% confidence interval for the predicted value.

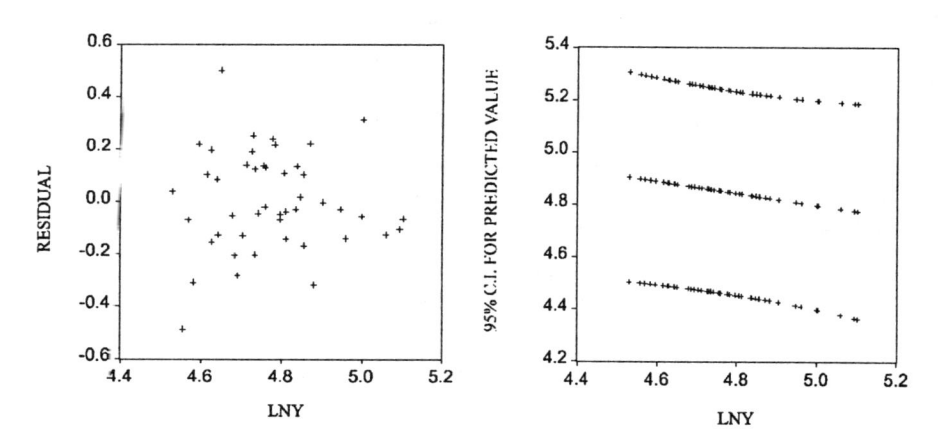

SAS Program for 3.14

```
Data CIGARETT;
  Input OBS STATE $ LNC LNP LNY;
  Cards;
  Proc reg data=cigarett;
      model lnc=lny;
      *plot residual.*lny='*';
      *plot (U95. L95.)*lny='-' p.*lny /overlay
       symbol='*';
  output out=out1 r=resid p=pred u95=upper95 l95=lower95;
  proc plot data=out1 vpercent=75 hpercent=100;
      plot resid*lny='*';
  proc plot data=out1 vpercent=95 hpercent=100;
      plot (Upper95 Lower95)*lny='-' Pred*lny='*'
/overlay;
  run;
```

3.15 Theil's *minimum mean square* estimator of β. $\tilde{\beta}$ can be written as:

a. $\tilde{\beta} = \sum_{i=1}^{n} X_i Y_i / [\sum_{i=1}^{n} X_i^2 + (\sigma^2/\beta^2)]$

Substituting $Y_i = \beta X_i + u_i$ we get

$$\tilde{\beta} = \frac{\beta \sum_{i=1}^{n} X_i^2 + \sum_{i=1}^{n} X_i u_i}{\sum_{i=1}^{n} X_i^2 + (\sigma^2/\beta^2)}$$

and since $E(X_i u_i) = 0$, we get

$$E(\tilde{\beta}) = \frac{\beta \sum_{i=1}^{n} X_i^2}{\sum_{i=1}^{n} X_i^2 + (\sigma^2/\beta^2)} = \beta\left(\frac{1}{1+c}\right)$$

where $c = \sigma^2/\beta^2 \sum_{i=1}^{n} X_i^2 > 0$.

b. Therefore, **Bias** $(\tilde{\beta}) = E(\tilde{\beta}) - \beta = \beta\left(\frac{1}{1+c}\right) - \beta = -[c/(1+c)]\beta$. This bias is positive (negative) when β is negative (positive). This also means that $\tilde{\beta}$ is biased towards zero.

c. **Bias**$^2(\tilde{\beta}) = [c^2/(1+c)^2]\beta^2$ and

$$\mathrm{var}(\tilde{\beta}) = E(\tilde{\beta} - E(\tilde{\beta}))^2 = E\left(\frac{\sum_{i=1}^{n} X_i u_i}{\sum_{i=1}^{n} X_i^2 + (\sigma^2/\beta^2)}\right)^2 = \frac{\sigma^2 \sum_{i=1}^{n} X_i^2}{\left[\sum_{i=1}^{n} X_i^2 + (\sigma^2/\beta^2)\right]^2}$$

using $\mathrm{var}(u_i) = \sigma^2$ and $\mathrm{cov}(u_i, u_j) = 0$ for $i \neq j$. This can also be written as

$$\mathrm{var}(\tilde{\beta}) = \frac{\sigma^2}{\sum_{i=1}^{n} X_i^2 (1+c)^2}. \quad \text{Therefore,}$$

$\mathrm{MSE}(\tilde{\beta}) = \mathrm{Bias}^2(\tilde{\beta}) + \mathrm{var}(\tilde{\beta})$

$$\mathrm{MSE}(\tilde{\beta}) = \frac{c^2}{(1+c)^2}\beta^2 + \frac{\sigma^2}{\sum_{i=1}^{n} X_i^2 (1+c)^2} = \frac{\beta^2 c^2 \sum_{i=1}^{n} X_i^2 + \sigma^2}{\sum_{i=1}^{n} X_i^2 (1+c)^2}$$

But $\beta^2 \sum_{i=1}^{n} X_i^2 c = \sigma^2$ from the definition of c. Hence,

$$\mathrm{MSE}(\tilde{\beta}) = \frac{\sigma^2(1+c)}{\sum_{i=1}^{n} X_i^2 (1+c)^2} = \frac{\sigma^2}{\sum_{i=1}^{n} X_i^2 (1+c)} = \frac{\sigma^2}{\sum_{i=1}^{n} X_i^2 + (\sigma^2/\beta^2)}$$

The **Bias** $(\hat{\beta}_{ols}) = 0$ and $\mathrm{var}(\hat{\beta}_{ols}) = \sigma^2 / \sum_{i=1}^{n} X_i^2$. Hence

$$\mathrm{MSE}(\hat{\beta}_{ols}) = \mathrm{var}(\hat{\beta}_{ols}) = \sigma^2 / \sum_{i=1}^{n} X_i^2.$$

This is larger than $MSE(\tilde{\beta})$ since the latter has a positive constant (σ^2/β^2) in the denominator.

3.16 a. Dependent Variable: LNEN

Analysis of Variance

Source	DF	Sum of Squares	Mean Square	F Value	Prob > F
Model	1	30.82384	30.82384	13.462	0.0018
Error	18	41.21502	2.28972		
C Total	19	72.03886			

Root MSE		1.51318	R-square	0.4279
Dep Mean		9.87616	Adj R-sq	0.3961
C.V.		15.32158		

Parameter Estimates

Variable	DF	Parameter Estimate	Standard Error	T for H0: Parameter=0	Prob > |T|
INTERCEP	1	1.988607	2.17622656	0.914	0.3729
LNRGDP	1	0.743950	0.20276460	3.669	0.0018

b. Plot of Residual *LNRGDP

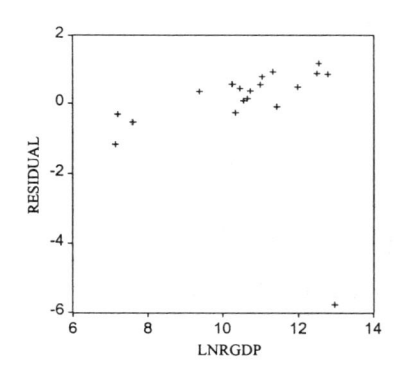

As clear from this plot, the W. Germany observation has a large residual.

c. For $H_o; \beta = 1$ we get $t = (\hat{\beta} - 1)/s.e.(\hat{\beta}) = (0.744 - 1)/0.203 = 1.26$. We do not reject H_o.

e. Dependent Variable: LNEN

Analysis of Variance

Source	DF	Sum of Squares	Mean Square	F Value	Prob > F
Model	1	59.94798	59.94798	535.903	0.0001
Error	18	2.01354	0.11186		
C Total	19	61.96153			

Root MSE		0.33446	R-square	0.9675
Dep Mean		10.22155	Adj R-sq	0.9657
C.V.		3.27211		

Parameter Estimates

Variable	DF	Parameter Estimate	Standard Error	T for H0: Parameter=0	Prob > |T|
INTERCEP	1	-0.778287	0.48101294	-1.618	0.1230
LNRGDP	1	1.037499	0.04481721	23.150	0.0001

Plot of Residual*LNRGDP

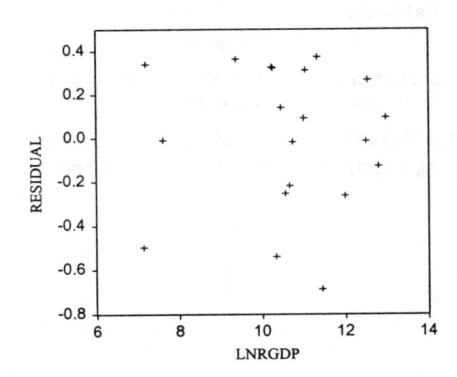

SAS PROGRAM

```
Data Rawdata;
  Input Country $ RGDP EN;
  Cards;
  Data Energy; Set Rawdata;
  LNRGDP=log(RGDP);   LNEN=log(EN);
  Proc reg data=energy;
```

```
         Model LNEN=LNRGDP;
         Output out=OUT1 R=RESID;
       Proc Plot data=OUT1 hpercent=85 vpercent=60;
         Plot RESID*LNRGDP='*';
       run;
```

3.17 b. Dependent Variable: LNRGDP

Analysis of Variance

Source	DF	Sum of Squares	Mean Square	F Value	Prob > F
Model	1	53.88294	53.88294	535.903	0.0001
Error	18	1.80983	0.10055		
C Total	19	55.69277			

Root MSE	0.31709	R-square	0.9675
Dep Mean	10.60225	Adj R-sq	0.9657
C.V.	2.99078		

Parameter Estimates

Variable	DF	Parameter Estimate	Standard Error	T for H0: Parameter=0	Prob > |T|
INTERCEP	1	1.070317	0.41781436	2.562	0.0196
LNEN	1	0.932534	0.04028297	23.150	0.0001

e. Log-log specification

Dependent Variable: LNEN1

Analysis of Variance

Source	DF	Sum of Squares	Mean Square	F Value	Prob > F
Model	1	59.94798	59.94798	535.903	0.0001
Error	18	2.01354	0.11186		
C Total	19	61.96153			

Chapter 3: Simple Linear Regression

```
Root MSE      0.33446      R-square    0.9675
Dep Mean     14.31589      Adj R-sq    0.9657
C.V.          2.33628
```

Parameter Estimates

Variable	DF	Parameter Estimate	Standard Error	T for H0: Parameter=0	Prob > |T|
INTERCEP	1	3.316057	0.48101294	6.894	0.0001
LNRGDP	1	1.037499	0.04481721	23.150	0.0001

Linear Specificiation
Dependent Variable: EN1

Analysis of Variance

Source	DF	Sum of Squares	Mean Square	F Value	Prob > F
Model	1	6.7345506E14	6.7345506E14	386.286	0.0001
Error	18	3.1381407E13	1.7434115E12		
C Total	19	7.0483646E14			

```
Root MSE     1320383.09457   R-square    0.9555
Dep Mean     4607256.0000    Adj R-sq    0.9530
C.V.           28.65877
```

Parameter Estimates

Variable	DF	Parameter Estimate	Standard Error	T for H0: Parameter=0	Prob > |T|
INTERCEP	1	-190151	383081.08995	-0.496	0.6256
RGDP	1	46.759427	2.37911166	19.654	0.0001

Linear Specification before the multiplication by 60
Dependent Variable: EN

Analysis of Variance

Source	DF	Sum of Squares	Mean Square	F Value	Prob > F
Model	1	187070848717	187070848717	386.286	0.0001
Error	18	8717057582.2	484280976.79		
C Total	19	195787906299			

```
Root MSE      22006.38491    R-square    0.9555
Dep Mean      76787.60000    Adj R-sq    0.9530
C.V.            28.6587
```

Parameter Estimates

Variable	DF	Parameter Estimate	Standard Error	T for H0: Parameter=0	Prob > |T|
INTERCEP	1	-3169.188324	6384.6848326	-0.496	0.6256
RGDP	1	0.779324	0.03965186	19.654	0.0001

What happens when we multiply our energy variable by 60? For the linear model specification, both $\hat{\alpha}^*$ and $\hat{\beta}^*$ are multiplied by 60, their standard errors are also multiplied by 60 and their t-statistics are the same.

For the log-log model specification, $\hat{\beta}$ is the same, but $\hat{\alpha}$ is equal to the old $\hat{\alpha}+\log 60$. The intercept therefore is affected but not the slope. Its standard error is the same, but its t-statistic is changed.

g. Plot of residuals for both linear and log-log models

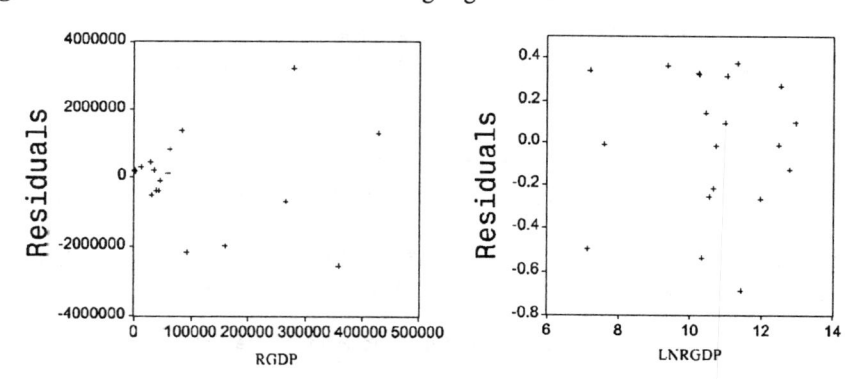

SAS PROGRAM

```
Data Rawdata;
  Input Country $ RGDP EN;
  Cards;
Data Energy; Set Rawdata;
LNRGDP=log(RGDP);   LNEN=log(EN);
EN1=EN*60;  LNEN1=log(EN1);
Proc reg data=energy; Model LNRGDP=LNEN;
Proc reg data=energy; Model LNEN1=LNRGDP/CLM, CLI;
    Output out=OUT1 R=LN_RESID;
Proc reg data=energy; Model EN1=RGDP;
```

```
          Output out=OUT2 R=RESID;
     Proc reg data=energy; Model EN=RGDP;
     data Resid; set out1(keep=lnrgdp ln_resid);
          set out2(keep=rgdp resid);
     Proc plot data=resid vpercent=60 hpercent=85;
          Plot ln_resid*lnrgdp='*';
          Plot resid*rgdp='*';
     run;
```

3.18 For parts (b) and (c), SAS will automatically compute confidence intervals for the mean (CLM option) and for a specific observation (CLI option), see the SAS program in 3.17.

95% CONFIDENCE PREDICTION INTERVAL

COUNTRY	Dep Var LNEN1	Predict Value	Std Err Predict	Lower95% Mean	Upper95% Mean	Lower95% Predict	Upper95% Predict
AUSTRIA	14.4242	14.4426	0.075	14.2851	14.6001	13.7225	15.1627
BELGIUM	15.0778	14.7656	0.077	14.6032	14.9279	14.0444	15.4868
CYPRUS	11.1935	11.2035	0.154	10.8803	11.5268	10.4301	11.9770
DENMARK	14.2997	14.1578	0.075	14.0000	14.3156	13.4376	14.8780
FINLAND	14.2757	13.9543	0.076	13.7938	14.1148	13.2335	14.6751
FRANCE	16.4570	16.5859	0.123	16.3268	16.8449	15.8369	17.3348
GREECE	14.0038	14.2579	0.075	14.1007	14.4151	13.5379	14.9780
ICELAND	11.1190	10.7795	0.170	10.4222	11.1368	9.9912	11.5678
IRELAND	13.4048	13.0424	0.093	12.8474	13.2375	12.3132	13.7717
ITALY	16.2620	16.2752	0.113	16.0379	16.5125	15.5335	17.0168
MALTA	10.2168	10.7152	0.173	10.3526	11.0778	9.9245	11.5059
NETHERLAND	15.4379	15.0649	0.081	14.8937	15.2361	14.3417	15.7882
NORWAY	14.2635	13.9368	0.077	13.7760	14.0977	13.2160	14.6577
PORTUGAL	13.4937	14.0336	0.076	13.8744	14.1928	13.3131	14.7540
SPAIN	15.4811	15.7458	0.097	15.5420	15.9495	15.0141	16.4774
SWEDEN	14.8117	14.7194	0.077	14.5581	14.8808	13.9985	15.4404
SWZERLAND	14.1477	14.3665	0.075	14.2094	14.5237	13.6465	15.0866
TURKEY	14.4870	15.1736	0.083	14.9982	15.3489	14.4494	15.8978
UK	16.5933	16.3259	0.115	16.0852	16.5667	15.5832	17.0687
W.GERMANY	16.8677	16.7713	0.130	16.4987	17.0440	16.0176	17.5251

References

Trenkler, G. (1996), "Optimal Weighting of Unbiased Estimators," *Econometric Theory*, Solution 95.3.1, 12: 585.

Zheng, J. X. (1994), "Efficiency as Correlation," *Econometric Theory*, Solution 93.1.3, 10:228.

CHAPTER 4
Multiple Regression Analysis

4.1 The regressions for parts (a), (b), (c), (d) and (e) are given below.

a. Regression of LNC on LNP and LNY

Dependent Variable: LNC

Analysis of Variance

Source	DF	Sum of Squares	Mean Square	F Value	Prob>F
Model	2	0.50098	0.25049	9.378	0.0004
Error	43	1.14854	0.02671		
C Total	45	1.64953			

Root MSE	0.16343	R-square	0.3037	
Dep Mean	4.84784	Adj R-sq	0.2713	
C.V.	3.37125			

Parameter Estimates

Variable	DF	Parameter Estimate	Standard Error	T for H0: Parameter=0	Prob>|T|
INTERCEP	1	4.299662	0.90892571	4.730	0.0001
LNP	1	-1.338335	0.32460147	-4.123	0.0002
LNY	1	0.172386	0.19675440	0.876	0.3858

b. Regression of LNC on LNP

Dependent Variable: LNC

Analysis of Variance

Source	DF	Sum of Squares	Mean Square	F Value	Prob>F
Model	1	0.48048	0.48048	18.084	0.0001
Error	44	1.16905	0.02657		
C Total	45	1.64953			

Root MSE	0.16300	R-square	0.2913	
Dep Mean	4.84784	Adj R-sq	0.2752	
C.V.	3.36234			

Parameter Estimates

Variable	DF	Parameter Estimate	Standard Error	T for H0: Parameter=0	Prob > |T|
INTERCEP	1	5.094108	0.06269897	81.247	0.0001
LNP	1	-1.198316	0.28178857	-4.253	0.0001

c. Regression of LNY on LNP

Dependent Variable: LNY

Analysis of Variance

Source	DF	Sum of Squares	Mean Square	F Value	Prob>F
Model	1	0.22075	0.22075	14.077	0.0005
Error	44	0.68997	0.01568		
C Total	45	0.91072			

Root MSE	0.12522	R-square	0.2424	
Dep Mean	4.77546	Adj R-sq	0.2252	
C.V.	2.62225			

Parameter Estimates

Variable	DF	Parameter Estimate	Standard Error	T for H0: Parameter=0	Prob> |T|
INTERCEP	1	4.608533	0.04816809	95.676	0.0001
LNP	1	0.812239	0.21648230	3.752	0.0005

d. Regression of LNC on the residuals of part (c).

Dependent Variable: LNC

Analysis of Variance

Source	DF	Sum of Squares	Mean Square	F Value	Prob>F
Model	1	0.02050	0.02050	0.554	0.4607
Error	44	1.62903	0.03702		
C Total	45	1.64953			

Root MSE	0.19241	R-square	0.0124	
Dep Mean	4.84784	Adj R-sq	-0.0100	
C.V.	3.96907			

Parameter Estimates

Variable	DF	Parameter Estimate	Standard Error	T for H0: Parameter=0	Prob > |T|
INTERCEP	1	4.847844	0.02836996	170.879	0.0001
RESID_C	1	0.172386	0.23164467	0.744	0.4607

e. Regression of Residuals from part (b) on those from part (c).

Dependent Variable: RESID_B

Analysis of Variance

Source	DF	Sum of Squares	Mean Square	F Value	Prob>F
Model	1	0.02050	0.02050	0.785	0.3803
Error	44	1.14854	0.02610		
C Total	45	1.16905			

Root MSE	0.16157	R-square	0.0175	
Dep Mean	-0.00000	Adj R-sq	-0.0048	
C.V.	-2.463347E16			

Parameter Estimates

Variable	DF	Parameter Estimate	Standard Error	T for H0: Parameter=0	Prob> |T|
INTERCEP	1	-6.84415E-16	0.02382148	-0.000	1.0000
RESID_C	1	0.172386	0.19450570	0.886	0.3803

f. The multiple regression coefficient estimate of real income in part (a) is equal to the slope coefficient of the regressions in parts (d) and (e). This demonstrates the residualing out interpretation of multiple regression coefficients.

4.2 *Simple Versus Multiple Regression Coefficients.* This is based on Baltagi (1987). The OLS residuals from $Y_i = \gamma + \delta_2 \hat{v}_{2i} + \delta_3 \hat{v}_{3i} + w_i$, say $\hat{w}_i$, satisfy the following conditions:

$$\sum_{i=1}^{n} \hat{w}_i = 0 \quad \sum_{i=1}^{n} \hat{w}_i \hat{v}_{2i} = 0 \quad \sum_{i=1}^{n} \hat{w}_i \hat{v}_{3i} = 0$$

with $Y_i = \hat{\gamma} + \hat{\delta}_2 \hat{v}_{2i} + \hat{\delta}_3 \hat{v}_{3i} + \hat{w}_i$.

Chapter 4: Multiple Regression Analysis

Multiply this last equation by $\hat{v}_{2i}$ and sum, we get $\sum_{i=1}^{n} Y_i \hat{v}_{2i} = \hat{\delta}_2 \sum_{i=1}^{n} \hat{v}_{2i}^2 + \hat{\delta}_3 \sum_{i=1}^{n} \hat{v}_{2i} \hat{v}_{3i}$

since $\sum_{i=1}^{n} \hat{v}_{2i} = 0$ and $\sum_{i=1}^{n} \hat{v}_{2i} \hat{w}_i = 0$. Substituting $\hat{v}_{3i} = X_{3i} - \hat{c} - \hat{d} X_{2i}$ and using the fact that $\sum_{i=1}^{n} \hat{v}_{2i} X_{3i} = 0$, we get $\sum_{i=1}^{n} Y_i \hat{v}_{2i} = \hat{\delta}_2 \sum_{i=1}^{n} \hat{v}_{2i}^2 - \hat{\delta}_3 \hat{d} \sum_{i=1}^{n} X_{2i} \hat{v}_{2i}$.

From $X_{2i} = \hat{a} + \hat{b} X_{3i} + \hat{v}_{2i}$, we get that $\sum_{i=1}^{n} X_{2i} \hat{v}_{2i} = \sum_{i=1}^{n} \hat{v}_{2i}^2$. Hence,

$\sum_{i=1}^{n} Y_i \hat{v}_{2i} = (\hat{\delta}_2 - \hat{\delta}_3 \hat{d}) \sum_{i=1}^{n} \hat{v}_{2i}^2$ and $\hat{\beta}_2 = \sum_{i=1}^{n} Y_i \hat{v}_{2i} / \sum_{i=1}^{n} \hat{v}_{2i}^2 = \hat{\delta}_2 - \hat{\delta}_3 \hat{d}$.

Similarly, one can show that $\hat{\beta}_3 = \hat{\delta}_3 - \hat{\delta}_2 \hat{b}$. Solving for $\hat{\delta}_2$ and $\hat{\delta}_3$ we get

$$\hat{\delta}_2 = (\hat{\beta}_2 + \hat{\beta}_3 \hat{d})/(1 - \hat{b}\hat{d}) \quad \text{and} \quad \hat{\delta}_3 = (\hat{\beta}_3 + \hat{b}\hat{\beta}_2)/(1 - \hat{b}\hat{d}).$$

4.3 a. Regressing X_i on a constant we get $\hat{X}_i = \bar{X}$ and $\hat{v}_i = X_i - \hat{X}_i = X_i - \bar{X} = x_i$.

Regressing Y_i on $\hat{v}_i$ we get $\hat{\beta}_{ols} = \sum_{i=1}^{n} Y_i \hat{v}_i / \sum_{i=1}^{n} \hat{v}_i^2 = \sum_{i=1}^{n} Y_i x_i / \sum_{i=1}^{n} x_i^2 = \sum_{i=1}^{n} x_i y_i / \sum_{i=1}^{n} x_i^2$.

b. Regressing a constant 1 on X_i we get

$$b = \sum_{i=1}^{n} X_i / \sum_{i=1}^{n} X_i^2 \quad \text{with residuals} \quad \hat{w}_i = 1 - (n\bar{X}/\sum_{i=1}^{n} X_i^2) X_i$$

so that, regressing Y_i on $\hat{w}_i$ yields $\hat{\alpha} = \sum_{i=1}^{n} \hat{w}_i Y_i / \sum_{i=1}^{n} \hat{w}_i^2$.

But $\sum_{i=1}^{n} \hat{w}_i Y_i = n\bar{Y} - n\bar{X} \sum_{i=1}^{n} X_i Y_i / \sum_{i=1}^{n} X_i^2 = \dfrac{n\bar{Y} \sum_{i=1}^{n} X_i^2 - n\bar{X} \sum_{i=1}^{n} X_i Y_i}{\sum_{i=1}^{n} X_i^2}$

and $\sum_{i=1}^{n} \hat{w}_i^2 = n + \dfrac{n^2 \bar{X}^2}{\sum_{i=1}^{n} X_i^2} - \dfrac{2n\bar{X} \sum_{i=1}^{n} X_i}{\sum_{i=1}^{n} X_i^2} = \dfrac{n \sum_{i=1}^{n} X_i^2 - n^2 \bar{X}^2}{\sum_{i=1}^{n} X_i^2} = \dfrac{n \sum_{i=1}^{n} x_i^2}{\sum_{i=1}^{n} X_i^2}$.

Therefore, $\hat{\alpha} = \dfrac{\bar{Y}\sum_{i=1}^{n}X_i^2 - \bar{X}\sum_{i=1}^{n}X_iY_i}{\sum_{i=1}^{n}x_i^2} = \dfrac{\bar{Y}\sum_{i=1}^{n}x_i^2 + n\bar{X}^2\bar{Y} - \bar{X}\sum_{i=1}^{n}X_iY_i}{\sum_{i=1}^{n}x_i^2}$

$= \bar{Y} - \dfrac{\bar{X}\left(\sum_{i=1}^{n}X_iY_i - n\bar{X}\bar{Y}\right)}{\sum_{i=1}^{n}x_i^2} = \bar{Y} - \hat{\beta}_{ols}\bar{X}$

where $\hat{\beta}_{ols} = \sum_{i=1}^{n}x_iy_i / \sum_{i=1}^{n}x_i^2$.

c. From part (a), $\text{var}(\hat{\beta}_{ols}) = \sigma^2 / \sum_{i=1}^{n}\hat{v}_i^2 = \sigma^2 / \sum_{i=1}^{n}x_i^2$ as it should be. From part (b),

$\text{var}(\hat{\alpha}_{ols}) = \sigma^2 / \sum_{i=1}^{n}\hat{w}_i^2 = \dfrac{\sigma^2 \sum_{i=1}^{n}X_i^2}{n\sum_{i=1}^{n}x_i^2}$ as it should be.

4.4 *Effect of Additional Regressors on R^2*

a. Least Squares on the $K = K_1 + K_2$ regressors minimizes the sum of squared error and yields $SSE_2 = \min \sum_{i=1}^{n}(Y_i - \alpha - \beta_2 X_{2i} - \ldots - \beta_{K_1}X_{K_1i} - \ldots - \beta_K X_{Ki})^2$ Let us denote the corresponding estimates by $(a, b_2, \ldots, b_{K_1}, \ldots, b_K)$. This implies that $SSE_2^* = \sum_{i=1}^{n}(Y_i - \alpha^* - \beta_2^* X_{2i} - \ldots - \beta_{K_1}^* X_{K_1i} - \ldots - \beta_K^* X_{Ki})^2$ based on arbitrary $(\alpha^*, \beta_2^*, \ldots, \beta_{K_1}^*, \ldots, \beta_K^*)$ satisfies $SSE_2^* \geq SSE_2$. In particular, substituting the least squares estimates using only K_1 regressors say $\hat{\alpha}, \hat{\beta}_2, \ldots, \hat{\beta}_{K_1}$ and $\hat{\beta}_{K_1+1} = 0, \ldots, \hat{\beta}_K = 0$ satisfy the above inequality. Hence, $SSE_1 \geq SSE_2$. Since $\sum_{i=1}^{n}y_i^2$ is fixed, this means that $R_2^2 \geq R_1^2$. This is based on the solution by Rao and White (1988).

b. From the definition of $\bar{R}^2$, we get $(1-\bar{R}^2) = [\sum_{i=1}^{n}e_i^2/(n-K)]/[\sum_{i=1}^{n}y_i^2/(n-1)]$. But $R^2 = 1 - (\sum_{i=1}^{n}e_i^2 / \sum_{i=1}^{n}y_i^2)$, hence $(1-\bar{R}^2) = \dfrac{\sum e_i^2}{\sum y_i^2} \cdot \dfrac{n-1}{n-k} = (1-R^2) \cdot \dfrac{n-1}{n-k}$ as required in (4.16).

4.5 This regression suffers from perfect multicollinearity. $X_2 + X_3$ is perfectly collinear with X_2 and X_3. Collecting terms in X_2 and X_3 we get

Chapter 4: Multiple Regression Analysis   45

$Y_i = \alpha + (\beta_2+\beta_4)X_{2i} + (\beta_3+\beta_4)X_{3i} + \beta_5 X_{2i}^2 + \beta_6 X_{3i}^2 + u_i$ so $(\beta_2+\beta_4)$, $(\beta_3+\beta_4)$, β_5 and β_6 are estimable by OLS.

4.6 a. If we regress e_t on X_{2t} and X_{3t} and a constant, we get zero regression coefficients and therefore zero predicted values, i.e., $\hat{e}_t = 0$. The residuals are therefore equal to $e_t - \hat{e}_t = e_t$ and the residual sum of squares is equal to the total sum of squares. Therefore, $R^2 = 1 - \dfrac{RSS}{TSS} = 1 - 1 = 0$.

b. For $Y_t = a + b\hat{Y}_t + v_t$, OLS yields $\hat{b} = \sum_{t=1}^{T}\hat{y}_t y_t / \sum_{t=1}^{T}\hat{y}_t^2$ and $\hat{a} = \overline{Y} - \hat{b}\overline{Y}$. Also, $Y_t = \hat{Y}_t + e_t$ which gives $\overline{Y} = \overline{\hat{Y}}$ since $\sum_{t=1}^{T} e_t = 0$. Also, $y_t = \hat{y}_t + e_t$. Therefore, $\sum_{t=1}^{T} y_t \hat{y}_t = \sum_{t=1}^{T} \hat{y}_t^2$ since $\sum_{t=1}^{T} \hat{y}_t e_t = 0$. Hence,

$\hat{b} = \sum_{t=1}^{T}\hat{y}_t^2 / \sum_{t=1}^{T}\hat{y}_t^2 = 1$ and $\hat{a} = \overline{Y} - 1 \cdot \overline{Y} = 0$

$\hat{Y}_t = 0 + 1 \cdot \hat{Y}_t = \hat{Y}_t$ and $Y_t - \hat{Y}_t = e_t$

$\sum_{t=1}^{T} e_t^2$ is the same as the original regression, $\sum_{t=1}^{T} y_t^2$ is still the same, therefore, R^2 is still the same.

c. For $Y_t = a + be_t + v_t$, OLS yields $\hat{b} = \sum_{t=1}^{T} e_t y_t / \sum_{t=1}^{T} e_t^2$ and $\hat{a} = \overline{Y} - \hat{b}\overline{e} = \overline{Y}$ since $\overline{e} = 0$. But $y_t = \hat{y}_t + e_t$, therefore, $\sum_{t=1}^{T} e_t y_t = \sum_{t=1}^{T} e_t \hat{y}_t + \sum_{t=1}^{T} e_t^2 = \sum_{t=1}^{T} e_t^2$ since $\sum_{t=1}^{T} e_t \hat{y}_t = 0$. Hence, $\hat{b} = \sum_{t=1}^{T} e_t^2 / \sum_{t=1}^{T} e_t^2 = 1$.

Also, the predicted value is now $\hat{Y}_t = \hat{a} + \hat{b}e_t = \overline{Y} + e_t$ and the new residual is

$= Y_t - \hat{Y}_t = y_t - e_t = \hat{y}_t$. Hence, the new RSS = old regression sum of squares

$= \sum_{t=1}^{T}\hat{y}_t^2$, and new $R^2 = 1 - \dfrac{\text{new RSS}}{\text{TSS}} = 1 - \dfrac{\text{Old Reg.SS}}{\text{TSS}} = \dfrac{\sum_{t=1}^{T} e_t^2}{\sum_{t=1}^{T} y_t^2} = (1-(\text{old } R^2))$.

4.7 For the Cobb-Douglas production given by equation (4.18) one can test H_o; $\alpha + \beta + \gamma + \delta = 1$ using the following t-statistic $t = \dfrac{(\hat{\alpha}+\hat{\beta}+\hat{\gamma}+\hat{\delta})-1}{\text{s.e.}(\hat{\alpha}+\hat{\beta}+\hat{\gamma}+\hat{\delta})}$ where the estimates are obtained from the unrestricted OLS regression given by (4.18). The $\text{var}(\hat{\alpha}+\hat{\beta}+\hat{\gamma}+\hat{\delta}) = \text{var}(\hat{\alpha}) + \text{var}(\hat{\beta}) + \text{var}(\hat{\gamma}) + \text{var}(\hat{\delta}) + 2\text{cov}(\hat{\alpha},\hat{\beta}) + 2\text{cov}(\hat{\alpha},\hat{\gamma}) + 2\text{cov}(\hat{\alpha},\hat{\delta}) + 2\text{cov}(\hat{\beta},\hat{\gamma}) + 2\text{cov}(\hat{\beta},\hat{\delta}) + 2\text{cov}(\hat{\gamma},\hat{\delta})$. These variance-covariance estimates are obtained from the unrestricted regression. The observed t-statistic is distributed as t_{n-5} under H_o.

4.8 a. The restricted regression for H_o; $\beta_2 = \beta_4 = \beta_6$ is given by

$Y_i = \alpha + \beta_2(X_{2i}+X_{4i}+X_{6i}) + \beta_3 X_{3i} + \beta_5 X_{5i} + \beta_7 X_{7i} + \beta_8 X_{8i} +..+ \beta_K X_{Ki} + u_i$

obtained by substituting $\beta_2 = \beta_4 = \beta_6$ in equation (4.1). The unrestricted regression is given by (4.1) and the F-statistic in (4.17) has two restrictions and is distributed $F_{2,n-K}$ under H_o.

b. The restricted regression for H_o; $\beta_2 = -\beta_3$ and $\beta_5 - \beta_6 = 1$ is given by

$Y_i + X_{6i} = \alpha + \beta_2(X_{2i}-X_{3i}) + \beta_4 X_{4i} + \beta_5(X_{5i}+X_{6i}) + \beta_7 X_{7i} +..+ \beta_K X_{Ki} + u_i$

obtained by substituting both restrictions in (4.1). The unrestricted regression is given by (4.1) and the F-statistic in (4.17) has two restrictions and is distributed $F_{2,n-K}$ under H_o.

4.10 a. For the data underlying Table 4.1, the following computer output gives the mean of log(wage) for females and for males. Out of 595 individuals observed, there were 528 Males and 67 Females. The corresponding means of log(wage) for Males and Females being $\overline{Y}_M = 7.004$ and $\overline{Y}_F = 6.530$, respectively. The regression of log(wage) on FEMALE and MALE without a constant yields coefficient estimates $\hat{\alpha}_F = \overline{Y}_F = 6.530$ and $\hat{\alpha}_M = \overline{Y}_M = 7.004$, as expected.

Chapter 4: Multiple Regression Analysis

```
Dependent Variable: LWAGE

                        Analysis of Variance

                        Sum of          Mean
     Source      DF     Squares         Square        F Value      Prob>F

     Model        2     28759.48792   14379.74396    84576.019     0.0001
     Error      593       100.82277       0.17002
     U Total    595     28860.31068

         Root MSE        0.41234     R-square        0.9965
         Dep Mean        6.95074     Adj R-sq        0.9965
         C.V.            5.93227

                          Parameter Estimates

                   Parameter      Standard       T for H0:
     Variable  DF  Estimate       Error          Parameter=0     Prob > |T|

     FEM       1   6.530366       0.05037494     129.635         0.0001
     M         1   7.004088       0.01794465     390.316         0.0001

     FEM=0
     -------------------------------------------------------------------
     Variable    N         Mean       Std Dev       Minimum       Maximum
     -------------------------------------------------------------------
     LWAGE     528    7.0040880     0.4160069     5.6767500     8.5370000
     -------------------------------------------------------------------

     FEM=1
     -------------------------------------------------------------------
     Variable    N         Mean       Std Dev       Minimum       Maximum
     -------------------------------------------------------------------
     LWAGE      67    6.5303664     0.3817668     5.7493900     7.2793200
     -------------------------------------------------------------------
```

b. Running log(wage) on a constant and the FEMALE dummy variable yields

$\hat{\alpha} = 7.004 = \overline{Y}_M = \hat{\alpha}_M$ and $\hat{\beta} = -0.474 = (\widehat{\alpha_F - \alpha_M})$. But $\hat{\alpha}_M = 7.004$.

Therefore, $\hat{\alpha}_F = \hat{\beta} + \hat{\alpha}_M = 7.004 - 0.474 = 6.530 = \overline{Y}_F = \hat{\alpha}_F$.

```
Dependent Variable: LWAGE

                        Analysis of Variance

                        Sum of          Mean
     Source      DF     Squares         Square        F Value      Prob>F

     Model        1     13.34252      13.34252       78.476        0.0001
     Error      593    100.82277       0.17002
     C Total    594    114.16529
```

```
              Root MSE           0.41234      R-square        0.1169
              Dep Mean           6.95074      Adj R-sq        0.1154
              C.V.               5.93227
```

```
                        Parameter Estimates

                        Parameter    Standard    T for HO:
         Variable  DF   Estimate     Error       Parameter=0   Prob > |T|

         INTERCEP  1     7.004088    0.01794465    390.316      0.0001
         FEM       1    -0.473722    0.05347565     -8.859      0.0001
```

4.12 a. The unrestricted regression is given by (4.28). This regression runs EARN on a constant, FEMALE, EDUCATION and (FEMALE x EDUCATION). The URSS = 76.63525. The restricted regression for equality of slopes and intercepts for Males and Females, tests the restriction $H_o; \alpha_F = \gamma = 0$. This regression runs EARN on a constant and EDUC. The RRSS = 90.36713. The SAS regression output is given below. There are two restrictions and the F-test given by (4.17) yields

$$F = \frac{(90.36713 - 76.63525)/2}{76.63525/591} = 52.94941.$$

This is distributed as $F_{2,591}$ under H_o. The null hypothesis is rejected.

```
Unrestricted Model (with FEM and FEM*EDUC)
Dependent Variable: LWAGE

                        Analysis of Variance

                        Sum of       Mean
         Source    DF   Squares      Square      F Value    Prob>F

         Model      3   37.53004     12.51001    96.475     0.0001
         Error    591   76.63525      0.12967
         C Total  594  114.16529


              Root MSE           0.36010      R-square        0.3287
              Dep Mean           6.95074      Adj R-sq        0.3253
              C.V.               5.18071
```

Parameter Estimates

Variable	DF	Parameter Estimate	Standard Error	T for H0: Parameter=0	Prob > |T|
INTERCEP	1	6.122535	0.07304328	83.821	0.0001
FEM	1	-0.905504	0.24132106	-3.752	0.0002
ED	1	0.068622	0.00555341	12.357	0.0001
F_EDC	1	0.033696	0.01844378	1.827	0.0682

Restricted Model (without FEM and FEM*EDUC)
Dependent Variable: LWAGE

Analysis of Variance

Source	DF	Sum of Squares	Mean Square	F Value	Prob>F
Model	1	23.79816	23.79816	156.166	0.0001
Error	593	90.36713	0.15239		
C Total	594	114.16529			

Root MSE	0.39037	R-square	0.2085	
Dep Mean	6.95074	Adj R-sq	0.2071	
C.V.	5.61625			

Parameter Estimates

Variable	DF	Parameter Estimate	Standard Error	T for H0: Parameter=0	Prob > |T|
INTERCEP	1	6.029192	0.07546051	79.899	0.0001
ED	1	0.071742	0.00574089	12.497	0.0001

b. The unrestricted regression is given by (4.27). This regression runs EARN on a constant, FEMALE and EDUCATION. The URSS = 77.06808. The restricted regression for the equality of intercepts given the same slopes for Males and Females, tests the restriction H_o; $\alpha_F = 0$ given that $\gamma = 0$. This is the same restricted regression given in part (a), running EARN on a constant and EDUC. The RRSS = 90.36713. The F-test given by (4.17) tests one restriction and yields

$$F = \frac{(90.36713 - 77.06808)/1}{77.06808/592} = 102.2.$$

This is distributed as $F_{1,592}$ under H_o. Note that this observed F-statistic is the square of the observed t-statistic of -10.107 for $\alpha_F = 0$ in the unrestricted regression. The SAS regression output is given below.

```
Unrestricted Model (with FEM)
Dependent Variable: LWAGE
```

Analysis of Variance

Source	DF	Sum of Squares	Mean Square	F Value	Prob>F
Model	2	37.09721	18.54861	142.482	0.0001
Error	592	77.06808	0.13018		
C Total	594	114.16529			

Root MSE	0.36081	R-square	0.3249	
Dep Mean	6.95074	Adj R-sq	0.3227	
C.V.	5.19093			

Parameter Estimates

Variable	DF	Parameter Estimate	Standard Error	T for H0: Parameter=0	Prob > |T|
INTERCEP	1	6.083290	0.06995090	86.965	0.0001
FEM	1	-0.472950	0.04679300	-10.107	0.0001
ED	1	0.071676	0.00530614	13.508	0.0001

c. The unrestricted regression is given by (4.28), see part (a). The restricted regression for the equality of intercepts allowing for different slopes for Males and Females, tests the restriction H_o; $\alpha_F = 0$ given that $\gamma \neq 0$. This regression runs EARN on a constant, EDUCATION and (FEMALE × EDUCATION). The RRSS = 78.46096. The SAS regression output is given below. The F-test given by (4.17), tests one restriction and yields:

$$F = \frac{(78.46096 - 76.63525)/1}{76.63525/591} = 14.0796.$$

This is distributed as $F_{1,591}$ under H_o. The null hypothesis is rejected. Note that this observed F-statistic is the square of the t-statistic (-3.752) on $\alpha_F = 0$ in the unrestricted regression.

Chapter 4: Multiple Regression Analysis

Restricted Model (without FEM)
Dependent Variable: LWAGE

Analysis of Variance

Source	DF	Sum of Squares	Mean Square	F Value	Prob>F
Model	2	35.70433	17.85216	134.697	0.0001
Error	592	78.46096	0.13254		
C Total	594	114.16529			

Root MSE	0.36405	R-square	0.3127	
Dep Mean	6.95074	Adj R-sq	0.3104	
C.V.	5.23763			

Parameter Estimates

Variable	DF	Parameter Estimate	Standard Error	T for H0: Parameter=0	Prob > |T|
INTERCEP	1	6.039577	0.07038181	85.812	0.0001
ED	1	0.074782	0.00536347	13.943	0.0001
F_EDC	1	-0.034202	0.00360849	-9.478	0.0001

4.13 a. Dependent Variable: LWAGE

Analysis of Variance

Source	DF	Sum of Squares	Mean Square	F Value	Prob>F
Model	12	52.48064	4.37339	41.263	0.0001
Error	582	61.68465	0.10599		
C Total	594	114.16529			

Root MSE	0.32556	R-square	0.4597	
Dep Mean	6.95074	Adj R-sq	0.4485	
C.V.	4.68377			

Parameter Estimates

Variable	DF	Parameter Estimate	Standard Error	T for H0: Parameter=0	Prob > |T|
INTERCEP	1	5.590093	0.19011263	29.404	0.0001
EXP	1	0.029380	0.00652410	4.503	0.0001
EXP2	1	-0.000486	0.00012680	-3.833	0.0001
WKS	1	0.003413	0.00267762	1.275	0.2030
OCC	1	-0.161522	0.03690729	-4.376	0.0001
IND	1	0.084663	0.02916370	2.903	0.0038
SOUTH	1	-0.058763	0.03090689	-1.901	0.0578

SMSA	1	0.166191	0.02955099	5.624	0.0001
MS	1	0.095237	0.04892770	1.946	0.0521
FEM	1	-0.324557	0.06072947	-5.344	0.0001
UNION	1	0.106278	0.03167547	3.355	0.0008
ED	1	0.057194	0.00659101	8.678	0.0001
BLK	1	-0.190422	0.05441180	-3.500	0.0005

b. H_o: EARN = α + u.

If you run EARN on an intercept only, you would get $\hat{\alpha}$ = 6.9507 which is average log wage or average earnings = $\bar{y}$. The total sum of squares = the residual sum of squares = $\sum_{i=1}^{n}(y_i-\bar{y})^2$ = 114.16529 and this is the restricted residual sum of squares (RRSS) needed for the F-test. The unrestricted model is given in Table 4.1 or part (a) and yields URSS = 61.68465. Hence, the joint significance for all slopes using (4.20) yields

$$F = \frac{(114.16529-61.68465)/12}{61.68465/582} = 41.26 \quad \text{also}$$

$$F = \frac{R^2}{1-R^2} \cdot \frac{n-K}{K-1} = \frac{0.4597}{1-0.4597} \cdot \frac{582}{12} = 41.26.$$

This F-statistic is distributed as $F_{12,582}$ under the null hypothesis. It has a p-value of 0.0001 as shown in Table 4.1 and we reject H_o. The Analysis of Variance table in the SAS output given in Table 4.1 always reports this F-statistic for the significance of all slopes for any regression.

c. The restricted model excludes FEM and BLACK. The SAS regression output is given below. The RRSS = 66.27893. The unrestricted model is given in Table 4.1 with URSS = 61.68465. The F-statistic given in (4.17) tests two restrictions and yields

$$F = \frac{(66.27893-61.68465)/2}{61.68465/582} = 21.6737.$$

This is distributed as $F_{2,582}$ under the null hypothesis. We reject H_o.

Model: Restricted Model (w/o FEMALE & BLACK)
Dependent Variable: LWAGE

Analysis of Variance

Source	DF	Sum of Squares	Mean Square	F Value	Prob>F
Model	10	47.88636	4.78864	42.194	0.0001
Error	584	66.27893	0.11349		
C Total	594	114.16529			

Root MSE	0.33688	R-square	0.4194	
Dep Mean	6.95074	Adj R-sq	0.4095	
C.V.	4.84674			

Parameter Estimates

Variable	DF	Parameter Estimate	Standard Error	T for H0: Parameter=0	Prob > |T|
INTERCEP	1	5.316110	0.19153698	27.755	0.0001
EXP	1	0.028108	0.00674771	4.165	0.0001
EXP2	1	-0.000468	0.00013117	-3.570	0.0004
WKS	1	0.004527	0.00276523	1.637	0.1022
OCC	1	-0.162382	0.03816211	-4.255	0.0001
IND	1	0.102697	0.03004143	3.419	0.0007
SOUTH	1	-0.073099	0.03175589	-2.302	0.0217
SMSA	1	0.142285	0.03022571	4.707	0.0001
MS	1	0.298940	0.03667049	8.152	0.0001
UNION	1	0.112941	0.03271187	3.453	0.0006
ED	1	0.059991	0.00680032	8.822	0.0001

d. The restricted model excludes MS and UNION. The SAS regression output is given below. This yields RRSS = 63.37107. The unrestricted model is given in Table 4.1 and yields URSS = 61.68465. The F-test given in (4.17) tests two restrictions and yields

$$F = \frac{(63.37107 - 61.68465)/2}{61.68465/582} = 7.9558.$$

This is distributed as $F_{2,582}$ under the null hypothesis. We reject H_o.

Restricted Model (without MS & UNION)
Dependent Variable: LWAGE

Analysis of Variance

Source	DF	Sum of Squares	Mean Square	F Value	Prob>F
Model	10	50.79422	5.07942	46.810	0.0001
Error	584	63.37107	0.10851		
C Total	594	114.16529			

Root MSE	0.32941	R-square	0.4449
Dep Mean	6.95074	Adj R-sq	0.4354
C.V.	4.73923		

Parameter Estimates

Variable	DF	Parameter Estimate	Standard Error	T for H0: Parameter=0	Prob > |T|
INTERCEP	1	5.766243	0.18704262	30.828	0.0001
EXP	1	0.031307	0.00657565	4.761	0.0001
EXP2	1	-0.000520	0.00012799	-4.064	0.0001
WKS	1	0.001782	0.00264789	0.673	0.5013
OCC	1	-0.127261	0.03591988	-3.543	0.0004
IND	1	0.089621	0.02948058	3.040	0.0025
SOUTH	1	-0.077250	0.03079302	-2.509	0.0124
SMSA	1	0.172674	0.02974798	5.805	0.0001
FEM	1	-0.425261	0.04498979	-9.452	0.0001
ED	1	0.056144	0.00664068	8.454	0.0001
BLK	1	-0.197010	0.05474680	-3.599	0.0003

e. From Table 4.1, using the coefficient estimate on Union, $\hat{\beta}_u = 0.106278$, we obtain $\hat{g}_u = e^{\hat{\beta}_u} - 1 = e^{0.106278} - 1 = 0.112131$ or (11.2131%). If the disturbances are log normal, Kennedy's (1981) suggestion yields

$\tilde{g}_u = e^{\hat{\beta}_u - 0.5 \cdot \text{var}(\hat{\beta}_u)} - 1 = e^{0.106278 - 0.5(0.001003335)} - 1 = 0.111573$ or (11.1573%).

f. From Table 4.1, using the coefficient estimate on MS, $\hat{\beta}_{MS} = 0.095237$, we obtain $\hat{g}_{MS} = e^{\hat{\beta}_{MS}} - 1 = e^{0.095237} - 1 = 0.09992$ or (9.992%).

4.14 Crude Quality

a. Regression of POIL on GRAVITY and SULPHUR.

Dependent Variable: POIL

Analysis of Variance

Source	DF	Sum of Squares	Mean Square	F Value	Prob>F
Model	2	249.21442	124.60721	532.364	0.0001
Error	96	22.47014	0.23406		
C Total	98	271.68456			

Root MSE	0.48380	R-square	0.9173	
Dep Mean	15.33727	Adj R-sq	0.9156	
C.V.	3.15442			

Parameter Estimates

Variable	DF	Parameter Estimate	Standard Error	T for H0: Parameter=0	Prob > |T|
INTERCEP	1	12.354268	0.23453113	52.676	0.0001
GRAVITY	1	0.146640	0.00759695	19.302	0.0001
SULPHUR	1	-0.414723	0.04462224	-9.294	0.0001

b. Regression of GRAVITY on SULPHUR.

Dependent Variable: GRAVITY

Analysis of Variance

Source	DF	Sum of Squares	Mean Square	F Value	Prob>F
Model	1	2333.89536	2333.89536	55.821	0.0001
Error	97	4055.61191	41.81043		
C Total	98	6389.50727			

Root MSE	6.46610	R-square	0.3653	
Dep Mean	24.38182	Adj R-sq	0.3587	
C.V.	26.52017			

Parameter Estimates

Variable	DF	Parameter Estimate	Standard Error	T for H0: Parameter=0	Prob>|T|
INTERCEP	1	29.452116	0.93961195	31.345	0.0001
SULPHUR	1	-3.549926	0.47513923	-7.471	0.0001

Regression of POIL on the Residuals from the Previous Regression

```
Dependent Variable: POIL
```

Analysis of Variance

Source	DF	Sum of Squares	Mean Square	F Value	Prob>F
Model	1	87.20885	87.20885	45.856	0.0001
Error	97	184.47571	1.90181		
C Total	98	271.68456			

Root MSE	1.37906	R-square	0.3210	
Dep Mean	15.33727	Adj R-sq	0.3140	
C.V.	8.99157			

Parameter Estimates

Variable	DF	Parameter Estimate	Standard Error	T for H0: Parameter=0	Prob > |T|
INTERCEP	1	15.337273	0.13860093	110.658	0.0001
RESID_V	1	0.146640	0.02165487	6.772	0.0001

c. Regression of POIL on SULPHUR

```
Dependent Variable: POIL
```

Analysis of Variance

Source	DF	Sum of Squares	Mean Square	F Value	Prob>F
Model	1	162.00557	162.00557	143.278	0.0001
Error	97	109.67900	1.13071		
C Total	98	271.68456			

Root MSE	1.06335	R-square	0.5963	
Dep Mean	15.33727	Adj R-sq	0.5921	
C.V.	6.93310			

Parameter Estimates

Variable	DF	Parameter Estimate	Standard Error	T for H0: Parameter=0	Prob > |T|
INTERCEP	1	16.673123	0.15451906	107.903	0.0001
SULPHUR	1	-0.935284	0.07813658	-11.970	0.0001

Chapter 4: Multiple Regression Analysis

Regression of Residuals in part (c) on those in part (b).

```
Dependent Variable: RESID_W
```

Analysis of Variance

Source	DF	Sum of Squares	Mean Square	F Value	Prob>F
Model	1	87.20885	87.20885	376.467	0.0001
Error	97	22.47014	0.23165		
C Total	98	109.67900			

Root MSE	0.48130	R-square	0.7951
Dep Mean	0.00000	Adj R-sq	0.7930
C.V.	2.127826E16		

Parameter Estimates

Variable	DF	Parameter Estimate	Standard Error	T for H0: Parameter=0	Prob > |T|
INTERCEP	1	3.082861E-15	0.04837260	0.000	1.0000
RESID_V	1	0.146640	0.00755769	19.403	0.0001

d. Regression based on the first 25 crudes.

```
Dependent Variable: POIL
```

Analysis of Variance

Source	DF	Sum of Squares	Mean Square	F Value	Prob>F
Model	2	37.68556	18.84278	169.640	0.0001
Error	22	2.44366	0.11108		
C Total	24	40.12922			

Root MSE	0.33328	R-square	0.9391
Dep Mean	15.65560	Adj R-sq	0.9336
C.V.	2.12882		

Parameter Estimates

Variable	DF	Parameter Estimate	Standard Error	T for H0: Parameter=0	Prob > |T|
INTERCEP	1	11.457899	0.34330283	33.375	0.0001
GRAVITY	1	0.166174	0.01048538	15.848	0.0001
SULPHUR	1	0.110178	0.09723998	1.133	0.2694

e. Deleting all crudes with sulphur content outside the range of 1 to 2 percent.

```
Dependent Variable: POIL

                          Analysis of Variance

                          Sum of          Mean
      Source      DF      Squares         Square       F Value      Prob>F

      Model        2      28.99180       14.49590      128.714      0.0001
      Error       25       2.81553        0.11262
      C Total     27      31.80732

              Root MSE         0.33559      R-square      0.9115
              Dep Mean        15.05250      Adj R-sq      0.9044
              C.V.             2.22947

                          Parameter Estimates

                       Parameter      Standard      T for H0:
      Variable  DF     Estimate       Error         Parameter=0    Prob > |T|

      INTERCEP   1     11.090789      0.37273724     29.755         0.0001
      GRAVITY    1      0.180260      0.01123651     16.042         0.0001
      SULPHUR    1      0.176138      0.18615220      0.946         0.3531
```

SAS PROGRAM

```
Data Crude;
Input POIL GRAVITY SULPHUR;
Cards;

Proc reg data=CRUDE;
    Model POIL=GRAVITY SULPHUR;

Proc reg data=CRUDE;
    Model GRAVITY=SULPHUR;
  Output out=OUT1 R=RESID_V;
run;

Data CRUDE1; set crude; set OUT1(keep=RESID_V);

Proc reg data=CRUDE1;
    Model POIL=RESID_V;

Proc reg data=CRUDE1;
    Model POIL=SULPHUR;
  Output out=OUT2 R=RESID_W;

Proc reg data=OUT2;
    Model RESID_W=RESID_V;

Data CRUDE2; set CRUDE(firstobs=1 obs=25);
```

Chapter 4: Multiple Regression Analysis

```
Proc reg data=CRUDE2;
     Model POIL=GRAVITY SULPHUR;
run;

data CRUDE3; set CRUDE;
     if SULPHUR < 1 then delete;
     if SULPHUR > 2 then delete;

Proc reg data=CRUDE3;
     Model POIL=GRAVITY SULPHUR;
run;
```

4.15 a. MODEL1 (1950-1972)

Dependent Variable: LNQMG

Analysis of Variance

Source	DF	Sum of Squares	Mean Square	F Value	Prob>F
Model	5	1.22628	0.24526	699.770	0.0001
Error	17	0.00596	0.00035		
C Total	22	1.23224			

Root MSE	0.01872	R-square	0.9952
Dep Mean	17.96942	Adj R-sq	0.9937
C.V.	0.10418		

Parameter Estimates

Variable	DF	Parameter Estimate	Standard Error	T for H0: Parameter=0	Prob > |T|
INTERCEP	1	1.680143	2.79355393	0.601	0.5555
LNCAR	1	0.363533	0.51515166	0.706	0.4899
LNPOP	1	1.053931	0.90483097	1.165	0.2602
LNRGNP	1	-0.311388	0.16250458	-1.916	0.0723
LNPGNP	1	0.124957	0.15802894	0.791	0.4400
LNPMG	1	1.048145	0.26824906	3.907	0.0011

MODEL2 (1950-1972)
Dependent Variable: QMG_CAR

Analysis of Variance

Source	DF	Sum of Squares	Mean Square	F Value	Prob>F
Model	3	0.01463	0.00488	4.032	0.0224
Error	19	0.02298	0.00121		
C Tota	22	0.03762			

Root MSE	0.03478	R-square	0.3890	
Dep Mean	-0.18682	Adj R-sq	0.2925	
C.V.	-18.61715			

Parameter Estimates

Variable	DF	Parameter Estimate	Standard Error	T for H0: Parameter=0	Prob > |T|
INTERCEP	1	-0.306528	2.37844176	-0.129	0.8988
RGNP_POP	1	-0.139715	0.23851425	-0.586	0.5649
CAR_POP	1	0.054462	0.28275915	0.193	0.8493
PMG_PGNP	1	0.185270	0.27881714	0.664	0.5144

b. From model (2) we have

$$\log QMG - \log CAR = \gamma_1 + \gamma_2 \log RGNP - \gamma_2 \log POP$$
$$+ \gamma_3 \log CAR - \gamma_3 \log POP + \gamma_4 \log PMG - \gamma_4 \log PGNP + v$$

For this to be the same as model (1), the following restrictions must hold:

$\beta_1 = \gamma_1$, $\beta_2 = \gamma_3 + 1$, $\beta_3 = -(\gamma_2 + \gamma_3)$, $\beta_4 = \gamma_2$, $\beta_5 = -\gamma_4$, and $\beta_6 = \gamma_4$.

d. Correlation Analysis

Variables: LNCAR   LNPOP   LNRGNP   LNPGNP   LNPMG

Simple Statistics

Variable	N	Mean	Std Dev	Sum	Minimum	Maximum
LNCAR	23	18.15623	0.26033	417.59339	17.71131	18.57813
LNPOP	23	12.10913	0.10056	278.50990	11.93342	12.25437
LNRGNP	23	7.40235	0.24814	170.25415	6.99430	7.81015
LNPGNP	23	3.52739	0.16277	81.12989	3.26194	3.86073
LNPMG	23	-1.15352	0.09203	-26.53096	-1.30195	-0.94675

Chapter 4: Multiple Regression Analysis

```
Pearson Correlation Coefficients / Prob > |R| under Ho: Rho=0 / N = 23
```

	LNCAR	LNPOP	LNRGNP	LNPGNP	LNPMG
LNCAR	1.00000	0.99588	0.99177	0.97686	0.94374
	0.0	0.0001	0.0001	0.0001	0.0001
LNPOP	0.99588	1.00000	0.98092	0.96281	0.91564
	0.0001	0.0	0.0001	0.0001	0.0001
LNRGNP	0.99177	0.98092	1.00000	0.97295	0.94983
	0.0001	0.0001	0.0	0.0001	0.0001
LNPGNP	0.97686	0.96281	0.97295	1.00000	0.97025
	0.0001	0.0001	0.0001	0.0	0.0001
LNPMG	0.94374	0.91564	0.94983	0.97025	1.00000
	0.0001	0.0001	0.0001	0.0001	0.0

This indicates the presence of multicollinearity among the regressors.

f. MODEL1 (1950-1987)

```
Dependent Variable: LNQMG
```

Analysis of Variance

Source	DF	Sum of Squares	Mean Square	F Value	Prob>F
Model	5	3.46566	0.69313	868.762	0.0001
Error	32	0.02553	0.00080		
C Total	37	3.49119			

Root MSE	0.02825	R-square	0.9927
Dep Mean	18.16523	Adj R-sq	0.9915
C.V.	0.15550		

Parameter Estimates

Variable	DF	Parameter Estimate	Standard Error	T for H0: Parameter=0	Prob > |T|
INTERCEP	1	9.986981	2.62061670	3.811	0.0006
LNCAR	1	2.559918	0.22812676	11.221	0.0001
LNPOP	1	-2.878083	0.45344340	-6.347	0.0001
LNRGNP	1	-0.429270	0.14837889	-2.893	0.0068
LNPGNP	1	-0.178866	0.06336001	-2.823	0.0081
LNPMG	1	-0.141105	0.04339646	-3.252	0.0027

MODEL2 (1950-1987)
Dependent Variable: QMG_CAR

Analysis of Variance

Source	DF	Sum of Squares	Mean Square	F Value	Prob>F
Model	3	0.35693	0.11898	32.508	0.0001
Error	34	0.12444	0.00366		
C Total	37	0.48137			

Root MSE	0.06050	R-square	0.7415	
Dep Mean	-0.25616	Adj R-sq	0.7187	
C.V.	-23.61671			

Parameter Estimates

Variable	DF	Parameter Estimate	Standard Error	T for H0: Parameter=0	Prob > |T|
INTERCEP	1	-5.853977	3.10247647	-1.887	0.0677
RGNP_POP	1	-0.690460	0.29336969	-2.354	0.0245
CAR_POP	1	0.288735	0.27723429	1.041	0.3050
PMG_PGNP	1	-0.143127	0.07487993	-1.911	0.0644

g. Model 2: SHOCK74 (=1 if year ≥ 1974)

Dependent Variable: QMG_CAR

Analysis of Variance

Source	DF	Sum of Squares	Mean Square	F Value	Prob>F
Model	4	0.36718	0.09179	26.527	0.0001
Error	33	0.11419	0.00346		
C Total	37	0.48137			

Root MSE	0.05883	R-square	0.7628	
Dep Mean	-0.25616	Adj R-sq	0.7340	
C.V.	-22.96396			

Parameter Estimates

Variable	DF	Parameter Estimate	Standard Error	for H0: Parameter=0	Prob > |T|
INTERCEP	1	-6.633509	3.05055820	-2.175	0.0369
SHOCK74	1	-0.073504	0.04272079	-1.721	0.0947
RGNP_POP	1	-0.733358	0.28634866	-2.561	0.0152
CAR_POP	1	0.441777	0.28386742	1.556	0.1292
PMG_PGNP	1	-0.069656	0.08440816	-0.825	0.4152

The t-statistic on SHOCK 74 yields -1.721 with a p-value of 0.0947. This is insignificant. Therefore, we cannot reject that gasoline demand per car did not permanently shift after 1973.

h. Model 2: DUMMY74 (=SCHOCK74 x PMG_PGNP)

```
Dependent Variable: QMG_CAR

               Analysis of Variance

                      Sum of        Mean
    Source     DF    Squares       Square      F Value      Prob>F

    Model       4    0.36706      0.09177       26.492      0.0001
    Error      33    0.11431      0.00346
    C Total    37    0.48137

        Root MSE      0.05886     R-square      0.7625
        Dep Mean     -0.25616     Adj R-sq      0.7337
        C.V.        -22.97560

                    Parameter Estimates

                  Parameter    Standard     T for H0:
    Variable DF   Estimate     Error        Parameter=0    Prob > |T|

    INTERCEP  1   -6.606761    3.05019318     -2.166        0.0376
    RGNP_POP  1   -0.727422    0.28622322     -2.541        0.0159
    CAR_POP   1    0.431492    0.28233413      1.528        0.1360
    PMG_PGNP  1   -0.083217    0.08083459     -1.029        0.3107
    DUMMY74   1    0.015283    0.00893783      1.710        0.0967
```

The interaction dummy named DUMMY74 has a t-statistic of 1.71 with a p-value of 0.0967. This is insignificant and we cannot reject that the price elasticity did not change after 1973.

SAS PROGRAM

```
Data RAWDATA;
Input Year CAR QMG PMG POP RGNP PGNP;
Cards;

Data USGAS; set RAWDATA;

LNQMG=LOG(QMG);
LNCAR=LOG(CAR);
LNPOP=LOG(POP);
LNRGNP=LOG(RGNP);
LNPGNP=LOG(PGNP);
LNPMG=LOG(PMG);
```

```
QMG_CAR=LOG(QMG/CAR);
RGNP_POP=LOG(RGNP/POP);
CAR_POP=LOG(CAR/POP);
PMG_PGNP=LOG(PMG/PGNP);

Data USGAS1; set USGAS;
If YEAR>1972 then delete;

Proc reg data=USGAS1;
    Model LNQMG=LNCAR LNPOP LNRGNP LNPGNP LNPMG;
    Model QMG_CAR=RGNP_POP CAR_POP PMG_PGNP;

Proc corr data=USGAS1;
    Var LNCAR LNPOP LNRGNP LNPGNP LNPMG;
run;

Proc reg data=USGAS;
    Model LNQMG=LNCAR LNPOP LNRGNP LNPGNP LNPMG;
    Model QMG_CAR=RGNP_POP CAR_POP PMG_PGNP;
run;

data DUMMY1; set USGAS;
If Year<1974 then SHOCK74=0; else SHOCK74=1;
DUMMY74=PMG_PGNP*SHOCK74;

Proc reg data=DUMMY1;
    Model QMG_CAR=SHOCK74 RGNP_POP CAR_POP PMG_PGNP;
    Model QMG_CAR=RGNP_POP CAR_POP PMG_PGNP DUMMY74;
run;
```

4.16

a. MODEL1

Dependent Variable: LNCONS

Analysis of Variance

Source	DF	Sum of Squares	Mean Square	F Value	Prob>F
Model	5	165.96803	33.19361	70.593	0.0001
Error	132	62.06757	0.47021		
C Total	137	228.03560			

Root MSE	0.68572	R-square	0.7278
Dep Mean	11.89979	Adj R-sq	0.7175
C.V.	5.76243		

Parameter Estimates

Variable	DF	Parameter Estimate	Standard Error	T for H0: Parameter=0	Prob > |T|
INTERCEP	1	-53.577951	4.53057139	-11.826	0.0001
LNPG	1	-1.425259	0.31539170	-4.519	0.0001
LNPE	1	0.131999	0.50531143	0.261	0.7943
LNPO	1	0.237464	0.22438605	1.058	0.2919
LNHDD	1	0.617801	0.10673360	5.788	0.0001
LNPI	1	6.554349	0.48246036	13.585	0.0001

b. The following plot show that states with low level of consumption are over predicted, while states with high level of consumption are under predicted. One can correct for this problem by either running a separate regression for each state, or use dummy variables for each state to allow for a varying intercept.

c. Model 2

Dependent Variable: LNCONS

Analysis of Variance

Source	DF	Sum of Squares	Mean Square	F Value	Prob>F
Model	10	226.75508	22.67551	2248.917	0.0001
Error	127	1.28052	0.01008		
C Total	137	228.03560			

Root MSE	0.10041	R-square	0.9944
Dep Mean	11.89979	Adj R-sq	0.9939
C.V.	0.84382		

Plot of PRED*LNCONS. Symbol is value of STATE.

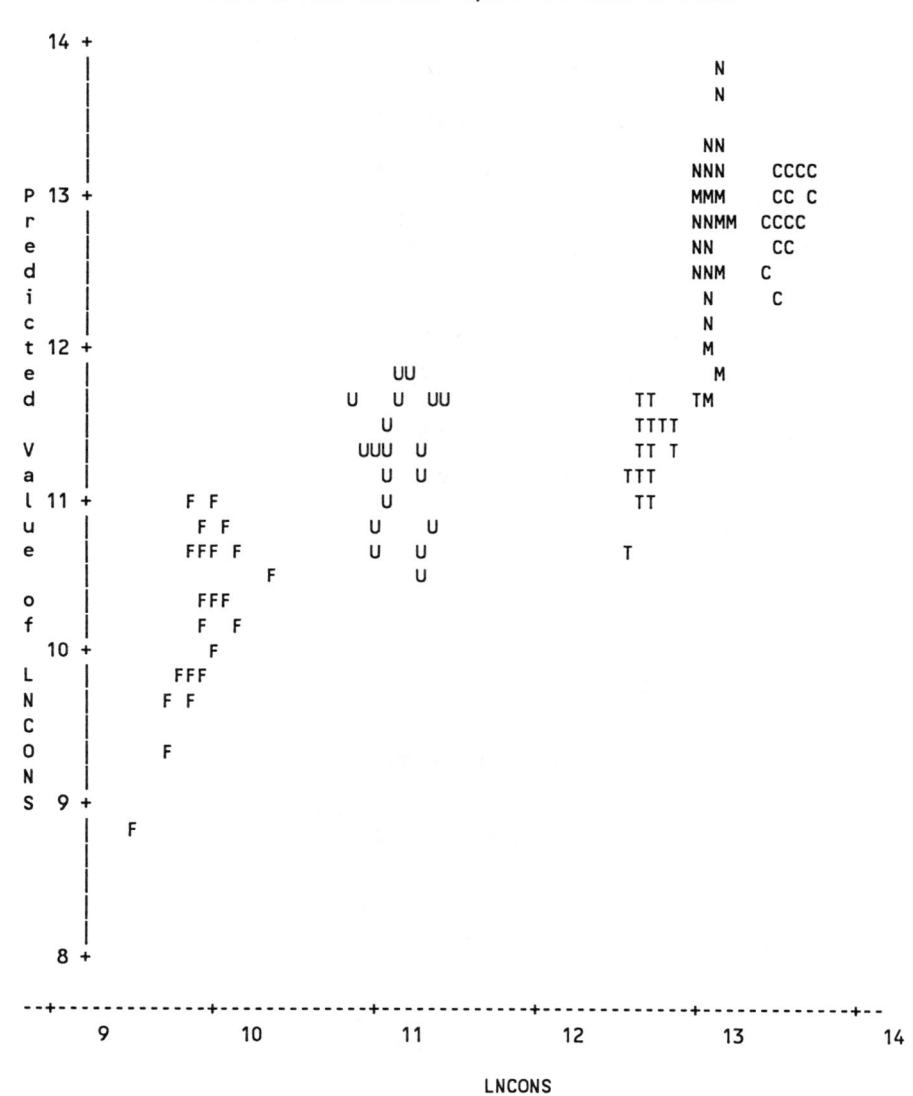

```
                                                                    N
      14 +                                                          N
         |
         |                                                         NN
         |                                                        NNN    CCCC
    P 13 +                                                        MMM    CC C
    r    |                                                        NNMM   CCCC
    e    |                                                         NN    CC
    d    |                                                        NNM    C
    i    |                                                         N     C
    c    |                                                         N
    t 12 +                                                         M
    e    |                    UU                                   M
    d    |              U   U  UU                         TT    TM
         |                  U                             TTTT
    V    |                UUU  U                          TT T
    a    |                 U  U                           TTT
    l 11 +        F F      U                              TT
    u    |         F F                                    
    e    |        FFF F          U   U                    
         |            F              U                 T
    o    |        FFF
    f    |        F  F
      10 +           F
    L    |        FFF
    N    |        F F
    C    |
    O    |        F
    N    |
    S  9 +
         |     F
         |
         |
         |
       8 +
       --+--------------+--------------+--------------+--------------+--------------+--
         9             10             11             12             13             14
                                          LNCONS
```

NOTE: 42 obs hidden.

Parameter Estimates

Variable	DF	Parameter Estimate	Standard Error	T for H0: Parameter=0	Prob > |T|
INTERCEP	1	6.260803	1.65972981	3.772	0.0002
LNPG	1	-0.125472	0.05512350	-2.276	0.0245
LNPE	1	-0.121120	0.08715359	-1.390	0.1670
LNPO	1	0.155036	0.03706820	4.182	0.0001
LNHDD	1	0.359612	0.07904527	4.549	0.0001
LNPI	1	0.416600	0.16626018	2.506	0.0135
DUMMY_NY	1	-0.702981	0.07640346	-9.201	0.0001
DUMMY_FL	1	-3.024007	0.11424754	-26.469	0.0001
DUMMY_MI	1	-0.766215	0.08491262	-9.024	0.0001
DUMMY_TX	1	-0.679327	0.04838414	-14.040	0.0001
DUMMY_UT	1	-2.597099	0.09925674	-26.165	0.0001

f. Dummy Variable Regression without an Intercept

NOTE: No intercept in model. R-square is redefined.
Dependent Variable: LNCONS

Analysis of Variance

Source	DF	Sum of Squares	Mean Square	F Value	Prob>F
Model	11	19768.26019	1797.11456	178234.700	0.0001
Error	127	1.28052	0.01008		
U Total	138	19769.54071			

Root MSE	0.10041	R-square	0.9999
Dep Mean	11.89979	Adj R-sq	0.9999
C.V.	0.84382		

Parameter Estimates

Variable	DF	Parameter Estimate	Standard Error	T for H0: Parameter=0	Prob > |T|
LNPG	1	-0.125472	0.05512350	-2.276	0.0245
LNPE	1	-0.121120	0.08715359	-1.390	0.1670
LNPO	1	0.155036	0.03706820	4.182	0.0001
LNHDD	1	0.359612	0.07904527	4.549	0.0001
LNPI	1	0.416600	0.16626018	2.506	0.0135
DUMMY_CA	1	6.260803	1.65972981	3.772	0.0002
DUMMY_NY	1	5.557822	1.67584667	3.316	0.0012
DUMMY_FL	1	3.236796	1.59445076	2.030	0.0444
DUMMY_MI	1	5.494588	1.67372513	3.283	0.0013
DUMMY_TX	1	5.581476	1.62224148	3.441	0.0008
DUMMY_UT	1	3.663703	1.63598515	2.239	0.0269

SAS PROGRAM

```
Data NATURAL;
Input STATE $ SCODE YEAR Cons Pg Pe Po LPgas HDD Pi;
Cards;
DATA NATURAL1; SET NATURAL;
LNCONS=LOG(CONS);
LNPG=LOG(PG);
LNPO=LOG(PO);
LNPE=LOG(PE);
LNHDD=LOG(HDD);
LNPI=LOG(PI);

******** PROB16.a **********;
*****************************;
Proc reg data=NATURAL1;
    Model LNCONS=LNPG LNPE LNPO LNHDD LNPI;
  Output out=OUT1 R=RESID P=PRED;
Proc plot data=OUT1 vpercent=75 hpercent=100;
    plot PRED*LNCONS=STATE;
Data NATURAL2; set NATURAL1;
If STATE='NY' THEN DUMMY_NY=1; ELSE DUMMY_NY=0;
If STATE='FL' THEN DUMMY_FL=1; ELSE DUMMY_FL=0;
If STATE='MI' THEN DUMMY_MI=1; ELSE DUMMY_MI=0;
If STATE='TX' THEN DUMMY_TX=1; ELSE DUMMY_TX=0;
If STATE='UT' THEN DUMMY_UT=1; ELSE DUMMY_UT=0;
If STATE='CA' THEN DUMMY_CA=1; ELSE DUMMY_CA=0;

******** PROB16.c **************;
********************************;
Proc reg data=NATURAL2;
    Model LNCONS=LNPG LNPE LNPO LNHDD LNPI DUMMY_NY DUMMY_FL DUMMY_MI DUMMY_TX
               DUMMY_UT;
******** PROB16.f ************;
*****************************;
Proc reg data=NATURAL2;
    Model LNCONS=LNPG LNPE LNPO LNHDD LNPI
              DUMMY_CA DUMMY_NY DUMMY_FL DUMMY_MI DUMMY_TX DUMMY_UT/NOINT;
run;
```

REFERENCES

Rao, U.L.G. and P.M. White (1988), "Effect of an Additional Regressor on R^2," *Econometric Theory*, Solution 86.3.1, 4: 352.

CHAPTER 5
Violations of the Classical Assumptions

5.1 From Chapter 3 we have shown that

$$e_i = Y_i - \hat{\alpha}_{ols} - \hat{\beta}_{ols}X_i = y_i - \hat{\beta}_{ols}x_i = (\beta - \hat{\beta}_{ols})x_i + (u_i - \bar{u})$$

for $i = 1, 2, \ldots, n$.

The second equality substitutes $\hat{\alpha}_{ols} = \bar{Y} - \hat{\beta}_{ols}\bar{X}$ and the third equality substitutes $y_i = \beta x_i + (u_i - \bar{u})$. Hence,

$$\sum_{i=1}^{n} e_i^2 = (\hat{\beta}_{ols} - \beta)^2 \sum_{i=1}^{n} x_i^2 + \sum_{i=1}^{n} (u_i - \bar{u})^2 - 2(\hat{\beta}_{ols} - \beta) \sum_{i=1}^{n} x_i(u_i - \bar{u}) \text{ and}$$

$$E(\sum_{i=1}^{n} e_i^2) = \sum_{i=1}^{n} x_i^2 \text{var}(\hat{\beta}_{ols}) + E[\sum_{t=1}^{T} (u_i - \bar{u})^2] - 2E(\sum_{i=1}^{n} x_i u_i)^2 / \sum_{i=1}^{n} x_i^2.$$

But $E(\sum_{i=1}^{n} x_i u_i)^2 = \sum_{i=1}^{n} x_i^2 \sigma_i^2$ since the u_i's are uncorrelated and heteroskedastic and

$$E[\sum_{i=1}^{n} (u_i - \bar{u})^2] = E(\sum_{i=1}^{n} u_i^2) + nE(\bar{u}^2) - 2E(\bar{u} \sum_{i=1}^{n} u_i) = \sum_{i=1}^{n} E(u_i^2) - nE(\bar{u}^2)$$

$$= \sum_{i=1}^{n} \sigma_i^2 - \frac{1}{n} E(\sum_{i=1}^{n} u_i^2) = \sum_{i=1}^{n} \sigma_i^2 - \frac{1}{n} \sum_{i=1}^{n} \sigma_i^2 = \left(\frac{n-1}{n}\right) \sum_{i=1}^{n} \sigma_i^2.$$

Hence, $E(s^2) = \frac{1}{n-2} E(\sum_{i=1}^{n} e_i^2)$

$$= \frac{1}{n-2} \left[\frac{\sum_{i=1}^{n} x_i^2 \sigma_i^2}{\sum_{i=1}^{n} x_i^2} + \left(\frac{n-1}{n}\right) \sum_{i=1}^{n} \sigma_i^2 - 2 \frac{\sum_{i=1}^{n} x_i^2 \sigma_i^2}{\sum_{i=1}^{n} x_i^2} \right]$$

$$= \frac{1}{n-2} \left[-\frac{\sum_{i=1}^{n} x_i^2 \sigma_i^2}{\sum_{i=1}^{n} x_i^2} + \left(\frac{n-1}{n}\right) \sum_{i=1}^{n} \sigma_i^2 \right].$$

Under homoskedasticity this reverts back to $E(s^2) = \sigma^2$.

5.2 $E[\widehat{var}(\hat{\beta}_{ols})] = E\left(\dfrac{s^2}{\sum_{i=1}^{n} x_i^2}\right) = \dfrac{E(s^2)}{\sum_{i=1}^{n} x_i^2}$. Using the results in problem 5.1, we

get $E[\widehat{var}(\hat{\beta}_{ols})] = \dfrac{1}{n-2}\left[-\dfrac{\sum_{i=1}^{n} x_i^2 \sigma_i^2}{\left(\sum_{i=1}^{n} x_i^2\right)^2} + \left(\dfrac{n-1}{n}\right)\dfrac{\sum_{i=1}^{n} \sigma_i^2}{\sum_{i=1}^{n} x_i^2}\right]$

$E[\widehat{var}(\hat{\beta}_{ols})] - var(\hat{\beta}_{ols}) = \dfrac{(n-1)}{n(n-2)}\dfrac{\sum_{i=1}^{n} \sigma_i^2}{\sum_{i=1}^{n} x_i^2} - \dfrac{\sum_{i=1}^{n} x_i^2 \sigma_i^2}{\left(\sum_{i=1}^{n} x_i^2\right)^2}\left(\dfrac{1}{n-2}+1\right)$

Substituting $\sigma_i^2 = bx_i^2$ where $b > 0$, we get

$E[\widehat{var}(\hat{\beta}_{ols})] - var(\hat{\beta}_{ols}) = \dfrac{b\sum_{i=1}^{n} x_i^2 \sum_{i=1}^{n} x_i^2 - nb\sum_{i=1}^{n} x_i^2 x_i^2}{n\left(\sum_{i=1}^{n} x_i^2\right)^2} \cdot \dfrac{n-1}{n-2}$

$= -\dfrac{n-1}{n-2} \cdot \dfrac{b\sum_{i=1}^{n}\left(x_i^2 - \sum_{i=1}^{n} x_i^2/n\right)^2}{\left(\sum_{i=1}^{n} x_i^2\right)^2} < 0 \quad \text{for} \quad b > 0.$

This means that, on the average, the estimated standard error of $\hat{\beta}_{ols}$ understates the true standard error. Hence, the t-statistic reported by the regression package for $H_o; \beta = 0$ is overblown.

5.3 *Weighted Least Squares.* This is based on Kmenta (1986).

a. From the first equation in (5.11), one could solve for $\tilde{\alpha}$

$\tilde{\alpha}\sum_{i=1}^{n}(1/\sigma_i^2) = \sum_{i=1}^{n}(Y_i/\sigma_i^2) - \tilde{\beta}\sum_{i=1}^{n}(X_i/\sigma_i^2).$

Dividing both sides by $\sum_{i=1}^{n}(1/\sigma_i^2)$ one gets

Chapter 5: Violations of the Classical Assumptions   71

$$\tilde{\alpha} = [\sum_{i=1}^{n}(Y_i/\sigma_i^2)/\sum_{i=1}^{n}(1/\sigma_i^2)] - \tilde{\beta}[\sum_{i=1}^{n}(X_i/\sigma_i^2)/\sum_{i=1}^{n}(1/\sigma_i^2)] = \overline{Y}^* - \tilde{\beta}\overline{X}^*.$$

Substituting $\tilde{\alpha}$ in the second equation of (5.11) one gets

$$\sum_{i=1}^{n} Y_i X_i/\sigma_i^2 = (\sum_{i=1}^{n} X_i/\sigma_i^2)[\sum_{i=1}^{n}(Y_i/\sigma_i^2)/\sum_{i=1}^{n}(1/\sigma_i^2)]$$

$$- \tilde{\beta}\left[\sum_{i=1}^{n}(X_i/\sigma_i^2)\right]^2 / \sum_{i=1}^{n}(1/\sigma_i^2) + \tilde{\beta}\sum_{i=1}^{n}(X_i^2/\sigma_i^2).$$

Multiplying both sides by $\sum_{i=1}^{n}(1/\sigma_i^2)$ and solving for $\tilde{\beta}$ one gets (5.12b)

$$\tilde{\beta} = \frac{[\sum_{i=1}^{n}(1/\sigma_i^2)][\sum_{i=1}^{n}(Y_i X_i/\sigma_i^2)] - [\sum_{i=1}^{n}(X_i/\sigma_i^2)][\sum_{i=1}^{n}(Y_i/\sigma_i^2)]}{[\sum_{i=1}^{n} X_i^2/\sigma_i^2][\sum_{i=1}^{n}(1/\sigma_i^2)] - [\sum_{i=1}^{n}(X_i/\sigma_i^2)]^2}$$

b. From the regression equation $Y_i = \alpha + \beta X_i + u_i$ one can multiply by w_i^* and sum to get $\sum_{i=1}^{n} w_i^* Y_i = \alpha \sum_{i=1}^{n} w_i^* + \beta \sum_{i=1}^{n} w_i^* X_i + \sum_{i=1}^{n} w_i^* u_i$. Now divide by $\sum_{i=1}^{n} w_i^*$ and use the definitions of $\overline{Y}^*$ and $\overline{X}^*$ to get $\overline{Y}^* = \alpha + \beta \overline{X}^* + \overline{u}^*$ where

$$\overline{u}^* = \sum_{i=1}^{n} w_i^* u_i / \sum_{i=1}^{n} w_i^*.$$

Subtract this equation from the original regression equation to get

$Y_i - \overline{Y}^* = \beta(X_i - \overline{X}^*) + (u_i - \overline{u}^*)$. Substitute this in the expression for $\tilde{\beta}$ in (5.12b), we get

$$\tilde{\beta} = \beta + \frac{\sum_{i=1}^{n} w_i^*(X_i - \overline{X}^*)(u_i - \overline{u}^*)}{\sum_{i=1}^{n} w_i^*(X_i - \overline{X}^*)^2} = \beta + \frac{\sum_{i=1}^{n} w_i^*(X_i - \overline{X}^*)u_i}{\sum_{i=1}^{n} w_i^*(X_i - \overline{X}^*)^2}$$

where the second equality uses the fact that

$$\sum_{i=1}^{n} w_i^*(X_i - \overline{X}^*) = \sum_{i=1}^{n} w_i^* X_i - (\sum_{i=1}^{n} w_i^*)(\sum_{i=1}^{n} w_i^* X_i)/\sum_{i=1}^{n} w_i^* = 0.$$

Therefore, $E(\tilde{\beta}) = \beta$ as expected, and

$$\text{var}(\tilde{\beta}) = E(\tilde{\beta}-\beta)^2 = E\left(\frac{\sum_{i=1}^{n} w_i^*(X_i-\bar{X}^*)u_i}{\sum_{i=1}^{n} w_i^*(X_i-\bar{X}^*)^2}\right)^2$$

$$= \frac{\sum_{i=1}^{n} w_i^{*2}(X_i-\bar{X}^*)^2 \sigma_i^2}{\left[\sum_{i=1}^{n} w_i^*(X_i-\bar{X}^*)^2\right]^2} = \frac{\sum_{i=1}^{n} w_i^*(X_i-\bar{X}^*)^2}{\left[\sum_{i=1}^{n} w_i^*(X_i-\bar{X}^*)^2\right]^2}$$

$$= \frac{1}{\sum_{i=1}^{n} w_i^*(X_i-\bar{X}^*)^2}$$

where the third equality uses the fact that the u_i's are not serially correlated and heteroskedastic and the fourth equality uses the fact that $w_i^* = (1/\sigma_i^2)$.

5.4 *Relative Efficiency of OLS Under Heteroskedasticity*

a. From equation (5.9) we have

$$\text{var}(\hat{\beta}_{ols}) = \sum_{i=1}^{n} x_i^2 \sigma_i^2 / (\sum_{i=1}^{n} x_i^2)^2 = \sigma^2 \sum_{i=1}^{n} x_i^2 X_i^\delta / (\sum_{i=1}^{n} x_i^2)^2$$

where $x_i = X_i - \bar{X}$. For $X_i = 1,2,..,10$ and $\delta = 0.5, 1, 1.5$ and 2. This is tabulated below.

b. From problem 3 part (b) we get

$$\text{var}(\tilde{\beta}_{BLUE}) = \frac{1}{\sum_{i=1}^{n} w_i^*(X_i-\bar{X}^*)^2}$$

where $w_i^* = (1/\sigma_i^2) = (1/\sigma^2 X_i^\delta)$ and $\bar{X}^* = \sum_{i=1}^{n} w_i^* X_i / \sum_{i=1}^{n} w_i^* = \dfrac{\sum_{i=1}^{n}(X_i/X_i^\delta)}{\sum_{i=1}^{n}(1/X_i^\delta)}$.

For $X_i = 1,2,..,10$ and $\delta = 0.5, 1, 1.5$ and 2, this variance is tabulated below.

c. The relative efficiency of OLS is given by the ratio of the two variances computed in parts (a) and (b). This is tabulated below for various values of δ.

δ	var($\hat{\beta}_{ols}$)	var($\tilde{\beta}_{BLUE}$)	var($\tilde{\beta}_{BLUE}$)/var($\hat{\beta}_{ols}$)
0.5	$0.0262\,\sigma^2$	$0.0237\,\sigma^2$	0.905
1	$0.0667\,\sigma^2$	$0.04794\,\sigma^2$	0.719
1.5	$0.1860\,\sigma^2$	$0.1017\,\sigma^2$	0.547
2	$0.5442\,\sigma^2$	$0.224\,\sigma^2$	0.412

As δ increases from 0.5 to 2, the relative efficiency of OLS with respect to BLUE decreases from 0.9 for mild heteroskedasticity to 0.4 for more serious heteroskedasticity.

5.6 *The AR(1) model.* From (5.26), by continuous substitution just like (5.29), one could stop at u_{t-s} to get

$$u_t = \rho^s u_{t-s} + \rho^{s-1}\epsilon_{t-s+1} + \rho^{s-2}\epsilon_{t-s+2} + .. + \rho\epsilon_{t-1} + \epsilon_t \qquad \text{for } t > s.$$

Note that the power of ρ and the subscript of ϵ always sum to t. Multiplying both sides by u_{t-s} and taking expected value, one gets

$$E(u_t u_{t-s}) = \rho^s E(u_{t-s}^2) + \rho^{s-1} E(\epsilon_{t-s+1} u_{t-s}) + .. + \rho E(\epsilon_{t-1} u_{t-s}) + E(\epsilon_t u_{t-s})$$

using (5.29), u_{t-s} is a function of ϵ_{t-s}, past values of ϵ_{t-s} and u_o. Since u_o is independent of the ϵ's, and the ϵ's themselves are not serially correlated, then u_{t-s} is independent of $\epsilon_t, \epsilon_{t-1},..., \epsilon_{t-s+1}$. Hence, all the terms on the right hand side of $E(u_t u_{t-s})$ except the first are zero. Therefore,

$$\text{cov}(u_t, u_{t-s}) = E(u_t u_{t-s}) = \rho^s \sigma_u^2 \text{ for } t > s.$$

5.7 *Relative Efficiency of OLS Under the AR(1) Model.*

a. $\hat{\beta}_{ols} = \sum_{t=1}^{T} x_t y_t / \sum_{t=1}^{T} x_t^2 = \beta + \sum_{t=1}^{T} x_t u_t / \sum_{t=1}^{T} x_t^2$

with $E(\hat{\beta}_{ols}) = \beta$ since x_t and u_t are independent. Also,

$$\text{var}(\hat{\beta}_{ols}) = E(\hat{\beta}_{ols} - \beta)^2 = E(\sum_{t=1}^{T} x_t u_t / \sum_{t=1}^{T} x_t^2)^2$$

$$= \sum_{t=1}^{T} x_t^2 E(u_t^2) / (\sum_{t=1}^{T} x_t^2)^2 + E[\sum\sum_{s \ne t} x_t x_s u_t u_s / (\sum_{t=1}^{T} x_t^2)^2]$$

$$= \frac{\sigma_u^2}{\sum_{t=1}^{T} x_t^2} \left[1 + 2\rho \frac{\sum_{t=1}^{T-1} x_t x_{t+1}}{\sum_{t=1}^{T} x_t^2} + 2\rho^2 \frac{\sum_{t=1}^{T-2} x_t x_{t+2}}{\sum_{t=1}^{T} x_t^2} + \ldots + 2\rho^{T-1} \frac{x_1 x_T}{\sum_{t=1}^{T} x_t^2} \right]$$

using the fact that $E(u_t u_s) = \rho^{|t-s|} \sigma_u^2$ as shown in problem 6.

Alternatively, one can use matrix algebra, see Chapter 9. For the AR(1) model, $\Omega = E(uu')$ is given by equation (9.9) of Chapter 9. So

$$x'\Omega x = (x_1, \ldots, x_T) \begin{bmatrix} 1 & \rho & \rho^2 & \ldots & \rho^{T-1} \\ \rho & 1 & \rho & \ldots & \rho^{T-2} \\ \vdots & \vdots & \vdots & & \vdots \\ \rho^{T-1} & \rho^{T-2} & \rho^{T-3} & \ldots & 1 \end{bmatrix} \begin{pmatrix} x_1 \\ x_2 \\ \vdots \\ x_T \end{pmatrix}$$

$$= (x_1^2 + \rho x_1 x_2 + \ldots + \rho^{T-1} x_T x_1) + (\rho x_1 x_2 + x_2^2 + \ldots + \rho^{T-2} x_T x_2)$$

$$+ (\rho^2 x_1 x_3 + \rho x_2 x_3 + \ldots + \rho^{T-3} x_T x_3) + \ldots + (\rho^{T-1} x_1 x_T + \rho^{T-2} x_2 x_T + \ldots + x_T^2)$$

collecting terms, we get

$$x'\Omega x = \sum_{t=1}^{T} x_t^2 + 2\rho \sum_{t=1}^{T-1} x_t x_{t+1} + 2\rho^2 \sum_{t=1}^{T-2} x_t x_{t+2} + \ldots + 2\rho^{T-1} x_1 x_T.$$

Using equation (9.5) of Chapter 9, we get $\text{var}(\hat{\beta}_{ols}) = \sigma_u^2 \dfrac{x'\Omega x}{(x'x)^2}$

$$= \frac{\sigma_u^2}{\sum_{t=1}^{T} x_t^2} \left[1 + 2\rho \frac{\sum_{t=1}^{T-1} x_t x_{t+1}}{\sum_{t=1}^{T} x_t^2} + 2\rho^2 \frac{\sum_{t=1}^{T-2} x_t x_{t+2}}{\sum_{t=1}^{T} x_t^2} + .. + 2\rho^{T-1} \frac{x_1 x_T}{\sum_{t=1}^{T} x_t^2} \right].$$

Similarly, Ω^{-1} for the AR(1) model is given by equation (9.10) of Chapter 9, and $\text{var}(\hat{\beta}_{BLUE}) = \sigma^2(X'\Omega^{-1}X)^{-1}$ below equation (9.4). In this case,

$$x'\Omega^{-1}x = \frac{1}{1-\rho^2}(x_1,\ldots,x_T)\begin{bmatrix} 1 & -\rho & 0 & \cdots & 0 & 0 & 0 \\ -\rho & 1+\rho^2 & -\rho & \cdots & 0 & 0 & 0 \\ \vdots & \vdots & \vdots & & \vdots & \vdots & \vdots \\ 0 & 0 & 0 & \cdots & -\rho & 1+\rho^2 & -\rho \\ 0 & 0 & 0 & \cdots & 0 & -\rho & 1 \end{bmatrix}\begin{pmatrix} x_1 \\ x_2 \\ \vdots \\ x_T \end{pmatrix}$$

$$x'\Omega^{-1}x = \frac{1}{1-\rho^2}[x_1^2 - \rho x_1 x_2 - \rho x_1 x_2 + (1+\rho^2)x_2^2 - \rho x_3 x_2 - \rho x_2 x_3 + (1+\rho^2)x_3^2$$

$$-\rho x_4 x_3 + .. + x_T^2 - \rho x_{T-1} x_T].$$

Collecting terms, we get

$$x'\Omega^{-1}x = \frac{1}{1-\rho^2}[\sum_{t=1}^{T} x_t^2 - 2\rho \sum_{t=1}^{T-1} x_t x_{t+1} + \rho^2 \sum_{t=1}^{T-1} x_t^2].$$

For $T \to \infty$, $\sum_{t=2}^{T-1} x_t^2$ is equivalent to $\sum_{t=1}^{T} x_t^2$, hence

$$\text{var}(\hat{\beta}_{PW}) = \sigma_u^2 (x'\Omega^{-1}x)^{-1} = \frac{\sigma_u^2}{\sum_{t=1}^{T} x_t^2} \left[\frac{1-\rho^2}{1+\rho^2 - 2\rho \sum_{t=1}^{T-1} x_t x_{t+1} / \sum_{t=1}^{T} x_t^2} \right].$$

b. For x_t following itself an AR(1) process: $x_t = \lambda x_{t-1} + v_t$, we know that λ is the correlation of x_t and x_{t-1}, and from problem 6, $\text{correl}(x_t, x_{t-s}) = \lambda^s$. As $T \to \infty$, λ is estimated well by $\sum_{t=1}^{T-1} x_t x_{t+1} / \sum_{t=1}^{T} x_t^2$. Also, λ^2 is estimated well by $\sum_{t=1}^{T-2} x_t x_{t+2} / \sum_{t=1}^{T} x_t^2$, etc. Hence

$$\text{asy eff}(\hat{\beta}_{ols}) = \lim_{T \to \infty} \frac{\text{var}(\hat{\beta}_{PW})}{\text{var}(\hat{\beta}_{ols})} = \frac{1-\rho^2}{(1+\rho^2-2\rho\lambda)(1+2\rho\lambda+2\rho^2\lambda^2+..)}$$

$$= \frac{(1-\rho^2)(1-\rho\lambda)}{(1+\rho^2-2\rho\lambda)(1+\rho\lambda)}$$

where the last equality uses the fact that $(1+\rho\lambda)/(1-\rho\lambda) = (1+2\rho\lambda+2\rho^2\lambda^2+..)$. For $\lambda = 0$, or $\rho = \lambda$, this asy eff($\hat{\beta}_{ols}$) is equal to $(1-\rho^2)/(1+\rho^2)$.

c. The asy eff($\hat{\beta}_{ols}$) derived in part (b) is tabulated below for various values of ρ and λ. A similar table is given in Johnston (1984, p. 312). For $\rho > 0$, loss in efficiency is big as ρ increases. For a fixed λ, this asymptotic efficiency drops from the 90 percent range to a 10 percent range as ρ increases from 0.2 to 0.9. Variation in λ has minor effects when $\rho > 0$. For $\rho < 0$, the efficiency loss is still big as the absolute value of ρ increases, for a fixed λ. However, now variation in λ has a much stronger effect. For a fixed negative ρ, the loss in efficiency decreases with λ. In fact, for $\lambda = 0.9$, the loss in efficiency drops from 99 percent to 53 percent as ρ goes from -0.2 to -0.9. This is in contrast to say $\lambda = 0.2$ where the loss in efficiency drops from 93 percent to 13 percent as ρ goes from -0.2 to -0.9.

(Asymptotic Relative Efficiency of $\hat{\beta}_{ols}$) × 100

ρ

λ	-0.9	-0.8	-0.7	-0.6	-0.5	-0.4	-0.3	-0.2	-0.1	0	0.1	0.2	0.3	0.4	0.5	0.6	0.7	0.8	0.9
0	10.5	22.0	34.2	47.1	60.0	72.4	83.5	92.3	98.0	100	98.0	92.3	83.5	72.4	60.0	47.1	34.2	22.0	10.5
0.1	11.4	23.5	36.0	48.8	61.4	73.4	84.0	92.5	98.1	100	98.0	92.2	83.2	71.8	59.0	45.8	32.8	20.7	9.7
0.2	12.6	25.4	38.2	50.9	63.2	74.7	84.8	92.9	98.1	100	98.1	92.3	83.2	71.6	58.4	44.9	31.8	19.8	9.1
0.3	14.1	27.7	40.9	53.5	65.5	76.4	85.8	93.3	98.2	100	98.1	92.5	83.5	71.7	58.4	44.5	31.1	19.0	8.6
0.4	16.0	30.7	44.2	56.8	68.2	78.4	87.1	93.9	98.4	100	98.3	92.9	84.1	72.4	58.8	44.6	30.8	18.5	8.2
0.5	18.5	34.4	48.4	60.6	71.4	80.8	88.6	94.6	98.6	100	98.4	93.5	85.1	73.7	60.0	45.3	31.1	18.4	7.9
0.6	22.0	39.4	53.6	65.4	75.3	83.6	90.3	95.5	98.8	100	98.6	94.3	86.6	75.7	62.1	47.1	32.0	18.6	7.8
0.7	27.3	46.2	60.3	71.2	79.9	86.8	92.3	96.4	99.0	100	98.9	95.3	88.7	78.8	65.7	50.3	34.2	19.5	7.8
0.8	35.9	56.2	69.3	78.5	85.4	90.6	94.6	97.5	99.3	100	99.2	96.6	91.4	83.2	71.4	56.2	38.9	22.0	8.4
0.9	52.8	71.8	81.7	87.8	92.0	94.9	97.1	98.7	99.6	100	99.6	98.1	95.1	89.8	81.3	68.3	50.3	29.3	10.5

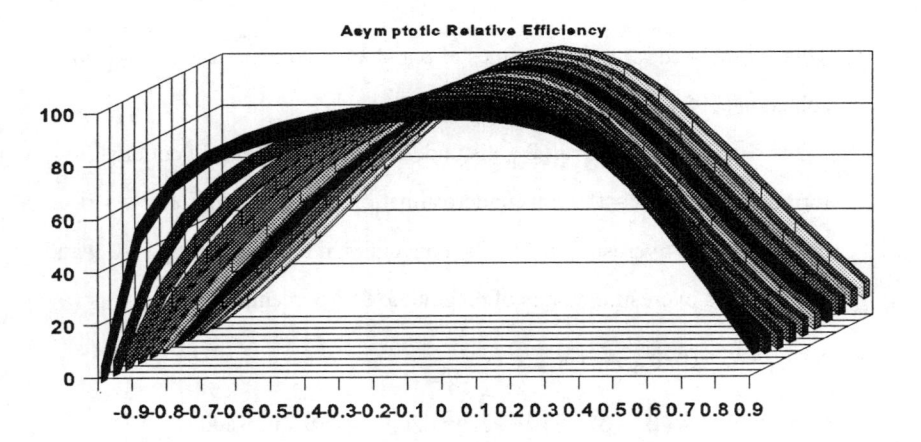

Asymptotic Relative Efficiency

d. Ignoring autocorrelation, $s^2/\sum_{t=1}^{T} x_t^2$ estimates $\sigma_u^2/\sum_{t=1}^{T} x_t^2$, but

$$\text{asy.var}(\hat{\beta}_{ols}) = (\sigma_u^2/\sum_{t=1}^{T} x_t^2)(1-\rho\lambda)/(1+\rho\lambda)$$

so the asy.bias in estimating the $\text{var}(\hat{\beta}_{ols})$ is

$$\text{asy.var}(\hat{\beta}_{ols}) - \sigma_u^2/\sum_{t=1}^{T} x_t^2 = \frac{\sigma_u^2}{\sum_{t=1}^{T} x_t^2}\left[\frac{1-\rho\lambda}{1+\rho\lambda} - 1\right] = \frac{\sigma_u^2}{\sum_{t=1}^{T} x_t^2}\left[\frac{-2\rho\lambda}{1+\rho\lambda}\right]$$

and asy.proportionate bias = $-2\rho\lambda/(1+\rho\lambda)$.

Percentage Bias in estimating var($\hat{\beta}_{ols}$)

ρ

λ	-0.9	-0.8	-0.7	-0.6	-0.5	-0.4	-0.3	-0.2	0	0.1	0.3	0.4	0.5	0.6	0.7	0.8	0.9
0	0	0	0	0	0	0	0	0	0	0	0	0	0	0	0	0	0
0.1	19.8	17.4	15.1	12.8	10.5	8.3	6.2	4.1	0	-2.0	-5.8	-7.7	-9.5	-11.3	-13.1	-14.8	-16.5
0.2	43.9	38.1	32.6	27.3	22.2	17.4	12.8	8.3	0	-3.9	-11.3	-14.8	-18.2	-21.4	-24.6	-27.6	-30.5
0.3	74.0	63.2	53.2	43.9	35.3	27.3	19.8	12.8	0	-5.8	-16.5	-21.4	-26.1	-30.5	-34.7	-38.7	-42.5
0.4	112.5	94.1	77.8	63.2	50.0	38.1	27.3	17.4	0	-7.7	-21.4	-27.6	-33.3	-38.7	-43.8	-48.5	-52.9
0.5	163.6	133.3	107.7	85.7	66.7	50.0	35.3	22.2	0	-9.5	-26.1	-33.3	-40.0	-46.2	-51.9	-57.1	-62.1
0.6	234.8	184.6	144.8	112.5	85.7	63.2	43.9	27.3	0	-11.3	-30.5	-38.7	-46.2	-52.9	-59.2	-64.9	-70.1
0.7	340.5	254.5	192.2	144.8	107.7	77.8	53.2	32.6	0	-13.1	-34.7	-43.8	-51.9	-59.2	-65.8	-71.8	-77.3
0.8	514.3	355.6	254.5	184.6	133.3	94.1	63.2	38.1	0	-14.8	-38.7	-48.5	-57.1	-64.9	-71.8	-78.0	-83.7
0.9	852.6	514.3	340.5	234.8	163.6	112.5	74.0	43.9	0	-16.5	-42.5	-52.9	-62.1	-70.1	-77.3	-83.7	-89.5

This is tabulated for various values of ρ and λ. A similar table is given in Johnston (1984, p. 312).

For ρ and λ positive, var($\hat{\beta}_{ols}$) is underestimated by the conventional formula. For $\rho = \lambda = 0.9$, this underestimation is almost 90%. For $\rho < 0$, the var($\hat{\beta}_{ols}$) is overestimated by the conventional formula. For $\rho = -0.9$ and $\lambda = 0.9$, this overestimation is of magnitude 853 percent.

e. $e_t = y_t - \hat{y}_t = (\beta - \hat{\beta}_{ols})x_t + u_t$

Hence, $\sum_{t=1}^{T} e_t^2 = (\hat{\beta}_{ols} - \beta)^2 \sum_{t=1}^{T} x_t^2 + \sum_{t=1}^{T} u_t^2 - 2(\hat{\beta}_{ols} - \beta) \sum_{t=1}^{T} x_t u_t$ and

$E(\sum_{t=1}^{T} e_t^2) = \sum_{t=1}^{T} x_t^2 \, var(\hat{\beta}_{ols}) + T\sigma_u^2 - 2(\sum_{s=1}^{T} x_s u_s)(\sum_{t=1}^{T} x_t u_t)/(\sum_{t=1}^{T} x_t^2)$

$= \sum_{t=1}^{T} x_t^2 \, var(\hat{\beta}_{ols}) + T\sigma_u^2$

$= -\sigma_u^2 \left[1 + 2\rho \frac{\sum_{t=1}^{T-1} x_t x_{t+1}}{\sum_{t=1}^{T} x_t^2} + 2\rho^2 \frac{\sum_{t=1}^{T-2} x_t x_{t+2}}{\sum_{t=1}^{T} x_t^2} + ... + 2\rho^{T-1} \frac{x_1 x_T}{\sum_{t=1}^{T} x_t^2} \right] + T\sigma_u^2$

Chapter 5: Violations of the Classical Assumptions   79

So that $E(s^2) = E(\sum_{t=1}^{T} e_t^2/(T-1))$

$$= \sigma_u^2 \left\{ T - \left(1 + 2\rho \frac{\sum_{t=1}^{T-1} x_t x_{t+1}}{\sum_{t=1}^{T} x_t^2} + 2\rho^2 \frac{\sum_{t=1}^{T-2} x_t x_{t+2}}{\sum_{t=1}^{T} x_t^2} + ... + 2\rho^{T-1} \frac{x_1 x_T}{\sum_{t=1}^{T} x_t^2}\right) \right\} / (T-1)$$

If $\rho = 0$, then $E(s^2) = \sigma_u^2$. If x_t follows an AR(1) model with parameter λ, then

for large T $E(s^2) = \sigma_u^2 \left(T - \frac{1+\rho\lambda}{1-\rho\lambda}\right) / (T-1)$.

For T = 101, $E(s^2) = \sigma_u^2 \left(101 - \frac{1+\rho\lambda}{1-\rho\lambda}\right) / 100$. This can be tabulated for various values of ρ and λ. For example, when $\rho=\lambda=0.9$, $E(s^2) = 0.915 \sigma_u^2$.

5.10 *Regressions with Non-Zero Mean Disturbances.*

a. For the gamma distribution, $E(u_i) = \theta$ and $var(u_i) = \theta$. Hence, the disturbances of the simple regression have non-zero mean but constant variance. Adding and subtracting θ on the right hand side of the regression we get $Y_i = (\alpha+\theta) + \beta X_i + (u_i - \theta) = \alpha^* + \beta X_i + u_i^*$ where $\alpha^* = \alpha + \theta$ and $u_i^* = u_i - \theta$ with $E(u_i^*) = 0$ and $var(u_i^*) = \theta$. OLS yields the BLU estimators of α^* and β since the u_i^* disturbances satisfy all the requirements of the Gauss-Markov Theorem, including the zero mean assumption. Hence, $E(\hat{\alpha}_{ols}) = \alpha^* = \alpha + \theta$ which is biased for α by the mean of the disturbances θ. But $E(s^2) = var(u_i^*) = var(u_i) = \theta$. Therefore,

$E(\hat{\alpha}_{ols} - s^2) = E(\hat{\alpha}_{ols}) - E(s^2) = \alpha + \theta - \theta = \alpha$.

b. Similarly, for the χ_v^2 distribution, we have $E(u_i) = v$ and $var(u_i) = 2v$. In this case, the same analysis applies except that $E(\hat{\alpha}_{ols}) = \alpha + v$ and $E(s^2) = 2v$. Hence, $E(\hat{\alpha}_{ols} - s^2/2) = \alpha$.

c. Finally, for the exponential distribution, we have $E(u_i) = \theta$ and $var(u_i) = \theta^2$. In this case, plim $\hat{\alpha}_{ols} = \alpha + \theta$ and plim $s^2 = \theta^2$. Hence,

plim($\hat{\alpha}_{ols} - s$) = plim $\hat{\alpha}_{ols}$ - plim $s = \alpha$.

5.11 *The Heteroskedastic Consequences of an Arbitrary Variance for the Initial Disturbance of an AR(1) Model.* Parts (a), (b), and (c) are based on the solution given by Kim (1991).

a. Continual substitution into the process $u_t = \rho u_{t-1} + \epsilon_t$ yields

$$u_t = \rho^t u_0 + \sum_{j=0}^{t-1} \rho^j \epsilon_{t-j}.$$

Exploiting the dependence conditions, $E(u_0 \epsilon_j) = 0$ for all j, and $E(\epsilon_i \epsilon_j) = 0$ for all $i \neq j$, the variance of u_t is

$$\sigma_t^2 = var(u_t) = \rho^{2t} var(u_0) + \sum_{j=0}^{t-1} \rho^{2j} var(\epsilon_{t-j}) = \rho^{2t}\sigma_\epsilon^2/\tau + \sigma_\epsilon^2 \sum_{j=0}^{t-1} \rho^{2j}$$

$$= \rho^{2t}\sigma_\epsilon^2/\tau + \sigma_\epsilon^2 [(1-\rho^{2t})/(1-\rho^2)] = [\{1/(1-\rho^2) + [1/\tau - 1/(1-\rho^2)]\rho^{2t}\}\sigma_\epsilon^2.$$

This expression for σ_t^2 depends on t. Hence, for an arbitrary value of τ, the disturbances are rendered heteroskedastic. If $\tau = 1 - \rho^2$ then σ_t^2 does not depend on t and σ_t^2 becomes homoskedastic.

b. Using the above equation for σ_t^2, define

$$a = \sigma_t^2 - \sigma_{t-1}^2 = [(1/\tau) - 1/(1-\rho^2)] \rho^{2t}(1-\rho^{-2}) \sigma_\epsilon^2 = bc$$

where $b = (1/\tau) - 1/(1-\rho^2)$ and $c = \rho^{2t}(1-\rho^{-2}) \sigma_\epsilon^2$. Note that $c < 0$. If $\tau > 1 - \rho^2$, then $b < 0$ implying that $a > 0$ and the variance is increasing. While if $\tau < 1 - \rho^2$, then $b > 0$ implying that $a < 0$ and the variance is decreasing. If $\tau = 1 - \rho^2$ then $a = 0$ and the process becomes homoskedastic.

Alternatively, one can show that $\dfrac{\partial \sigma_t^2}{\partial t} = \dfrac{2\sigma_\epsilon^2 \rho^{2t} \log \rho}{\tau(1-\rho^2)} [(1-\rho^2) - \tau]$

where $\partial a^{bx}/\partial x = b a^{bx} \log a$, has been used. Since $\log \rho < 0$, the right hand side of the above equation is positive (σ_t^2 is increasing) if $\tau > 1 - \rho^2$, and is

Chapter 5: Violations of the Classical Assumptions   81

negative otherwise.

c. Using $u_t = \rho^t u_0 + \sum_{j=0}^{t-1} \rho^j \epsilon_{t-j}$, and noting that $E(u_t) = 0$ for all t, finds, for $t \geq s$,

$$\text{cov}(u_t, u_{t-s}) = E(u_t u_{t-s}) = \left(\rho^t u_0 + \sum_{i=0}^{t-1} \rho^i \epsilon_{t-i}\right)\left(\rho^{t-s} u_0 + \sum_{j=0}^{t-s-1} \rho^j \epsilon_{t-s-j}\right)$$

$$= \rho^{2t-s} \text{var}(u_0) + \rho^s \sum_{i=0}^{t-s-1} \rho^{2i} \text{var}(\epsilon_{t-s-i})$$

$$= \rho^{2t-s} \sigma_\epsilon^2 / \tau + \sigma_\epsilon^2 [\rho^s (1-\rho^{2(t-s)})/(1-\rho^2)]$$

$$= \rho^s \{1/(1-\rho^2) + [(1/\tau)-1/(1-\rho^2)]\rho^{2(t-s)}\} \sigma_\epsilon^2 = \rho^s \sigma_{t-s}^2$$

d. *The Bias of the Standard Errors of OLS Process with an Arbitrary Variance on the Initial Observation.* This is based on the solution by Koning (1992). Consider the time-series model

$$y_t = \beta x_t + u_t$$

$$u_t = \rho u_{t-1} + \epsilon_t$$

with $0 < \rho < 1$, $\epsilon_t \sim \text{IIN}(0, \sigma_\epsilon^2)$, $u_0 \sim \text{IIN}(0, (\sigma_\epsilon^2/\tau))$, $\tau > 0$. The x's are positively autocorrelated. Note that

$$\text{var}(u_t) = \sigma_t^2 = \sigma_\epsilon^2 \left(\frac{\rho^{2t}}{\tau} + \sum_{s=1}^{t-1} \rho^{2s}\right) = \sigma_\epsilon^2 \left(\frac{\rho^{2t}}{\tau} + \frac{1-\rho^{2t}}{1-\rho^2}\right).$$

The population variance of the OLS estimator for β is

$$\text{var}(\hat{\beta}_{ols}) = \frac{\sum_{t=1}^{T}\sum_{s=1}^{T} x_s x_t \text{cov}(u_t, u_s)}{\left(\sum_{t=1}^{T} x_t^2\right)^2} = \frac{\sum_{t=1}^{T} \sigma_t^2 x_t^2 + 2\sum_{t=1}^{T}\sum_{s>t} x_s x_t \rho^{s-t} \sigma_t^2}{\left(\sum_{t=1}^{T} x_t^2\right)^2}.$$

Also, note that in the stationary case ($\tau = 1-\rho^2$), we have:

$$\text{var}(\hat{\beta}_{ols}) = \frac{\sigma_\epsilon^2/(1-\rho^2)}{\sum_{t=1}^{T} x_t^2} + \frac{(2\sigma_\epsilon^2/(1-\rho^2)) \sum_{t=1}^{T}\sum_{s>t} x_s x_t \rho^{s-t}}{\left(\sum_{t=1}^{T} x_t^2\right)^2} \geq \frac{\sigma_\epsilon^2/(1-\rho^2)}{\sum_{t=1}^{T} x_t^2},$$

since each term in the double summation is non-negative. Hence, the true variance of $\hat{\beta}_{ols}$ (the left hand side of the last equation) is greater than the estimated variance of $\hat{\beta}_{ols}$ (the right hand side of the last equation). Therefore, the true t-ratio corresponding to H_o; $\beta = 0$ versus H_1; $\beta \neq 0$ is lower than the estimated t-ratio. In other words, OLS rejects too often.

From the equation for σ_t^2 it is easily seen that $\partial \sigma_t^2/\partial \tau < 0$, and hence, that the true variance of $\hat{\beta}_{ols}$ is larger than the estimated var($\hat{\beta}_{ols}$) if $\tau < 1 - \rho^2$. Hence, the true t-ratio decreases and compared to the stationary case, OLS rejects more often. The opposite holds for $\tau > 1 - \rho^2$.

5.12 *ML Estimation of Linear Regression Model with AR(1) Errors and Two Observations.* This is based on Baltagi and Li (1995).

a. The OLS estimator of β is given by

$$\hat{\beta}_{ols} = \sum_{i=1}^{2} x_i y_i / \sum_{i=1}^{2} x_i^2 = (y_1 x_1 + y_2 x_2)/(x_1^2 + x_2^2).$$

b. The log-likelihood function is given by

$\log L = -\log 2\pi - \log \sigma^2 - (1/2)\log(1-\rho^2) - (u_2^2 - 2\rho u_1 u_2 + u_1^2)/2\sigma^2(1-\rho^2);$

setting $\partial \log L/\partial \sigma^2 = 0$ gives $\hat{\sigma}^2 = (u_2^2 - 2\rho u_1 u_2 + u_1^2)/2(1-\rho^2);$

setting $\partial \log L/\partial \rho = 0$ gives $\rho\sigma^2(1-\rho^2) + u_1 u_2 + \rho^2 u_1 u_2 - \rho u_2^2 - \rho u_1^2 = 0;$

substituting $\hat{\sigma}^2$ in this last equation, we get $\hat{\rho} = 2u_1 u_2/(u_1^2 + u_2^2);$

setting $\partial \log L/\partial \beta = 0$ gives $u_2(x_2 - \rho x_1) + u_1(x_1 - \rho x_2) = 0;$

substituting $\hat{\rho}$ in this last equation, we get $(u_1 x_1 - u_2 x_2)(u_1^2 - u_2^2) = 0$. Note that $u_1^2 = u_2^2$ implies a $\hat{\rho}$ of ± 1, and this is ruled out by the stationarity of the AR(1) process. Solving $(u_1 x_1 - u_2 x_2) = 0$ gives the required MLE of β:

$\hat{\beta}_{mle} = (y_1 x_1 - y_2 x_2)/(x_1^2 + x_2^2).$

Chapter 5: Violations of the Classical Assumptions

c. By substituting $\hat{\beta}_{mle}$ into u_1 and u_2, one gets $\hat{u}_1 = x_2(x_1y_2-x_2y_1)/(x_1^2-x_2^2)$ and $\hat{u}_2 = x_1(x_1y_2 - x_2y_1)/(x_1^2 - x_2^2)$, which upon substitution in the expression for $\hat{\rho}$ give the required estimate of ρ: $\quad \hat{\rho} = 2x_1x_2/(x_1^2+x_2^2)$.

d. If $x_1 \to x_2$, with $x_2 \neq 0$, then $\hat{\rho} \to 1$. For $\hat{\beta}_{mle}$, we distinguish between two cases.

   (i) For $y_1 = y_2$, $\hat{\beta}_{mle} \to y_2/(2x_2)$, which is half the limit of $\hat{\beta}_{ols} \to y_2 + x_2$. The latter is the slope of the line connecting the origin to the observation (x_2,y_2).

   (ii) For $y_1 \neq y_2$, $\hat{\beta}_{mle} \to \pm\infty$, with the sign depending on the sign of x_2, (y_1-y_2), and the direction from which x_1 approaches x_2. In this case, $\hat{\beta}_{ols} \to \bar{y}/x_2$, where $\bar{y} = (y_1+y_2)/2$. This is the slope of the line connecting the origin to $(x_2, \bar{y})$.

   Similarly, if $x_1 \to -x_2$, with $x_2 \neq 0$, then $\hat{\rho} \to -1$. For $\hat{\beta}_{mle}$, we distinguish between two cases:

   (i) For $y_1 = -y_2$, $\hat{\beta}_{mle} \to y_2$, $\hat{\beta}_{mle} \to y_2/(2x_2)$, which is half the limit of $\hat{\beta}_{ols} \to y_2/x_2$.

   (ii) For $y_1 \neq -y_2$, $\hat{\beta}_{mle} \to \pm\infty$, with the sign depending on the sign of x_2, (y_1+y_2), and the direction from which x_1 approaches $-x_2$. In this case, $\hat{\beta}_{ols} \to (y_2-y_1)/2x_2 = (y_2-y_1)/(x_2-x_1)$, which is the standard formula for the slope of a straight line based on two observations. In conclusion, $\hat{\beta}_{ols}$ is a more reasonable estimator of β than $\hat{\beta}_{mle}$ for this two-observation example.

5.13 The backup regressions are given below: These are performed using SAS.

```
OLS REGRESSION OF LNC ON CONSTANT, LNP, AND LNY

Dependent Variable: LNC
```

Analysis of Variance

Source	DF	Sum of Squares	Mean Square	F Value	Prob>F
Model	2	0.50098	0.25049	9.378	0.0004
Error	43	1.14854	0.02671		
C Total	45	1.64953			

Root MSE	0.16343	R-square	0.3037		
Dep Mean	4.84784	Adj R-sq	0.2713		
C.V.	3.37125				

Parameter Estimates

Variable	DF	Parameter Estimate	Standard Error	T for H0: Parameter=0	Prob > |T|
INTERCEP	1	4.299662	0.90892571	4.730	0.0001
LNP	1	-1.338335	0.32460147	-4.123	0.0002
LNY	1	0.172386	0.19675440	0.876	0.3858

PLOT OF RESIDUAL VS. LNY

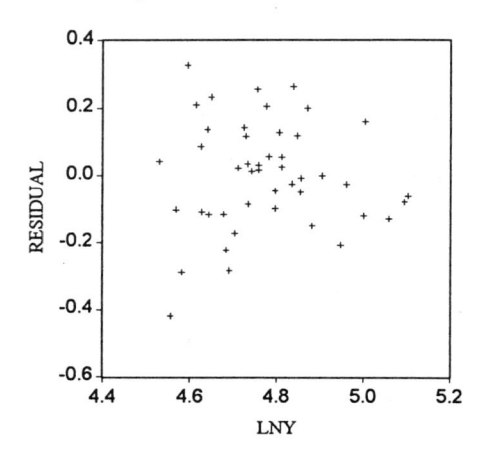

a. Regression for Glejser Test (1969)

```
Dependent Variable: ABS_E
MODEL: Z1=LNY^(-1)
```

Analysis of Variance

Source	DF	Sum of Squares	Mean Square	F Value	Prob>F
Model	1	0.04501	0.04501	5.300	0.0261
Error	44	0.37364	0.00849		
C Total	45	0.41865			

Root MSE	0.09215	R-square	0.1075
Dep Mean	0.12597	Adj R-sq	0.0872
C.V.	73.15601		

Parameter Estimates

Variable	DF	Parameter Estimate	Standard Error	T for H0: Parameter=0	Prob > |T|
INTERCEP	1	-0.948925	0.46709261	-2.032	0.0483
Z1(LNY^-1)	1	5.128691	2.22772532	2.302	0.0261

```
MODEL: Z2=LNY^(-0.5)
Dependent Variable: ABS_E
```

Analysis of Variance

Source	DF	Sum of Squares	Mean Square	F Value	Prob>F
Model	1	0.04447	0.04447	5.229	0.0271
Error	44	0.37418	0.00850		
C Total	45	0.41865			

Root MSE	0.09222	R-square	0.1062
Dep Mean	0.12597	Adj R-sq	0.0859
C.V.	73.20853		

Parameter Estimates

Variable	DF	Parameter Estimate	Standard Error	T for H0: Parameter=0	Prob > |T|
INTERCEP	1	-2.004483	0.93172690	-2.151	0.0370
Z2(LNY^-.5)	1	4.654129	2.03521298	2.287	0.0271

MODEL: Z3=LNY^(0.5)
Dependent Variable: ABS_E

Analysis of Variance

Source	DF	Sum of Squares	Mean Square	F Value	Prob>F
Model	1	0.04339	0.04339	5.087	0.0291
Error	44	0.37526	0.00853		
C Total	45	0.41865			

Root MSE	0.09235	R-square	0.1036	
Dep Mean	0.12597	Adj R-sq	0.0833	
C.V.	73.31455			

Parameter Estimates

Variable	DF	Parameter Estimate	Standard Error	T for H0: Parameter=0	Prob > |T|
INTERCEP	1	2.217240	0.92729847	2.391	0.0211
Z3(LNY^.5)	1	-0.957085	0.42433823	-2.255	0.0291

MODEL: Z4=LNY^1
Dependent Variable: ABS_E

Analysis of Variance

Source	DF	Sum of Squares	Mean Square	F Value	Prob>F
Model	1	0.04284	0.04284	5.016	0.0302
Error	44	0.37581	0.00854		
C Total	45	0.41865			

Root MSE	0.09242	R-square	0.1023	
Dep Mean	0.12597	Adj R-sq	0.0819	
C.V.	73.36798			

Parameter Estimates

Variable	DF	Parameter Estimate	Standard Error	T for H0: Parameter=0	Prob > |T|
INTERCEP	1	1.161694	0.46266689	2.511	0.0158
Z4(LNY^1)	1	-0.216886	0.09684233	-2.240	0.0302

Chapter 5: Violations of the Classical Assumptions

c. Regression for Goldfeld and Quandt Test (1965) with first 17 obervations

Dependent Variable: LNC

Analysis of Variance

Source	DF	Sum of Squares	Mean Square	F Value	Prob>F
Model	2	0.22975	0.11488	2.354	0.1315
Error	14	0.68330	0.04881		
C Total	16	0.91305			

Root MSE	0.22092	R-square	0.2516	
Dep Mean	4.85806	Adj R-sq	0.1447	
C.V.	4.54756			

Parameter Estimates

Variable	DF	Parameter Estimate	Standard Error	T for H0: Parameter=0	Prob > $|T|$
INTERCEP	1	3.983911	4.51225092	0.883	0.3922
LNP	1	-1.817254	0.86970957	-2.089	0.0554
LNY	1	0.248409	0.96827122	0.257	0.8013

REGRESSION FOR GOLDFELD AND QUANDT TEST (1965) w/ last 17 obs

Dependent Variable: LNC

Analysis of Variance

Source	DF	Sum of Squares	Mean Square	F Value	Prob>F
Model	2	0.17042	0.08521	5.482	0.0174
Error	14	0.21760	0.01554		
C Total	16	0.38803			

Root MSE	0.12467	R-square	0.4392	
Dep Mean	4.78796	Adj R-sq	0.3591	
C.V.	2.60387			

Parameter Estimates

Variable	DF	Parameter Estimate	Standard Error	T for H0: Parameter=0	Prob > $|T|$
INTERCEP	1	6.912881	1.57090447	4.401	0.0006
LNP	1	-1.248584	0.39773565	-3.139	0.0072
LNY	1	-0.363625	0.31771223	-1.145	0.2716

d. Data for Spearman Rank Correlation Test

OBS	STATE	LNC	RANKY	ABS_E	RANKE	D
1	MS	4.93990	1	0.04196	11	10
2	UT	4.40859	2	0.41867	46	44
3	WV	4.82454	3	0.10198	21	18
4	NM	4.58107	4	0.28820	44	40
5	AR	5.10709	5	0.32868	45	40
6	LA	4.98602	6	0.21014	38	32
7	SC	5.07801	7	0.08730	19	12
8	OK	4.72720	8	0.10845	22	14
9	AL	4.96213	9	0.13671	30	21
10	ID	4.74902	10	0.11628	24	14
11	KY	5.37906	11	0.23428	40	29
12	SD	4.81545	12	0.11470	23	11
13	AZ	4.66312	13	0.22128	39	26
14	ND	4.58237	14	0.28253	43	29
15	MT	4.73313	15	0.17266	34	19
16	WY	5.00087	16	0.02320	5	-11
17	TN	5.04939	17	0.14323	31	14
18	IN	5.11129	18	0.11673	25	7
19	GA	4.97974	19	0.03583	10	-9
20	TX	4.65398	20	0.08446	18	-2
21	IA	4.80857	21	0.01372	3	-18
22	ME	4.98722	22	0.25740	41	19
23	WI	4.83026	23	0.01754	4	-19
24	OH	4.97952	24	0.03201	9	-15
25	VT	5.08799	25	0.20619	36	11
26	MO	5.06430	26	0.05716	15	-11
27	KS	4.79263	27	0.04417	12	-15
28	NE	4.77558	28	0.09793	20	-8
29	MI	4.94744	29	0.12797	28	-1
30	FL	4.80081	30	0.05625	14	-16
31	MN	4.69589	31	0.02570	8	-23
32	PA	4.80363	32	0.02462	6	-26
33	NV	4.96642	33	0.26506	42	9
34	RI	4.84693	34	0.11760	26	-8
35	VA	4.93065	35	0.04776	13	-22
36	WA	4.66134	36	0.00638	2	-34
37	DE	5.04705	37	0.20120	35	-2
38	CA	4.50449	38	0.14953	32	-6
39	IL	4.81445	39	0.00142	1	-38
40	MD	4.77751	40	0.20664	37	-3
41	NY	4.66496	41	0.02545	7	-34
42	MA	4.73877	42	0.12018	27	-15
43	NH	5.10990	43	0.15991	33	-10
44	DC	4.65637	44	0.12810	29	-15
45	CT	4.66983	45	0.07783	17	-28
46	NJ	4.70633	46	0.05940	16	-30

SPEARMAN RANK CORRELATION TEST

OBS	R	T
1	-0.28178	1.94803

Chapter 5: Violations of the Classical Assumptions

e. Harvey's Multiplicative Heteroskedasticity Test (1976)

Dependent Variable: LNE_SQ

Analysis of Variance

Source	DF	Sum of Squares	Mean Square	F Value	Prob>F
Model	1	14.36012	14.36012	2.992	0.0907
Error	44	211.14516	4.79875		
C Total	45	225.50528			

Root MSE	2.19061	R-square	0.0637	
Dep Mean	-4.97462	Adj R-sq	0.0424	
C.V.	-44.03568			

Parameter Estimates

Variable	DF	Parameter Estimate	Standard Error	T for H0: Parameter=0	Prob > |T|
INTERCEP	1	24.852752	17.24551850	1.441	0.1566
LLNY	1	-19.082690	11.03125044	-1.730	0.0907

Variable	N	Mean	Std Dev	Minimum	Maximum
HV_TEMP	46	25.0589810	22.8713299	-2.4583422	113.8840204
LNE_SQ	46	-4.9746160	2.2385773	-13.1180671	-1.7413218

HARVEY'S MULTIPLICATIVE HETEROSKEDASTICITY TEST (1976)

OBS	HV_TEST
1	2.90997

f. Regression for Breusch & Pagan Test (1979)

Dependent Variable: X2

Analysis of Variance

Source	DF	Sum of Squares	Mean Square	F Value	Prob>F
Model	1	10.97070	10.97070	6.412	0.0150
Error	44	75.28273	1.71097		
C Total	45	86.25344			

Root MSE	1.30804	R-square	0.1272	
Dep Mean	1.00001	Adj R-sq	0.1074	
C.V.	130.80220			

Parameter Estimates

Variable	DF	Parameter Estimate	Standard Error	T for H0: Parameter=0	Prob > |T|
INTERCEP	1	17.574476	6.54835118	2.684	0.0102
LNY	1	-3.470761	1.37065694	-2.532	0.0150

BREUSCH & PAGAN TEST (1979)

OBS	RSSBP
1	5.48535

g. Regression for WhiteTest (1979)

Dependent Variable: E_SQ

Analysis of Variance

Source	DF	Sum of Squares	Mean Square	F Value	Prob>F
Model	5	0.01830	0.00366	4.128	0.0041
Error	40	0.03547	0.00089		
C Total	45	0.05377			

Root MSE	0.02978	R-square	0.3404
Dep Mean	0.02497	Adj R-sq	0.2579
C.V.	119.26315		

Parameter Estimates

Variable	DF	Parameter Estimate	Standard Error	T for H0: Parameter=0	Prob > |T|
INTERCEP	1	18.221989	5.37406002	3.391	0.0016
LNP	1	9.506059	3.30257013	2.878	0.0064
LNY	1	-7.893179	2.32938645	-3.389	0.0016
LNP_SQ	1	1.281141	0.65620773	1.952	0.0579
LNPY	1	-2.078635	0.72752332	-2.857	0.0068
LNY_SQ	1	0.855726	0.25304827	3.382	0.0016

Normality Test (Jarque-Bera)
This chart was done with EViews.

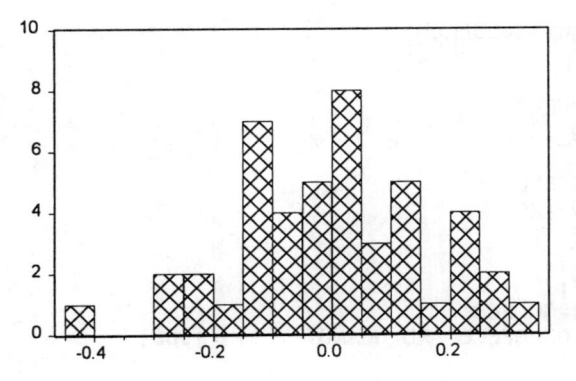

Series: RESID	
Sample 1 46	
Observations 46	
Mean	-9.95E-16
Median	0.007568
Maximum	0.328677
Minimum	-0.418675
Std. Dev.	0.159760
Skewness	-0.181935
Kurtosis	2.812520
Jarque-Bera	0.321137
Probability	0.851659

SAS PROGRAM

```
Data CIGAR;
Input OBS STATE $ LNC LNP LNY;
CARDS;

Proc reg data=CIGAR;
     Model LNC=LNP LNY;
  Output OUT=OUT1 R=RESID;

Proc Plot data=OUT1 hpercent=85 vpercent=60;
     Plot RESID*LNY='*';
run;

***** GLEJSER'S TEST(1969) *****;
*********************************;

Data GLEJSER; set OUT1;
  ABS_E=ABS(RESID);
  Z1=LNY**-1;
  Z2=LNY**-.5;
  Z3=LNY**.5;
  Z4=LNY;

Proc reg data=GLEJSER;
     Model ABS_E=Z1;
     Model ABS_E=Z2;
     Model ABS_E=Z3;
     Model ABS_E=Z4;
TITLE 'REGRESSION FOR GLEJSER TEST (1969)';
LABEL Z1='LNY^(-1)'
      Z2='LNY^(-0.5)'
      Z3='LNY^(0.5)'
      Z4='LNY^(1)';
run;
```

```
***** GOLDFELD & QUANDT TEST(1965) *****;
*****************************************;

Proc sort data=CIGAR out=GOLDFELD;
    By LNY;

Data GQTEST1; set GOLDFELD;
    If _N_<18; OBS=_N_;

Data GQTEST2; set GOLDFELD;
    If _N_>29; OBS=_N_-29;

Proc reg data=GQTEST1;
    Model LNC=LNP LNY;
  Output out=GQ_OUT1 R=GQ_RES1;
TITLE 'REGRESSION FOR GOLDFELD AND QUANDT TEST (1965)
        w/ first 17 obs';

Proc reg data=GQTEST2;
    Model LNC=LNP LNY;
  Output out=GQ_OUT2 R=GQ_RES2;
TITLE 'REGRESSION FOR GOLDFELD AND QUANDT TEST (1965)
        w/ last 17 obs';
run;

***** SPEARMAN'S RANK CORRELATION TEST *****;
**********************************************;

Data SPEARMN1; set GOLDFELD;
  RANKY=_N_;

Proc sort data=GLEJSER out=OUT2;
  By ABS_E;

Data TEMP1; set OUT2;
  RANKE=_N_;

Proc sort data=TEMP1 out=SPEARMN2;
  By LNY;

Data SPEARMAN;
  Merge SPEARMN1 SPEARMN2;
        By LNY;
  D=RANKE-RANKY;
* Difference b/w Ranking of |RES| and Ranking of LNY;
  D_SQ=D**2;

Proc means data=SPEARMAN NOPRINT;
  Var D_SQ;
  Output out=OUT3 SUM=SUM_DSQ;

Data SPTEST;  Set OUT3;
  R=1-((6*SUM_DSQ)/(46**3-46));
  T=SQRT(R**2*(46-2)/(1-R**2));
```

Chapter 5: Violations of the Classical Assumptions

```sas
Proc print data=SPEARMAN;
  Var STATE LNC RANKY ABS_E RANKE D;
TITLE 'DATA FOR SPEARMAN RANK CORRELATION TEST';

Proc print data=SPTEST;
  Var R T;
TITLE 'SPEARMAN RANK CORRELATION TEST';
run;

*HARVEY'S MULTIPLICATIVE HETEROSKEDASTICITY TEST (1976) **;
**************************************************;

Data HARVEY; set OUT1;
  E_SQ=RESID**2;
  LNE_SQ=LOG(E_SQ);
  LLNY=LOG(LNY);

Proc reg data=HARVEY;
     Model LNE_SQ=LLNY;
  Output out=OUT4 R=RLNESQ;
TITLE 'HARVEY``S MULTIPLICATIVE HETEROSKEDASTICITY TEST
       (1976)';

Data HARVEY1; set OUT4;
  HV_TEMP=LNE_SQ**2-RLNESQ**2;

Proc means data=HARVEY1;
  Var HV_TEMP LNE_SQ;
  Output out=HARVEY2 SUM=SUMTMP SUMLNESQ;

Data HARVEY3; set HARVEY2;
  HV_TEST=(SUMTMP-46*(SUMLNESQ/46)**2)/4.9348;

Proc print data=HARVEY3;
  Var HV_TEST;
TITLE 'HARVEY``S MULTIPLICATIVE HETEROSKEDASTICITY TEST
       (1976)';

***** BREUSCH & PAGAN TEST (1979) *****;
**************************************;

Proc means data=HARVEY;
  Var E_SQ;
  Output out=OUT5 MEAN=S2HAT;
  TITLE 'BREUSCH & PAGAN TEST (1979)';

Proc print data=OUT5;
  Var S2HAT;

Data BPTEST; set HARVEY;
  X2=E_SQ/0.024968;

Proc reg data=BPTEST;
     Model X2=LNY;
```

```
  Output out=OUT6 R=BP_RES;
  TITLE 'REGRESSION FOR BREUSCH & PAGAN TEST (1979)';

Data BPTEST1; set OUT6;
  BP_TMP=X2**2-BP_RES**2;

Proc means data=BPTEST1;
  Var BP_TMP X2;
  Output out=OUT7 SUM=SUMBPTMP SUMX2;

Data BPTEST2; set OUT7;
  RSSBP=(SUMBPTMP-SUMX2**2/46)/2;

Proc print data=BPTEST2;
  Var RSSBP;

***** WHITE'S TEST (1980) *****;
*******************************;

Data WHITE; set HARVEY;
  LNP_SQ=LNP**2;
  LNY_SQ=LNY**2;
  LNPY=LNP*LNY;

Proc reg data=WHITE;
    Model E_SQ=LNP LNY LNP_SQ LNPY LNY_SQ;
  TITLE 'REGRESSION FOR WHITE TEST (1979)';
run;
```

5.15 The backup regressions are given below. These are performed using SAS.

    a. OLS Regression of consumption on constant and income

```
Dependent Variable: C
```

Analysis of Variance

Source	DF	Sum of Squares	Mean Square	F Value	Prob>F
Model	1	264463459.78	264463459.78	11209.210	0.0001
Error	42	990923.13097	23593.40788		
C Total	43	265454382.91			

Root MSE	153.60146	R-square	0.9963
Dep Mean	9250.54545	Adj R-sq	0.9962
C.V.	1.66046		

Chapter 5: Violations of the Classical Assumptions

Parameter Estimates

Variable	DF	Parameter Estimate	Standard Error	T for H0: Parameter=0	Prob > |T|
INTERCEP	1	-65.795821	90.99082402	-0.723	0.4736
Y	1	0.915623	0.00864827	105.874	0.0001

Durbin-Watson D 0.461
(For Number of Obs.) 44
1st Order Autocorrelation 0.707

b. Cochrane-Orcutt (1949) Method

NOTE: No intercept in model. R-square is redefined.
Dependent Variable: RESID   Residual

Analysis of Variance

Source	DF	Sum of Squares	Mean Square	F Value	Prob>F
Model	1	555133.91473	555133.91473	55.731	0.0001
Error	42	418360.04492	9960.95345		
U Total	43	973493.95965			

Root MSE	99.80458	R-square	0.5702	
Dep Mean	-3.07022	Adj R-sq	0.5600	
C.V.	-3250.72732			

Parameter Estimates

Variable	DF	Parameter Estimate	Standard Error	T for H0: Parameter=0	Prob > |T|
RHO	1	0.792116	0.10610610	7.465	0.0001

REGRESSION WITH COCHRANE-ORCUTT(1949) METHOD

NOTE: No intercept in model. R-square is redefined.
Dependent Variable: CO_C

Analysis of Variance

Source	DF	Sum of Squares	Mean Square	F Value	Prob>F
Model	2	198151710.54	99075855.272	9755.993	0.0001
Error	41	416370.73707	10155.38383		
U Total	43	198568081.28			

```
              Root MSE       100.77392    R-square      0.9979
              Dep Mean      2079.09308    Adj R-sq      0.9978
              C.V.             4.84701
```

Parameter Estimates

Variable	DF	Parameter Estimate	Standard Error	T for H0: Parameter=0	Prob > |T|
CO_INT	1	-168.488581	301.66743280	-0.559	0.5795
CO_Y	1	0.926261	0.02663823	34.772	0.0001

c. Prais-Winsten (1954, Yule-Walker) 2-Step Method

```
Autoreg Procedure
Dependent Variable = C
```

Ordinary Least Squares Estimates

```
              SSE           990923.1    DFE                  42
              MSE           23593.41    Root MSE        153.6015
              SBC           573.4119    AIC             569.8435
              Reg Rsq         0.9963    Total Rsq         0.9963
              Durbin's t    7.422247    PROB>t            0.0001
              Durbin-Watson   0.4608
```

Variable	DF	B Value	Std Error	t Ratio	Approx Prob
Intercept	1	-65.7958208	90.991	-0.723	0.4736
Y	1	0.9156232	0.008648	105.874	0.0001

Estimates of Autocorrelations

Lag	Covariance	Correlation	-1 9 8 7 6 5 4 3 2 1 0 1 2 3 4 5 6 7 8 9 1
0	22520.98	1.000000	|********************|
1	15927.82	0.707244	|************** |

```
                  Preliminary MSE = 11256.13
```

Estimates of the Autoregressive Parameters

Lag	Coefficient	Std Error	t Ratio
1	-0.70724357	0.11041016	-6.405602

Chapter 5: Violations of the Classical Assumptions

Yule-Walker Estimates

SSE	431266	DFE	41
MSE	10518.68	Root MSE	102.5606
SBC	541.2855	AIC	535.9329
Reg Rsq	0.9858	Total Rsq	0.9984
Durbin-Watson	1.7576		

Variable	DF	B Value	Std Error	t Ratio	Approx Prob
Intercept	1	-52.6348364	181.93	-0.289	0.7738
Y	1	0.9165547	0.017165	53.398	0.0001

d. Prais-Winsten(1954, Yule-Walker) Iterative Method

Yule-Walker Estimates

SSE	428239.5	DFE	41
MSE	10444.86	Root MSE	102.2001
SBC	541.0303	AIC	535.6778
Reg Rsq	0.9844	Total Rsq	0.9984
Durbin-Watson	1.7996		

Variable	DF	B Value	Std Error	t Ratio	Approx Prob
Intercept	1	-48.4072690	191.13	-0.253	0.8013
Y	1	0.9163298	0.018019	50.853	0.0001

Maximum Likelihood Estimates

SSE	422433.3	DFE	41
MSE	10303.25	Root MSE	101.5049
SBC	540.652	AIC	535.2994
Reg Rsq	0.9773	Total Rsq	0.9984
Durbin-Watson	1.9271		

Variable	DF	B Value	Std Error	t Ratio	Approx Prob
Intercept	1	-24.8162978	233.50	-0.106	0.9159
Y	1	0.9148510	0.022159	41.285	0.0001
A(1)	1	-0.7881142	0.11	-7.347	0.0001

Autoregressive parameters assumed given.

Autoreg Procedure

Variable	DF	B Value	Std Error	t Ratio	Approx Prob
Intercept	1	-24.8162978	231.74	-0.107	0.9152
Y	1	0.9148510	0.021766	42.031	0.0001

e. Durbin (1960) Method

Dependent Variable: C

Analysis of Variance

Source	DF	Sum of Squares	Mean Square	F Value	Prob>F
Model	3	253053288.45	84351096.151	9169.543	0.0001
Error	39	358762.98953	9199.05101		
C Total	42	253412051.44			

Root MSE	95.91168	R-square	0.9986	
Dep Mean	9330.32558	Adj R-sq	0.9985	
C.V.	1.02796			

Parameter Estimates

Variable	DF	Parameter Estimate	Standard Error	T for H0: Parameter=0	Prob > |T|
INTERCEP	1	-40.791743	59.78575886	-0.682	0.4991
CLAG	1	0.800000	0.10249693	7.805	0.0001
Y	1	0.719519	0.08649875	8.318	0.0001
YLAG	1	-0.529709	0.12837635	-4.126	0.0002

DURBIN(1960) METHOD

NOTE: No intercept in model. R-square is redefined.
Dependent Variable: D_C

Analysis of Variance

Source	DF	Sum of Squares	Mean Square	F Value	Prob>F
Model	2	184638724.05	92319362.023	9086.413	0.0001
Error	41	416566.35382	10160.15497		
U Total	43	185055290.4			

Root MSE	100.79759	R-square	0.9977	
Dep Mean	2006.92093	Adj R-sq	0.9976	
C.V.	5.02250			

Parameter Estimates

Variable	DF	Parameter Estimate	Standard Error	T for H0: Parameter=0	Prob > |T|
D_INT	1	-166.190749	313.48280828	-0.530	0.5989
D_Y	1	0.926096	0.02759143	33.565	0.0001

f. Breusch and Godfrey (1978) LM Test

```
Dependent Variable: RESID      Residual
```

Analysis of Variance

Source	DF	Sum of Squares	Mean Square	F Value	Prob>F
Model	2	564141.24338	282070.62169	27.590	0.0001
Error	40	408947.38671	10223.68467		
C Total	42	973088.63008			

Root MSE	101.11224	R-square	0.5797
Dep Mean	-3.07022	Adj R-sq	0.5587
C.V.	-3293.31907		

Parameter Estimates

Variable	DF	Parameter Estimate	Standard Error	T for H0: Parameter=0	Prob > |T|
INTERCEP	1	-53.564768	62.03682611	-0.863	0.3930
RHO	1	0.801980	0.10805085	7.422	0.0001
Y	1	0.005511	0.00586003	0.940	0.3526

SAS PROGRAMS

```
Data CONS;
Input YEAR Y C;
cards;

Data CONSUMP; set CONS;
YLAG=LAG(Y);
CLAG=LAG(C);

Proc reg data=CONSUMP;
     Model C=Y/DW;
  Output out=OUT1 R=RESID;
Title 'REGRESSION OF C ON CONSTANT AND Y';

Data TEMP1; set OUT1;
RHO=LAG(RESID);

Proc reg data=TEMP1;
     Model RESID=RHO/noint;
Title 'REGRESSION OF RESIDUAL FOR THE 1st ORDER
AUTOCORRELATION';
```

```
***** COCHRANE-ORCUTT(1949) METHOD *****;
**********************************************;

Data CO_DATA; set CONSUMP;
CO_Y=Y-0.792116*YLAG;      *** RHO=0.792116 ***;
CO_C=C-0.792116*CLAG;
CO_INT=1-0.792116;

Proc reg data=CO_DATA;
    Model CO_C=CO_INT CO_Y/noint;

TITLE 'COCHRANE-ORCUTT(1949) METHOD';

***** PRAIS-WINSTEN(1954, YULE-WALKER) METHOD *****;
****************************************************;

Proc autoreg data=CONSUMP;
    Model C=Y /DW=1 DWPROB LAGDEP NLAG=1 METHOD=YW;
TITLE 'PRAIS-WINSTEN(1954, YULE-WALKER) 2-STEP METHOD';

Proc autoreg data=CONSUMP;
    Model C=Y /DW=1 DWPROB LAGDEP NLAG=1 ITER METHOD=YW;
TITLE 'PRAIS-WINSTEN(1954, YULE-WALKER) ITERATIVE METHOD';

Proc autoreg data=CONSUMP;
    Model C=Y /DW=1 DWPROB LAGDEP NLAG=1 METHOD=ML;
TITLE 'MAXIMUM LIKELIHOOD METHOD';

***** DURBIN'S METHOD (1960) *****;
************************************;

Proc reg data=CONSUMP;
    Model C=CLAG Y YLAG;
Title 'DURBIN(1960) METHOD';

Data DURBIN; set CONSUMP;
  D_C=C-0.8*CLAG;
  D_Y=Y-0.8*YLAG;
D_INT=1-0.8;

Proc reg data=DURBIN;
    Model D_C=D_INT D_Y/NOINT;
Title 'DURBIN(1960) METHOD';


*BREUSCH AND GODFREY (1978) LM TEST FOR AUTOCORRELATION ;
**********************************************************;

Proc reg data=TEMP1;
    Model RESID=RHO Y;
Title 'BREUSCH AND GODFREY (1978) LM TEST';
run;
```

5.16 Using EViews, Q_{t+1} is simply Q(1) and one can set the sample range from 1954-1976.

a. The OLS regression over the period 1954-1976 yields

$$RS_t = \underset{(8.53)}{-6.14} + \underset{(1.44)}{6.33} Q_{t+1} - \underset{(1.37)}{1.67} P_t$$

with $R^2 = 0.62$ and D.W. = 1.07. The t-statistic for $\gamma = 0$ yields t = -1.67/1.37 = -1.21 which is insignificant with a p-value of 0.24. Therefore, the inflation rate is insignificant in explaining real stock returns.

```
LS // Dependent Variable is RS
Sample: 1954 1976
Included observations: 23
```

Variable	Coefficient	Std. Error	t-Statistic	Prob.
C	-6.137282	8.528957	-0.719582	0.4801
Q(1)	6.329580	1.439842	4.396024	0.0003
P	-1.665309	1.370766	-1.214875	0.2386

R-squared	0.616110	Mean dependent var	8.900000
Adjusted R-squared	0.577721	S.D. dependent var	21.37086
S.E. of regression	13.88743	Akaike info criterion	5.383075
Sum squared resid	3857.212	Schwarz criterion	5.531183
Log likelihood	-91.54095	F-statistic	16.04912
Durbin-Watson stat	1.066618	Prob(F-statistic)	0.000070

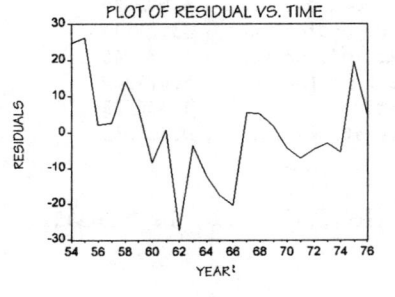

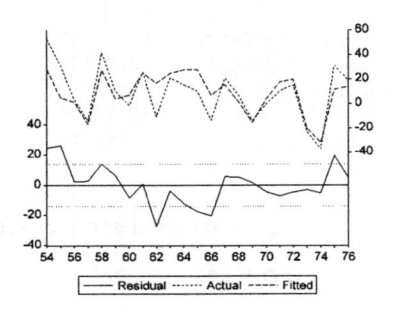

b. The D.W. = 1.07. for n = 23 and two slope coefficients, the 5 percent critical values of the D.W. are $d_L = 1.17$ and $d_U = 1.54$. Since $1.07 < d_L$, this indicates the presence of positive serial correlation.

c. The Breusch and Godfrey test for first-order serial correlation runs the regression of OLS residuals e_t on the regressors in the model and e_{t-1}. This yields

$$e_t = -4.95 + 1.03\ Q_{t+1} + 0.49\ P_t + 0.45\ e_{t-1}$$
$$\quad\ (8.35)\quad (1.44)\qquad\ (1.30)\quad\ (0.22)$$

with $R^2 = 0.171$ and $n = 23$. The LM statistic is nR^2 which yields 3.94. This distributed as χ_1^2 under the null hypothesis and has a p-value of 0.047. This is significant at the 5 percent level and indicates the presence of first-order serial correlation.

```
Breusch-Godfrey Serial Correlation LM Test:

F-statistic       3.922666    Probability    0.062305
Obs*R-squared     3.935900    Probability    0.047266

Test Equation:
LS // Dependent Variable is RESID

Variable   Coefficient   Std. Error   t-Statistic   Prob.

C           -4.953720     8.350094    -0.593253    0.5600
Q(1)         1.030343     1.442030     0.714509    0.4836
P            0.487511     1.303845     0.373903    0.7126
RESID(-1)    0.445119     0.224743     1.980572    0.0623

R-squared             0.171126    Mean dependent var    4.98E-15
Adjusted R-squared    0.040251    S.D. dependent var   13.24114
S.E. of regression   12.97192     Akaike info criterion 5.282345
Sum squared resid  3197.143       Schwarz criterion     5.479822
Log likelihood      -89.38255     F-statistic           1.307555
Durbin-Watson stat    1.818515    Prob(F-statistic)     0.301033
```

d. The Cochrane-Orcutt option CO on HUMMER yields the following regression

$$RS_t^* = -14.19 + 7.47\ Q_{t+1}^* - 0.92\ P_t^*$$
$$\qquad\ \ (9.34)\quad (1.20)\qquad (1.63)$$

where $RS_t^* = RS_t - \hat{\rho} RS_{t-1}$, $Q_{t+1}^* = Q_{t+1} - \hat{\rho} Q_t$ and $P_t^* = P_t - \hat{\rho} P_{t-1}$ with $\hat{\rho}_{CO}$

$= 0.387$.

Chapter 5: Violations of the Classical Assumptions   103

e. The AR(1) options on EViews yields the following results:

$$RS_t = \underset{(7.92)}{-7.32} + \underset{(1.36)}{5.91}\ Q_{t+1} - \underset{(1.28)}{1.25}\ P_t$$

with $R^2 = 0.68$. The estimate of ρ is $\hat{\rho} = -0.027$ which is a standard error of 0.014 and a t-statistic for $\rho = 0$ of -1.92. This has a p-value of 0.07. Note that even after correcting for serial correlation, P_t remains insignificant while Q_{t+1} remains significant. The estimates as well as their standard errors are affected by the correction for serial correlation. Compare with part (a).

PRAIS-WINSTEN PROCEDURE

```
LS // Dependent Variable is RS
Sample: 1954 1976
Included observations: 23
Convergence achieved after 4 iterations
```

Variable	Coefficient	Std. Error	t-Statistic	Prob.
C	-7.315299	7.921839	-0.923435	0.3674
Q(1)	5.905362	1.362572	4.333981	0.0004
P	-1.246799	1.277783	-0.975752	0.3414
AR(1)	-0.027115	0.014118	-1.920591	0.0699

R-squared	0.677654	Mean dependent var	8.900000
Adjusted R-squared	0.626757	S.D. dependent var	21.37086
S.E. of regression	13.05623	Akaike info criterion	5.295301
Sum squared resid	3238.837	Schwarz criterion	5.492779
Log likelihood	-89.53155	F-statistic	13.31429
Durbin-Watson stat	1.609639	Prob(F-statistic)	0.000065
Inverted AR Roots	-.03		

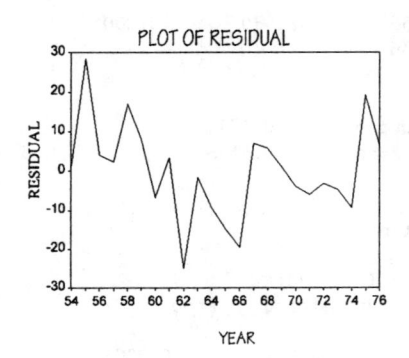

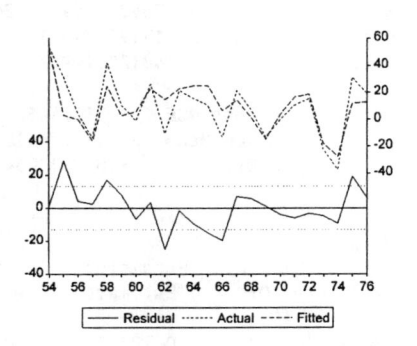

5. 18 The back up regressions are given below. These are performed using SAS.

a. Dependent Variable: EMP

Analysis of Variance

Source	DF	Sum of Squares	Mean Square	F Value	Prob>F
Model	1	2770902.9483	2770902.9483	4686.549	0.0001
Error	73	43160.95304	591.24593		
C Total	74	2814063.9013			

Root MSE	24.31555	R-square	0.9847	
Dep Mean	587.94141	Adj R-sq	0.9845	
C.V.	4.13571			

Parameter Estimates

Variable	DF	Parameter Estimate	Standard Error	T for H0: Parameter=0	Prob > |T|
INTERCEP	1	-670.591096	18.59708022	-36.059	0.0001
RGNP_1	1	1.008467	0.01473110	68.458	0.0001

Durbin-Watson D	0.349
(For Number of Obs.)	75
1st Order Autocorrelation	0.788

c. Dependent Variable: RESID

Analysis of Variance

Source	DF	Sum of Squares	Mean Square	F Value	Prob>F
Model	1	26929.25956	26929.25956	148.930	0.0001
Error	73	13199.74001	180.81836		
U Total	74	40128.99957			

Root MSE	13.44687	R-square	0.6711	
Dep Mean	-0.74410	Adj R-sq	0.6666	
C.V.	-1807.13964			

Parameter Estimates

Variable	DF	Parameter Estimate	Standard Error	T for H0: Parameter=0	Prob > |T|
RESID_1	1	0.791403	0.06484952	12.204	0.0001

Chapter 5: Violations of the Classical Assumptions

COCHRANE-ORCUTT(1949) METHOD

Dependent Variable: EMP_STAR

Analysis of Variance

Source	DF	Sum of Squares	Mean Square	F Value	Prob>F
Model	2	1358958.4052	679479.20261	3767.845	0.0001
Error	72	12984.21373	180.33630		
U Total	74	1371942.619			

Root MSE	13.42894	R-square	0.9905	
Dep Mean	129.70776	Adj R-sq	0.9903	
C.V.	10.35322			

Parameter Estimates

Variable	DF	Parameter Estimate	Standard Error	T for H0: Parameter=0	Prob > |T|
C_STAR	1	-628.267447	50.28495095	-12.494	0.0001
RGNPSTAR	1	0.972263	0.03867418	25.140	0.0001

d. Prais-Winsten(1954, Yule-Walker) 2-Step Method

Autoreg Procedure
Dependent Variable = EMP

Ordinary Least Squares Estimates

SSE	53280.82	DFE	74
MSE	720.0111	Root MSE	26.83302
SBC	722.3376	AIC	717.6761
Reg Rsq	0.9816	Total Rsq	0.9816
Durbin's t	13.43293	PROB>t	0.0001
Durbin-Watson	0.3023		

Variable	DF	B Value	Std Error	t Ratio	Approx Prob
Intercept	1	-672.969905	20.216	-33.289	0.0001
RGNP	1	1.003782	0.016	62.910	0.0001

Estimates of Autocorrelations

Lag Covariance Correlation -1 9 8 7 6 5 4 3 2 1 0 1 2 3 4 5 6 7 8 9 1

| 0 | 701.0635 | 1.000000 | \| \|******************\| |
| 1 | 574.9163 | 0.820063 | \| \|****************   \| |

Preliminary MSE = 229.5957

Estimates of the Autoregressive Parameters

Lag	Coefficient	Std Error	t Ratio
1	-0.82006318	0.06697949	-12.243497

Yule-Walker Estimates

SSE	14259.84	DFE	73
MSE	195.3403	Root MSE	13.97642
SBC	627.6068	AIC	620.6146
Reg Rsq	0.8919	Total Rsq	0.9951
Durbin-Watson	2.2216		

Variable	DF	B Value	Std Error	t Ratio	Approx Prob
Intercept	1	-559.809933	47.373	-11.817	0.0001
RGNP	1	0.914564	0.037	24.539	0.0001

e. Breusch and Godfrey (1978) LM Test

Dependent Variable: RESID   Residual

Analysis of Variance

Source	DF	Sum of Squares	Mean Square	F Value	Prob>F
Model	2	27011.59888	13505.79944	73.331	0.0001
Error	71	13076.42834	184.17505		
C Total	73	40088.02723			

Root MSE	13.57111	R-square	0.6738	
Dep Mean	-0.74410	Adj R-sq	0.6646	
C.V.	-1823.83627			

Parameter Estimates

Variable	DF	Parameter Estimate	Standard Error	T for H0: Parameter=0	Prob > |T|
INTERCEP	1	-7.177291	10.64909636	-0.674	0.5025
RESID_1	1	0.790787	0.06546691	12.079	0.0001
RGNP_1	1	0.005026	0.00840813	0.598	0.5519

SAS PROGRAM

```
Data ORANGE;
Input DATE EMP RGNP;
Cards;

Data ORANGE1; set ORANGE;
RGNP_1=LAG(RGNP);
```

Chapter 5: Violations of the Classical Assumptions 107

```
RGNP_2=LAG2(RGNP);
EMP_1=LAG(EMP);

Proc reg data=ORANGE1;
     Model EMP=RGNP_1/DW;
  Output out=OUT1 R=RESID;

Data TEMP; set OUT1;
RESID_1=LAG(RESID);

Proc reg data=TEMP;
     Model RESID=RESID_1/noint;
run;

***** COCHRANE-ORCUTT(1949) METHOD *****;
****************************************;

Data CO_DATA; set ORANGE1;
EMP_STAR=EMP-0.791403*EMP_1;        *** RHO=0.791403 ***;

RGNPSTAR=RGNP_1-0.791403*RGNP_2;
C_STAR=1-0.791403;

Proc reg data=CO_DATA;
     Model EMP_STAR=C_STAR RGNPSTAR/noint;
TITLE 'COCHRANE-ORCUTT(1949) METHOD';

***** PRAIS-WINSTEN(1954, YULE-WALKER) METHOD *****;
****************************************************;

Proc autoreg data=ORANGE1;
     Model EMP=RGNP /DW=1 DWPROB LAGDEP NLAG=1 METHOD=YW;
TITLE 'PRAIS-WINSTEN(1954, YULE-WALKER) 2-STEP METHOD';

*BREUSCH & GODFREY (1978) LM TEST FOR AUTOCORRELATION***;
*********************************************************;

Proc reg data=TEMP;
     Model RESID=RESID_1 RGNP_1;
Title 'BREUSCH AND GODFREY (1978) LM TEST';
run;
```

CHAPTER 6
Distributed Lags and Dynamic Models

6.1 a. Using the Linear Arithmetic lag given in equation (6.2), a six year lag on income gives a regression of consumption on a constant and $Z_t = \sum_{i=0}^{6}(7-i)X_{t-i}$ where X_t denotes income. In this case,

$$Z_t = 7X_t + 6X_{t-1} + .. + X_{t-6}.$$

The SAS regression output is given below. It yields

$$\hat{C}_t = 0.102 + 0.035\ Z_t$$
$$\ \ \ \ \ \ (0.12)\ \ \ (0.0005)$$

Linear Arithmetic Lag

Dependent Variable: C

Analysis of Variance

Source	DF	Sum of Squares	Mean Square	F Value	Prob>F
Model	1	2.14863	2.14863	5709.490	0.0001
Error	36	0.01355	0.00038		
C Total	37	2.16218			

Root MSE	0.01940	R-square	0.9937	
Dep Mean	9.15840	Adj R-sq	0.9936	
C.V.	0.21182			

Parameter Estimates

Variable	DF	Parameter Estimate	Standard Error	T for H0: Parameter=0	Prob > |T|
INTERCEP	1	0.102387	0.11989134	0.854	0.3988
ZT	1	0.035096	0.00046447	75.561	0.0001

From equation (6.2)

$\beta_i = [(s+1)-i]\beta$ for $i = 0,..,6$

Chapter 6: Distributed Lags and Dynamic Models

one can retrieve the $\hat{\beta}_i$'s:

$\hat{\beta}_0 = 7\hat{\beta} = 0.245672$ $\hat{\beta}_4 = 3\hat{\beta} = 0.105288$

$\hat{\beta}_1 = 6\hat{\beta} = 0.210576$ $\hat{\beta}_5 = 2\hat{\beta} = 0.070192$

$\hat{\beta}_2 = 5\hat{\beta} = 0.175480$ $\hat{\beta}_6 = \hat{\beta} = 0.035096$

$\hat{\beta}_3 = 4\hat{\beta} = 0.140384$

see also the restricted SAS regression imposing these restrictions and testing each restriction separately.

```
                Linear Arithmetic model restrictions

NOTE: Restrictions have been applied to parameter estimates.
Dependent Variable: C

                        Analysis of Variance

                            Sum of          Mean
    Source          DF    Squares         Square        F Value       Prob>F

    Model            1    2.14863        2.14863       5709.490       0.0001
    Error           36    0.01355        0.00038
    C Total         37    2.16218

         Root MSE       0.01940     R-square      0.9937
         Dep Mean       9.15840     Adj R-sq      0.9936
         C.V.           0.21182

                        Parameter Estimates

                    Parameter       Standard     T for H0:
    Variable   DF    Estimate          Error    Parameter=0    Prob > |T|

    INTERCEP    1    0.102387     0.11989134        0.854        0.3988
    Y           1    0.245671     0.00325128       75.561        0.0001
    YLAG1       1    0.210575     0.00278681       75.561        0.0001
    YLAG2       1    0.175479     0.00232234       75.561        0.0001
    YLAG3       1    0.140383     0.00185787       75.561        0.0001
    YLAG4       1    0.105287     0.00139341       75.561        0.0001
    YLAG5       1    0.070192     0.00092894       75.561        0.0001
    YLAG6       1    0.035096     0.00046447       75.561        0.0001
    RESTRICT   -1    0.009251     0.00218154        4.241        0.0001
    RESTRICT   -1    0.002527     0.00148714        1.699        0.0980
    RESTRICT   -1   -0.003272     0.00136217       -2.402        0.0216
    RESTRICT   -1   -0.005693     0.00180090       -3.161        0.0032
```

| RESTRICT | -1 | -0.006562 | 0.00248232 | -2.644 | 0.0121 |
| RESTRICT | -1 | -0.007161 | 0.00314571 | -2.276 | 0.0289 |

b. Using an Almon-lag second degree polynomial described in equation (6.4), a six year lag on income gives a regression of consumption on a constant, $Z_0 = \sum_{i=0}^{6} X_{t-i}$, $Z_1 = \sum_{i=0}^{6} i X_{t-i}$ and $Z_2 = \sum_{i=0}^{6} i^2 X_{t-i}$. This yields the Almon-lag without near or far end-point constraints. A near end-point constraint imposes $\beta_{-1} = 0$ in equation (6.1) which yields $a_0 - a_1 + a_2 = 0$ in equation (6.4). Substituting for a_0 in equation (6.4) yields the regression in (6.5). The corresponding SAS regression output is given below.

ALMON LAG(S=6,P=2) Near End-point Restriction

PDLREG Procedure
Dependent Variable = C

Ordinary Least Squares Estimates

SSE	0.016433	DFE	35
MSE	0.00047	Root MSE	0.021668
SBC	-175.597	AIC	-180.51
Reg Rsq	0.9924	Total Rsq	0.9924
Durbin-Watson	0.6491		

Variable	DF	B Value	Std Error	t Ratio	Approx Prob
Intercept	1	0.104806316	0.13705	0.765	0.4496
Y**0	1	0.370825802	0.00555	66.859	0.0001
Y**1	1	-0.294792129	0.08325	-3.541	0.0011
Y**2	1	-0.277246503	0.04841	-5.727	0.0001

Restriction	DF	L Value	Std Error	t Ratio	Approx Prob
Y(-1)	-1	-0.008519656	0.00197	-4.331	0.0001

Variable	Parameter Value	Std Error	t Ratio	Approx Prob
Y(0)	0.15604	0.021	7.34	0.0001
Y(1)	0.25158	0.032	7.87	0.0001
Y(2)	0.28662	0.032	8.93	0.0001

```
         Y(3)         0.26116        0.022    12.03   0.0001
         Y(4)         0.17520        0.003    66.86   0.0001
         Y(5)         0.02874        0.031     0.92   0.3620
         Y(6)        -0.17822        0.073    -2.44   0.0201

                           Estimate of Lag Distribution
         Variable         -0.178           0                    0.2866

         Y(0)                          |*************                     |
         Y(1)                          |*********************             |
         Y(2)                          |**************************        |
         Y(3)                          |************************          |
         Y(4)                          |***************                   |
         Y(5)                          |**                                |
         Y(6)         |***************|                                   |
```

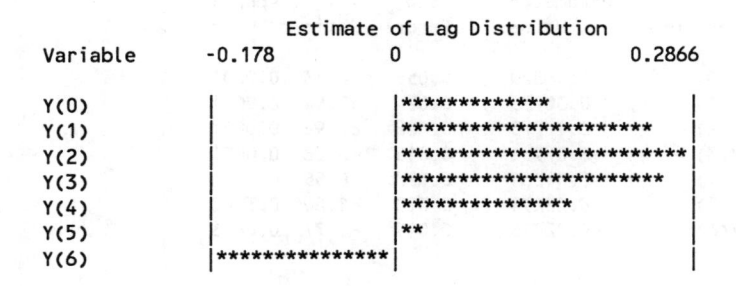

c. The far end-point constraint imposes $\beta_7 = 0$. This translates into the following restriction $a_0 + 7a_1 + 49a_2 = 0$. Substituting for a_0 in (6.4) yields the regression in (6.6) with s = 6, i.e., the regression of consumption on a constant, $(Z_1\text{-}7Z_0)$ and $(Z_2\text{-}49Z_0)$. The corresponding SAS regression output is given below.

```
              ALMON LAG(S=6,P=2) Far End-point Restriction

PDLREG Procedure
Dependent Variable = C

                    Ordinary Least Squares Estimates

              SSE             0.009208     DFE                 35
              MSE             0.000263     Root MSE       0.01622
              SBC            -197.609      AIC           -202.521
              Reg Rsq           0.9957     Total Rsq       0.9957
              Durbin-Watson     0.6475

       Variable     DF       B Value     Std Error   t Ratio  Approx Prob

       Intercept    1     -0.041355729    0.10631     -0.389    0.6996
       Y**0         1      0.376231142    0.00428     87.961    0.0001
       Y**1         1     -0.436935994    0.06189     -7.060    0.0001
       Y**2         1      0.143656538    0.03537      4.062    0.0003
```

Restriction	DF	L Value	Std Error	t Ratio	Approx Prob
Y(7)	-1	-0.003584133	0.00146	-2.451	0.0194

Variable	Parameter Value	Std Error	t Ratio	Approx Prob
Y(0)	0.46829	0.055	8.53	0.0001
Y(1)	0.30735	0.024	12.84	0.0001
Y(2)	0.17775	0.002	87.96	0.0001
Y(3)	0.07951	0.015	5.28	0.0001
Y(4)	0.01261	0.023	0.55	0.5846
Y(5)	-0.02294	0.023	-1.00	0.3242
Y(6)	-0.02715	0.015	-1.77	0.0853

Estimate of Lag Distribution

Variable	-0.027		0.4683
Y(0)		|**|	
Y(1)		|**************************|	
Y(2)		|****************|	
Y(3)		|*******|	
Y(4)		|*|	
Y(5)	|*|		
Y(6)	|*|		

d. The RRSS is given by the residual sum of squares of the regression in part (a). The URSS is obtained from running consumption on a constant and six lags on income. The corresponding SAS regression is given below. The number of restrictions given in (6.2) is 6. Hence, the Chow F-statistic yields

$$F = \frac{(0.01355 - 0.00627)/6}{0.00627/30} = 5.8054$$

and this is distributed as $F_{6,30}$ under the null hypothesis. This rejects the arithmetic lag restriction.

Chapter 6: Distributed Lags and Dynamic Models

UNRESTRICTED MODEL

Dependent Variable: C

Analysis of Variance

Source	DF	Sum of Squares	Mean Square	F Value	Prob>F
Model	7	2.15591	0.30799	1473.031	0.0001
Error	30	0.00627	0.00021		
C Total	37	2.16218			

Root MSE	0.01446	R-square	0.9971	
Dep Mean	9.15840	Adj R-sq	0.9964	
C.V.	0.15788			

Parameter Estimates

Variable	DF	Parameter Estimate	Standard Error	T for H0: Parameter=0	Prob > |T|
INTERCEP	1	-0.092849	0.09577016	-0.969	0.3400
Y	1	0.865012	0.14795354	5.847	0.0001
YLAG1	1	0.063999	0.22303900	0.287	0.7761
YLAG2	1	-0.136316	0.21443891	-0.636	0.5298
YLAG3	1	0.009104	0.21183864	0.043	0.9660
YLAG4	1	0.192795	0.21343826	0.903	0.3736
YLAG5	1	-0.067975	0.21517360	-0.316	0.7543
YLAG6	1	0.074452	0.14613719	0.509	0.6142

e. The RRSS is given by the second degree Almon-lag model with far end-point constraint described in part (c). The URSS is the same as that obtained in part (d). The number of restrictions is 5. Therefore,

$$F = \frac{(0.009208 - 0.00627)/5}{0.00627/30} = 2.8115$$

and this is distributed as $F_{5,30}$ under the null hypothesis that the restrictions in (c) are true. This rejects H_o.

6.2 a. For the Almon-lag third degree polynomial

$$\beta_i = a_o + a_1 i + a_2 i^2 + a_3 i^3 \text{ for } i = 0,1,\ldots,5.$$

In this case, (6.1) reduces to

$$Y_t = \alpha + \sum_{i=0}^{5} (a_o + a_1 i + a_2 i^2 + a_3 i^3) X_{t-i} + u_t$$

$$= \alpha + a_o \sum_{i=0}^{5} X_{t-i} + a_1 \sum_{i=0}^{5} i X_{t-i} + a_2 \sum_{i=0}^{5} i^2 X_{t-i} + a_3 \sum_{i=0}^{5} i^3 X_{t-i} + u_t.$$

Now α, a_o, a_1, a_2 and a_3 can be estimated from the regression of Y_t on a constant, $Z_0 = \sum_{i=0}^{5} X_{t-i}, Z_1 = \sum_{i=0}^{5} i X_{t-i}, Z_2 = \sum_{i=0}^{5} i^2 X_{t-i}$ and $Z_3 = \sum_{i=0}^{5} i^3 X_{t-i}$. The corresponding SAS regression for the Consumption-Income data is given below.

Almon lag (S=5, P=3)

Dependent Variable: C

Analysis of Variance

Source	DF	Sum of Squares	Mean Square	F Value	Prob>F
Model	4	2.31126	0.57782	2939.828	0.0001
Error	34	0.00668	0.00020		
C Total	38	2.31794			

Root MSE	0.01402	R-square	0.9971	
Dep Mean	9.14814	Adj R-sq	0.9968	
C.V.	0.15325			

Parameter Estimates

Variable	DF	Parameter Estimate	Standard Error	T for H0: Parameter=0	Prob > |T|
INTERCEP	1	-0.057499	0.08894219	-0.646	0.5223
Z0	1	0.870710	0.12271595	7.095	0.0001
Z1	1	-1.272007	0.37956328	-3.351	0.0020
Z2	1	0.493755	0.19743864	2.501	0.0174
Z3	1	-0.054682	0.02616943	-2.090	0.0442

Covariance of Estimates

COVB	INTERCEP	Z0	Z1	Z2	Z3
INTERCEP	0.007911	-0.003052	0.005238	-0.002063	0.000233
Z0	-0.003052	0.015059	-0.041566	0.019446	-0.002383
Z1	0.005238	-0.041566	0.144068	-0.073668	0.009509
Z2	-0.002063	0.019446	-0.073668	0.038982	-0.005136
Z3	0.000233	-0.002383	0.009509	-0.005136	0.000685

Chapter 6: Distributed Lags and Dynamic Models

b. The estimate of β_3 is $\hat{\beta}_3 = \hat{a}_0 + 3\hat{a}_1 + 9\hat{a}_2 + 27\hat{a}_3$ with

$\text{var}(\hat{\beta}_3) = \text{var}(\hat{a}_0) + 9\,\text{var}(\hat{a}_1) + 81\,\text{var}(\hat{a}_2) + 27^2\,\text{var}(\hat{a}_3)$

$\qquad + 2.3.\,\text{cov}(\hat{a}_0, \hat{a}_1) + 2.9.\,\text{cov}(\hat{a}_0, \hat{a}_2) + 2.27.\,\text{cov}(\hat{a}_0, \hat{a}_3)$

$\qquad + 2.3.9.\,\text{cov}(\hat{a}_1, \hat{a}_2) + 2.3.27\,\text{cov}(\hat{a}_1, \hat{a}_3) + 2.9.27\,\text{cov}(\hat{a}_2, \hat{a}_3).$

From the above regression in part (a), this yields

$\text{var}(\hat{\beta}_3) = (0.01506) + 9(0.14407) + 81(0.03898) + 27^2(0.00068)$

$\qquad + 6(-0.04157) + 18(0.01945) + 54(-0.00238) + 54(-0.07367)$

$\qquad + 162(0.00951) + 486(-0.00514) = 0.00135.$

c. Imposing the near end-point constraint $\beta_{-1} = 0$ yields the following restriction on the third degree polynomial in a's:

$a_0 - a_1 + a_2 - a_3 = 0.$

solving for a_0 and substituting above yields the following constrained regression:

$Y_t = \alpha + a_1(Z_1 + Z_0) + a_2(Z_2 - Z_0) + a_3(Z_1 + Z_3) + u_t$

The corresponding SAS regression is reported below.

Alman Lag (S=5, P=3) Near End-point Constraint

Dependent Variable: C

Analysis of Variance

Source	DF	Sum of Squares	Mean Square	F Value	Prob>F
Model	3	2.30824	0.76941	2776.062	0.0001
Error	35	0.00970	0.00028		
C Total	38	2.31794			

Root MSE	0.01665	R-square	0.9958	
Dep Mean	9.14814	Adj R-sq	0.9955	
C.V.	0.18198			

Parameter Estimates

Variable	DF	Parameter Estimate	Standard Error	T for H0: Parameter=0	Prob > |T|
INTERCEP	1	0.002691	0.10403149	0.026	0.9795
ZZ1	1	0.171110	0.01635695	10.461	0.0001
ZZ2	1	-0.261338	0.05107648	-5.117	0.0001
ZZ3	1	0.042516	0.00990505	4.292	0.0001

d. The near end-point restriction can be tested using the Chow F-statistic. The URSS is the residuals sum of squares obtained from (a) without imposing the near end-point constraint. The RRSS is that from part (c) imposing the near end-point constraint. One restriction is involved. Therefore,

$$F = \frac{(0.00970 - 0.00668)/1}{0.00668/34} = 15.3713$$

and this is distributed as $F_{1,34}$ under the null hypothesis that the near end-point constraint is true. This restriction is rejected.

e. The URSS is obtained from regressing consumption on a constant and five year lags on income. This was done in the text and yielded URSS = 0.00667. The RRSS is given in part (a). The number of restrictions is 3. Therefore,

$$F = \frac{(0.00668 - 0.00667)/3}{0.00667/32} = 0.01599$$

and this is distributed as $F_{3,32}$ under the null hypothesis. We do not reject H_o.

SAS PROGRAM (for 6.1 and 6.2)

```
data AAA;
input year income consmp;
cards;

data CONSUM; set AAA;
Y=log(income);
C=log(consmp);
Clag1=LAG(C);

Ylag1=LAG(Y);
Ylag2=LAG2(Y);
Ylag3=LAG3(Y);
Ylag4=LAG4(Y);
Ylag5=LAG5(Y);
```

```
Ylag6=LAG6(Y);

Zt=7*Y+6*Ylag1+5*Ylag2+4*Ylag3+3*Ylag4+2*Ylag5+Ylag6;

Z0=Y+Ylag1+Ylag2+Ylag3+Ylag4+Ylag5;
Z1=Ylag1+2*Ylag2+3*Ylag3+4*Ylag4+5*Ylag5;
Z2=Ylag1+4*Ylag2+9*Ylag3+16*Ylag4+25*Ylag5;
Z3=Ylag1+8*Ylag2+27*Ylag3+64*Ylag4+125*Ylag5;

ZZ1=Z1+Z0;
ZZ2=Z2-Z0;
ZZ3=Z1+Z3;
/*********************************************/
/*              PROBLEM 6.1 (a)             */
/*********************************************/
PROC REG DATA=CONSUM;
     MODEL C=Zt;
TITLE 'Linear Arithmetic Lag';
RUN;

PROC REG DATA=CONSUM;
     MODEL C=Y Ylag1 Ylag2 Ylag3 Ylag4 Ylag5 Ylag6;
     RESTRICT
                    Y=7*Ylag6,
                 Ylag1=6*Ylag6,
                 Ylag2=5*Ylag6,
                 Ylag3=4*Ylag6,
                 Ylag4=3*Ylag6,
                 Ylag5=2*Ylag6;
TITLE 'Linear Arithmetic Model Restrictions';
RUN;

/*********************************************/
/*      PROBLEM 6.1 parts(b) and (c)        */
/*********************************************/

PROC PDLREG DATA=CONSUM;
     MODEL C=Y(6,2,,FIRST);
TITLE 'Almon Lag (S=6,P=2) Near End-point Restriction';
RUN;

PROC PDLREG DATA=CONSUM;
     MODEL C=Y(6,2,,LAST);
TITLE 'Almon Lag (S=6,P=2) Far End-point Restriction';
RUN;

/*********************************************/
/*            PROBLEM 6.1 part (d)          */
/*********************************************/
```

```
PROC REG DATA=CONSUM;
     MODEL C=Y Ylag1 Ylag2 Ylag3 Ylag4 Ylag5 Ylag6;
TITLE 'Unrestricted Model';
RUN;
/**********************************************/
/*              PROBLEM 6.2 part (a)         */
/**********************************************/
PROC REG DATA=CONSUM;
     MODEL C=Z0 Z1 Z2 Z3/COVB;
TITLE 'Almon Lag (S=5, P=3)';
RUN;
/**********************************************/
/*              PROBLEM 6.2 (c)              */
/**********************************************/
PROC REG DATA=CONSUM;
     MODEL C=ZZ1 ZZ2 ZZ3;
TITLE 'Almon Lag (S=5, P=3) Near End-point Constraint';
RUN;
```

6.3 a. From (6.18), $Y_t = \beta Y_{t-1} + v_t$. Therefore, $Y_{t-1} = \beta Y_{t-2} + v_{t-1}$ and

$$\rho Y_{t-1} = \rho\beta Y_{t-2} + \rho v_{t-1}.$$

Subtracting this last equation from (6.18) and re-arranging terms, one gets

$$Y_t = (\beta+\rho)Y_{t-1} - \rho\beta Y_{t-2} + \epsilon_t.$$

Multiply both sides by Y_{t-1} and sum, we get

$$\sum_{t=2}^{T} Y_t Y_{t-1} = (\beta+\rho)\sum_{t=2}^{T} Y_{t-1}^2 - \rho\beta \sum_{t=2}^{T} Y_{t-1}Y_{t-2} + \sum_{t=2}^{T} Y_{t-1}\epsilon_t.$$

Divide by $\sum_{t=2}^{T} Y_{t-1}^2$ and take probability limits, we get

$$\text{plim} \sum_{t=2}^{T} Y_t Y_{t-1}/\sum_{t=2}^{T} Y_{t-1}^2 = (\beta+\rho) - \rho\beta\, \text{plim}(\sum_{t=2}^{T} Y_{t-1}Y_{t-2}/\sum_{t=2}^{T} Y_{t-1}^2)$$

$$+ \text{plim}(\sum_{t=2}^{T} Y_{t-1}\epsilon_t/\sum_{t=2}^{T} Y_{t-1}^2).$$

The last term has zero probability limits since it is the $\text{cov}(Y_{t-1},\epsilon_t)/\text{var}(Y_{t-1})$. The numerator is zero whereas the denominator is finite and positive. As

Chapter 6: Distributed Lags and Dynamic Models

$T\to\infty$, $\text{plim}\sum_{t=2}^{T} Y_{t-1}Y_{t-2}$ is almost identical to $\text{plim}\sum_{t=2}^{T} Y_t Y_{t-1}$. Hence, as $T\to\infty$, one can solve for

$$\text{plim}\sum_{t=2}^{T} Y_t Y_{t-1}/\sum_{t=2}^{T} Y_{t-1}^2 = (\beta+\rho)/(1+\rho\beta).$$

Therefore,

$$\text{plim}\,\hat{\beta}_{ols} = (\beta+\rho)/(1+\rho\beta)$$

and

$$\text{plim}(\hat{\beta}_{ols}-\beta) = \rho(1-\beta^2)/(1+\rho\beta).$$

b. One can tabulate the asymptotic bias of $\hat{\beta}_{ols}$ derived in (a) for various values of $|\rho| < 1$ and $|\beta| < 1$.

c. For $\hat{\rho} = \sum_{t=2}^{T}\hat{v}_t\hat{v}_{t-1}/\sum_{t=2}^{T}\hat{v}_{t-1}^2$ with $\hat{v}_t = Y_t - \hat{\beta}_{ols} Y_{t-1}$ we compute

$$\sum_{t=2}^{T}\hat{v}_{t-1}^2/T = \sum_{t=2}^{T} Y_t^2/T - 2\hat{\beta}_{ols}(\sum_{t=2}^{T} Y_t Y_{t-1}/T) + \hat{\beta}_{ols}^2(\sum_{t=2}^{T} Y_{t-1}^2/T).$$

Defining the covariances of Y_t by γ_s, and using the fact that

$$\text{plim}(\sum Y_t Y_{t-s}/T) = \gamma_s \qquad \text{for } s = 0,1,2,...,$$

we get

$$\text{plim}\sum_{t=2}^{T}\hat{v}_{t-1}^2/T = \gamma_0 - 2(\text{plim}\,\hat{\beta}_{ols})\gamma_1 + (\text{plim}\,\hat{\beta}_{ols})^2\gamma_0.$$

But

$$\text{plim}\,\hat{\beta}_{ols} = \text{plim}\left[\frac{\sum_{t=2}^{T} Y_t Y_{t-1}/T}{\sum_{t=2}^{T} Y_{t-1}^2/T}\right] = \gamma_1/\gamma_0.$$

Hence,

$$\text{plim}\sum_{t=2}^{T}\hat{v}_{t-1}^2/T = \gamma_0 - 2\gamma_1^2/\gamma_0 + \gamma_1^2/\gamma_0 = \gamma_0 - (\gamma_1^2/\gamma_0) = (\gamma_0^2 - \gamma_1^2)/\gamma_0.$$

Also, $\hat{v}_{t-1} = Y_{t-1} - \hat{\beta}_{ols} Y_{t-2}$. Multiply this equation by $\hat{v}_t$ and sum, we get

$$\sum_{t=2}^{T}\hat{v}_t\hat{v}_{t-1} = \sum_{t=2}^{T} Y_{t-1}\hat{v}_t - \hat{\beta}_{ols}\sum_{t=2}^{T} Y_{t-2}\hat{v}_{t-1}.$$

But, by the property of least squares $\sum_{t=2}^{T} Y_{t-1} \hat{v}_t = 0$, hence

$$\sum_{t=2}^{T} \hat{v}_t \hat{v}_{t-1}/T = -\hat{\beta}_{ols} \sum_{t=2}^{T} Y_{t-2} \hat{v}_t/T = -\hat{\beta}_{ols} \sum_{t=2}^{T} Y_{t-2} Y_t/T + \hat{\beta}_{ols}^2 \sum_{t=2}^{T} Y_{t-2} Y_{t-1}/T$$

and

$$\text{plim} \sum_{t=2}^{T} \hat{v}_t \hat{v}_{t-1}/T = -\frac{\gamma_1}{\gamma_0} \cdot \gamma_2 + \frac{\gamma_1^2}{\gamma_0^2} \cdot \gamma_1 = \frac{\gamma_1(\gamma_1^2 - \gamma_0 \gamma_2)}{\gamma_0^2}.$$

Hence,

$$\text{plim} \, \hat{\rho} = \text{plim}(\sum_{t=2}^{T} \hat{v}_t \hat{v}_{t-1}/T)/\text{plim}(\sum_{t=2}^{T} \hat{v}_{t-1}^2/T) = \frac{\gamma_1}{\gamma_0} \cdot \frac{(\gamma_1^2 - \gamma_0 \gamma_2)}{(\gamma_0^2 - \gamma_1^2)}.$$

From part (a), we know that

$$Y_t = (\beta+\rho)Y_{t-1} - \rho\beta Y_{t-2} + \epsilon_t$$

multiply both sides by Y_{t-2} and sum and take plim after dividing by T, we get

$$\gamma_2 = (\beta+\rho)\gamma_1 - \rho\beta\gamma_0$$

so that

$$\gamma_0 \gamma_2 = (\beta+\rho)\gamma_1 \gamma_0 + \rho\beta\gamma_0^2$$

and

$$\gamma_1^2 - \gamma_0 \gamma_2 = \gamma_1^2 - (\beta+\rho)\gamma_1 \gamma_0 + \rho\beta\gamma_0^2.$$

But from part (a), $\text{plim} \, \hat{\beta}_{ols} = \gamma_1/\gamma_0 = (\beta+\rho)/(1+\rho\beta)$. Substituting

$$(\beta+\rho)\gamma_0 = (1+\rho\beta)\gamma_1$$

above we get

$$\gamma_1^2 - \gamma_0 \gamma_2 = \gamma_1^2 - (1+\rho\beta)\gamma_1^2 + \rho\beta\gamma_0^2 = \rho\beta(\gamma_0^2 - \gamma_1^2).$$

Hence,

$$\text{plim} \, \hat{\rho} = \frac{\gamma_1}{\gamma_0} \cdot \rho\beta = \rho\beta(\beta+\rho)/(1+\rho\beta)$$

and

Chapter 6: Distributed Lags and Dynamic Models

$\text{plim}(\hat{\rho}-\rho) = (\rho\beta^2+\rho^2\beta-\rho-\rho^2\beta)/(1+\rho\beta) = \rho(\beta^2-1)/(1+\rho\beta) = -\text{plim}(\hat{\beta}_{ols}-\beta)$.

The asymptotic bias of $\hat{\rho}$ is negative that of $\hat{\beta}_{ols}$.

d. Since $d = \sum_{t=2}^{T}(\hat{v}_t - \hat{v}_{t-1})^2 / \sum_{t=1}^{T} \hat{v}_t^2$ and as $T\to\infty$, $\sum_{t=2}^{T} \hat{v}_t^2$ is almost identical to $\sum_{t=2}^{T} \hat{v}_{t-1}^2$,

then plim $d \approx 2(1-\text{plim}\,\hat{\rho})$

where $\hat{\rho}$ was defined in (c). But

$\text{plim}\,\hat{\rho} = \rho - \rho(1-\beta^2)/(1+\rho\beta) = (\rho^2\beta+\rho\beta^2)/(1+\rho\beta) = \rho\beta(\rho+\beta)/(1+\rho\beta)$.

Hence, plim $d = 2[1 - \dfrac{\rho\beta(\beta+\rho)}{1+\beta\rho}]$.

e. Knowing the true disturbances, the Durbin-Watson statistic would be

$d^* = \sum_{t=2}^{T}(v_t - v_{t-1})^2 / \sum_{t=1}^{T} v_t^2$

and its plim $d^* = 2(1-\rho)$. This means that from part (d)

$\text{plim}(d-d^*) = 2(1-\text{plim}\,\hat{\rho}) - 2(1-\rho) = 2[\rho-\beta\rho(\rho+\beta)/(1+\rho\beta)]$

$= 2[\rho+\rho^2\beta-\beta\rho^2-\beta^2\rho]/(1+\rho\beta) = 2\rho(1-\beta^2)/(1+\rho\beta) = 2\,\text{plim}(\hat{\beta}_{ols}-\beta)$

from part (a). The asymptotic bias of the D.W. statistic is twice that of $\hat{\beta}_{ols}$.

The plim d and plim d^* and the asymptotic bias in d can be tabulated for various values of ρ and β.

6.4 a. From (6.18) with MA(1) disturbances, we get

$Y_t = \beta Y_{t-1} + \epsilon_t + \theta\epsilon_{t-1}$ with $|\beta| < 1$.

In this case,

$\hat{\beta}_{ols} = \sum_{t=2}^{T} Y_t Y_{t-1} / \sum_{t=2}^{T} Y_{t-1}^2 = \beta + \sum_{t=2}^{T} Y_{t-1}\epsilon_t / \sum_{t=2}^{T} Y_{t-1}^2 + \theta \sum_{t=2}^{T} Y_{t-1}\epsilon_t / \sum_{t=2}^{T} Y_{t-1}^2$

so that

$$\text{plim}(\hat{\beta}_{ols}-\beta) = \text{plim}(\sum_{t=2}^{T} Y_{t-1}\epsilon_t/T)/\text{plim}(\sum_{t=2}^{T} Y_{t-1}^2/T)$$

$$+ \theta\, \text{plim}(\sum_{t=2}^{T} Y_{t-1}\epsilon_{t-1}/T)/\text{plim}(\sum_{t=2}^{T} Y_{t-1}^2/T).$$

Now the above model can be written as

$$(1-\beta L)Y_t = (1+\theta L)\epsilon_t$$

or

$$Y_t = (1+\theta L)\sum_{i=0}^{\infty} \beta^i L^i \epsilon_t$$

$$Y_t = (1+\theta L)(\epsilon_t + \beta\epsilon_{t-1} + \beta^2\epsilon_{t-2} + ..)$$

$$Y_t = \epsilon_t + (\theta+\beta)[\epsilon_{t-1} + \beta\epsilon_{t-2} + \beta^2\epsilon_{t-3} + ..]$$

From the last expression, it is clear that $E(Y_t) = 0$ and

$$\text{var}(Y_t) = \sigma_\epsilon^2 [1+(\theta+\beta)^2/(1-\beta^2)] = \text{plim} \sum_{t=2}^{T} Y_{t-1}^2/T.$$

Also,

$$\sum_{t=2}^{T} Y_{t-1}\epsilon_t/T = \sum_{t=2}^{T} \epsilon_t[\epsilon_{t-1}+(\theta+\beta)(\epsilon_{t-2}+\beta\epsilon_{t-3}+..)]/T$$

Since the ϵ_t's are not serially correlated, each term on the right hand side has zero plim. Hence, $\text{plim} \sum_{t=2}^{T} Y_{t-1}\epsilon_t/T = 0$ and the first term on the right hand side of $\text{plim}(\hat{\beta}_{ols}-\beta)$ is zero. Similarly,

$$\sum_{t=2}^{T} Y_{t-1}\epsilon_{t-1}/T = \sum_{t=2}^{T} \epsilon_{t-1}^2/T + (\theta+\beta)[\sum_{t=2}^{T} \epsilon_{t-1}\epsilon_{t-2}/T + \beta\sum_{t=2}^{T}\epsilon_{t-1}\epsilon_{t-3}/T+..]$$

which yields $\text{plim} \sum_{t=2}^{T} Y_{t-1}\epsilon_{t-1}/T = \sigma_\epsilon^2$ since the second term on the right hand side has zero plim. Therefore,

$$\text{plim}(\hat{\beta}_{ols}-\beta) = \theta\sigma_\epsilon^2/\sigma_\epsilon^2 [1+(\theta+\beta)^2/(1-\beta^2)] = \theta(1-\beta^2)/(1-\beta^2+\theta^2+\beta^2+2\theta\beta)$$

$$= \theta(1-\beta^2)/(1+\theta^2+2\theta\beta) = \delta(1-\beta^2)/(1+2\beta\delta)$$

Chapter 6: Distributed Lags and Dynamic Models 123

where $\delta = \theta/(1+\theta^2)$.

b. The asymptotic bias of $\hat{\beta}_{ols}$ derived in part (a) can be tabulated for various values of β and $0 < \theta < 1$.

c. Let $\hat{v}_t = Y_t - \hat{\beta}_{ols}Y_{t-1} = \beta Y_{t-1} - \hat{\beta}_{ols}Y_{t-1} + v_t = v_t - (\hat{\beta}_{ols} - \beta)Y_{t-1}$. But

$$\hat{\beta}_{ols} - \beta = \sum_{t=2}^{T} Y_{t-1}v_t / \sum_{t=2}^{T} Y_{t-1}^2.$$

Therefore,

$$\sum_{t=2}^{T} \hat{v}_t^2 = \sum_{t=2}^{T} v_t^2 + (\sum_{t=2}^{T} Y_{t-1}v_t)^2 / \sum_{t=2}^{T} Y_{t-1}^2 - 2(\Sigma Y_{t-1}v_t)^2 / \sum_{t=2}^{T} Y_{t-1}^2$$

$$= \sum_{t=2}^{T} v_t^2 - (\sum_{t=2}^{T} Y_{t-1}v_t)^2 / \sum_{t=2}^{T} Y_{t-1}^2.$$

But, $v_t = \epsilon_t + \theta\epsilon_{t-1}$ with $var(v_t) = \sigma_\epsilon^2(1+\theta^2)$. Hence,

$$\text{plim } \sum_{t=2}^{T} v_t^2/T = \sigma_\epsilon^2 + \theta^2\sigma_\epsilon^2 = \sigma_\epsilon^2(1+\theta^2).$$

Also,

$$\text{plim } \sum_{t=2}^{T} Y_{t-1}v_t/T = \text{plim } \sum_{t=2}^{T} Y_{t-1}\epsilon_t/T + \theta\text{plim } \sum_{t=2}^{T} Y_{t-1}\epsilon_t/T = 0 + \theta\sigma_\epsilon^2 = \theta\sigma_\epsilon^2$$

from part (a). Therefore,

$$\text{plim } \sum_{t=2}^{T} \hat{v}_t^2/T = \sigma_\epsilon^2(1+\theta^2) - \theta^2\sigma_\epsilon^4/\sigma_\epsilon^2[1+(\theta+\beta)^2/(1-\beta^2)]$$

$$= \sigma_\epsilon^2[1+\theta^2-\theta^2(1-\beta^2)/(1-\theta^2+2\theta\beta)]$$

$$= \sigma_\epsilon^2[1+\theta^2-\theta\delta(1-\beta^2)/(1+2\delta\beta)] = \sigma_\epsilon^2[1+\theta^2-\theta\theta^*]$$

where $\delta = \dfrac{\theta}{(1+\theta^2)}$ and $\theta^* = \delta(1-\beta^2)/(1+2\beta\delta)$.

6.5 a. The regression given in (6.10) yields the $\widehat{var}$(coeff. of Y_{t-1}) = 0.08269. The regression of the resulting OLS residuals on their lagged values, yields

$\hat{\rho} = 0.43146$.

These SAS regressions are given below. Durbin's h given by (6.19) yields

$h = (0.43146) [43/(1-43(0.08269))]^{1/2} = 3.367$.

This is asymptotically distributed as $N(0,1)$ under the null hypothesis of $\rho = 0$. This rejects H_o indicating the presence of serial correlation.

Durbin-h test

Dependent Variable: C

Analysis of Variance

Source	DF	Sum of Squares	Mean Square	F Value	Prob>F
Model	2	3.05661	1.52831	12079.251	0.0001
Error	40	0.00506	0.00013		
C Total	42	3.06167			

Root MSE	0.01125	R-square	0.9983	
Dep Mean	9.10606	Adj R-sq	0.9983	
C.V.	0.12352			

Parameter Estimates

Variable	DF	Parameter Estimate	Standard Error	T for H0: Parameter=0	Prob > |T|
INTERCEP	1	-0.028071	0.05971057	-0.470	0.6408
CLAG1	1	0.425049	0.08269017	5.140	0.0001
Y	1	0.572918	0.08282744	6.917	0.0001

b. The Breusch and Godfrey regression to test for first-order serial correlation is given below. This regresses the OLS residuals on their lagged value and the regressors in the model including the lagged dependent variable. This yields an $R^2 = 0.1898$. The number of observations is 42. Therefore, the

LM statistic $= TR^2 = 42(0.1898) = 7.972$.

This is asymptotically distributed as χ_1^2 under the null hypothesis of $\rho = 0$. We reject the null at the 5% significance level.

Chapter 6: Distributed Lags and Dynamic Models

```
                    LM Test for AR1 by BREUSCH & GODFREY

Dependent Variable: RES         Residual

                        Analysis of Variance

                            Sum of         Mean
        Source      DF      Squares        Square      F Value      Prob>F

        Model        3      0.00096        0.00032       2.967       0.0440
        Error       38      0.00409        0.00011
        C Total     41      0.00505


        Root MSE         0.01037      R-square      0.1898
        Dep Mean         0.00009      Adj R-sq      0.1258
        C.V.          11202.01501

                          Parameter Estimates

                       Parameter       Standard     T for H0:
   Variable   DF       Estimate         Error       Parameter=0    Prob > |T|

   INTERCEP    1      -0.009418       0.05758575      -0.164         0.8710
   RLAG1       1       0.468006       0.15700139       2.981         0.0050
   CLAG1       1      -0.063769       0.08063434      -0.791         0.4339
   Y           1       0.064022       0.08109334       0.789         0.4347
```

c. The Breusch and Godfrey regression to test for second-order serial correlation is given below. This regresses the OLS residuals e_t on e_{t-1} and e_{t-2} and the regressors in the model including the lagged dependent variable. This yields an $R^2 = 0.2139$. The number of observations is 41. Therefore, the LM statistic $= TR^2 = 41(0.2139) = 8.7699$.

This is asymptotically distributed as χ_2^2 under the null hypothesis of no second-order serial correlation. We reject the null at the 5% significance level.

LM Test for AR2 by BREUSCH & GODFREY

Dependent Variable: RES   Residual

Analysis of Variance

Source	DF	Sum of Squares	Mean Square	F Value	Prob>F
Model	4	0.00108	0.00027	2.448	0.0638
Error	36	0.00397	0.00011		
C Total	40	0.00504			

Root MSE	0.01050	R-square	0.2139	
Dep Mean	0.00011	Adj R-sq	0.1265	
C.V.	9503.70002			

Parameter Estimates

Variable	DF	Parameter Estimate	Standard Error	T for H0: Parameter=0	Prob > |T|
INTERCEP	1	-0.004200	0.06137905	-0.068	0.9458
RLAG1	1	0.528816	0.16915130	3.126	0.0035
RLAG2	1	-0.185773	0.17923858	-1.036	0.3069
CLAG1	1	-0.033806	0.08786935	-0.385	0.7027
Y	1	0.033856	0.08870382	0.382	0.7049

SAS PROGRAM

```
PROC REG DATA=CONSUM;
    MODEL C=Clag1 Y;
OUTPUT OUT=OUT1 R=RES;
TITLE 'Durbin-h Test';
RUN;

DATA DW;  SET OUT1;
Rlag1=LAG(RES);
Rlag2=LAG2(RES);

PROC REG DATA=DW;
    MODEL RES=Rlag1 Clag1 Y;
TITLE 'LM Test for AR1 by BREUSCH & GODFREY';
RUN;
```

Chapter 6: Distributed Lags and Dynamic Models

```
PROC REG DATA=DW;
    MODEL RES=Rlag1 Rlag2 Clag1 Y;
TITLE 'LM Test for AR2 by BREUSCH & GODFREY';
RUN;
```

6.6 U.S. Gasoline Data, Table 4.2.

   a. STATIC MODEL

Dependent Variable: QMG_CAR

Analysis of Variance

Source	DF	Sum of Squares	Mean Square	F Value	Prob>F
Model	3	0.35693	0.11898	32.508	0.0001
Error	34	0.12444	0.00366		
C Total	37	0.48137			

Root MSE	0.06050	R-square	0.7415
Dep Mean	-0.25616	Adj R-sq	0.7187
C.V.	-23.61671		

Parameter Estimates

Variable	DF	Parameter Estimate	Standard Error	T for H0: Parameter=0	Prob > |T|
INTERCEP	1	-5.853977	3.10247647	-1.887	0.0677
RGNP_POP	1	-0.690460	0.29336969	-2.354	0.0245
CAR_POP	1	0.288735	0.27723429	1.041	0.3050
PMG_PGNP	1	-0.143127	0.07487993	-1.911	0.0644

DYNAMIC MODEL

                           Autoreg Procedure

Dependent Variable = QMG_CAR

                     Ordinary Least Squares Estimates

SSE	0.014269	DFE	32
MSE	0.000446	Root MSE	0.021117
SBC	-167.784	AIC	-175.839
Reg Rsq	0.9701	Total Rsq	0.9701
Durbin h	2.147687	PROB>h	0.0159

Godfrey's Serial Correlation Test

Alternative	LM	Prob>LM
AR(+ 1)	4.6989	0.0302
AR(+ 2)	5.1570	0.0759

Variable	DF	B Value	Std Error	t Ratio	Approx Prob
Intercept	1	0.523006	1.1594	0.451	0.6550
RGNP_POP	1	0.050519	0.1127	0.448	0.6571
CAR_POP	1	-0.106323	0.1005	-1.058	0.2981
PMG_PGNP	1	-0.072884	0.0267	-2.733	0.0101
LAG_DEP	1	0.907674	0.0593	15.315	0.0001

c. LM Test for AR1 by BREUSCH & GODFREY

Dependent Variable: RESID

Analysis of Variance

Source	DF	Sum of Squares	Mean Square	F Value	Prob>F
Model	5	0.00178	0.00036	0.863	0.5172
Error	30	0.01237	0.00041		
C Total	35	0.01415			

Root MSE	0.02031	R-square	0.1257	
Dep Mean	-0.00030	Adj R-sq	-0.0200	
C.V.	-6774.59933			

Parameter Estimates

Variable	DF	Parameter Estimate	Standard Error	T for H0: Parameter=0	Prob > |T|
INTERCEP	1	-0.377403	1.15207945	-0.328	0.7455
RESID_1	1	0.380254	0.18431559	2.063	0.0479
RGNP_POP	1	-0.045964	0.11158093	-0.412	0.6833
CAR_POP	1	0.031417	0.10061545	0.312	0.7570
PMG_PGNP	1	0.007723	0.02612020	0.296	0.7695
LAG_DEP	1	-0.034217	0.06069319	-0.564	0.5771

SAS PROGRAM

```
Data RAWDATA;
Input Year CAR QMG PMG POP RGNP PGNP;
```

Chapter 6: Distributed Lags and Dynamic Models 129

```
Cards;

Data USGAS; set RAWDATA;
LNQMG=LOG(QMG);
LNCAR=LOG(CAR);
LNPOP=LOG(POP);
LNRGNP=LOG(RGNP);
LNPGNP=LOG(PGNP);
LNPMG=LOG(PMG);
QMG_CAR=LOG(QMG/CAR);
RGNP_POP=LOG(RGNP/POP);
CAR_POP=LOG(CAR/POP);
PMG_PGNP=LOG(PMG/PGNP);
LAG_DEP=LAG(QMG_CAR);

Proc reg data=USGAS;
    Model QMG_CAR=RGNP_POP CAR_POP PMG_PGNP;
  TITLE ' STATIC MODEL';

Proc autoreg data=USGAS;
    Model QMG_CAR=RGNP_POP CAR_POP PMG_PGNP
                    LAG_DEP/LAGDEP=LAG_DEP godfrey=2;
  OUTPUT OUT=MODEL2 R=RESID;
  TITLE ' DYNAMIC MODEL';
RUN;

DATA DW_DATA;  SET MODEL2;
RESID_1=LAG(RESID);
PROC REG DATA=DW_DATA;
    MODEL RESID=RESID_1 RGNP_POP CAR_POP PMG_PGNP LAG_DEP;
TITLE 'LM Test for AR1 by BREUSCH & GODFREY';
RUN;
```

6.7
  a. Unrestricted Model

```
Autoreg Procedure
Dependent Variable = QMG_CAR
```

                Ordinary Least Squares Estimates

              SSE   0.031403   DFE   22
              MSE   0.001427   Root MSE   0.037781
              SBC   -96.1812   AIC   -110.839
              Reg Rsq   0.9216   Total Rsq   0.9216
              Durbin-Watson   0.5683

Variable	DF	B Value	Std Error	t Ratio	Approx Prob
Intercept	1	-7.46541713	3.0889	-2.417	0.0244
RGNP_POP	1	-0.58684334	0.2831	-2.073	0.0501
CAR_POP	1	0.24215182	0.2850	0.850	0.4046
PMG_PGNP	1	-0.02611161	0.0896	-0.291	0.7734
PMPG_1	1	-0.15248735	0.1429	-1.067	0.2975
PMPG_2	1	-0.13752842	0.1882	-0.731	0.4726
PMPG_3	1	0.05906629	0.2164	0.273	0.7875
PMPG_4	1	-0.21264747	0.2184	-0.974	0.3408
PMPG_5	1	0.22649780	0.1963	1.154	0.2609
PMPG_6	1	-0.41142284	0.1181	-3.483	0.0021

b. Almon Lag (S=6,P=2)

PDLREG Procedure
Dependent Variable = QMG_CAR

Ordinary Least Squares Estimates

SSE	0.04017	DFE		26
MSE	0.001545	Root MSE	0.039306
SBC	-102.165	AIC	-110.96
Reg Rsq	0.8998	Total Rsq	0.8998
Durbin-Watson	0.5094		

Variable	DF	B Value	Std Error	t Ratio	Approx Prob
Intercept	1	-5.06184299	2.9928	-1.691	0.1027
RGNP_POP	1	-0.35769028	0.2724	-1.313	0.2006
CAR_POP	1	0.02394559	0.2756	0.087	0.9314
PMG_PGNP**0	1	-0.24718333	0.0340	-7.278	0.0001
PMG_PGNP**1	1	-0.05979404	0.0439	-1.363	0.1847
PMG_PGNP**2	1	-0.10450923	0.0674	-1.551	0.1331

Variable	Parameter Value	Std Error	t Ratio	Approx Prob
PMG_PGNP(0)	-0.11654	0.045	-2.58	0.0159
PMG_PGNP(1)	-0.07083	0.020	-3.46	0.0019
PMG_PGNP(2)	-0.04792	0.024	-1.98	0.0584
PMG_PGNP(3)	-0.04781	0.028	-1.74	0.0944
PMG_PGNP(4)	-0.07052	0.021	-3.33	0.0026
PMG_PGNP(5)	-0.11603	0.021	-5.40	0.0001
PMG_PGNP(6)	-0.18434	0.054	-3.42	0.0021

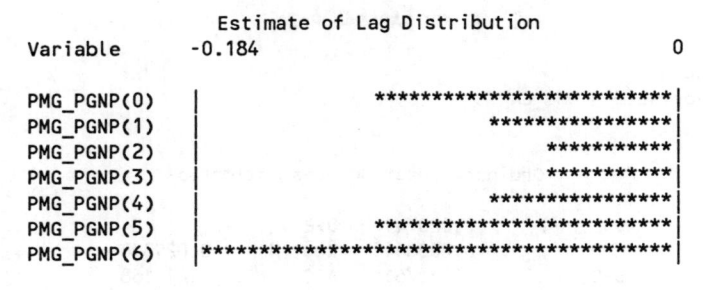

```
                  Estimate of Lag Distribution
Variable          -0.184                                      0

PMG_PGNP(0)   |              ***************************|
PMG_PGNP(1)   |                  ****************|
PMG_PGNP(2)   |                       **********|
PMG_PGNP(3)   |                       **********|
PMG_PGNP(4)   |                  ****************|
PMG_PGNP(5)   |              ***************************|
PMG_PGNP(6)   |*******************************************|
```

c. Almon Lag (S=4,P=2)

PDLREG Procedure
Dependent Variable = QMG_CAR

Ordinary Least Squares Estimates

SSE	0.065767	DFE	28
MSE	0.002349	Root MSE	0.048464
SBC	-94.7861	AIC	-103.944
Reg Rsq	0.8490	Total Rsq	0.8490
Durbin-Watson	0.5046		

Variable	DF	B Value	Std Error	t Ratio	Approx Prob
Intercept	1	-6.19647990	3.6920	-1.678	0.1044
RGNP_POP	1	-0.57281368	0.3422	-1.674	0.1053
CAR_POP	1	0.21338192	0.3397	0.628	0.5351
PMG_PGNP**0	1	-0.19423745	0.0414	-4.687	0.0001
PMG_PGNP**1	1	-0.06534647	0.0637	-1.026	0.3138
PMG_PGNP**2	1	0.03085234	0.1188	0.260	0.7970

Variable	Parameter Value	Std Error	t Ratio	Approx Prob
PMG_PGNP(0)	-0.02905	0.070	-0.41	0.6834
PMG_PGNP(1)	-0.07445	0.042	-1.78	0.0851
PMG_PGNP(2)	-0.10336	0.061	-1.70	0.0999
PMG_PGNP(3)	-0.11578	0.033	-3.51	0.0015
PMG_PGNP(4)	-0.11170	0.092	-1.22	0.2329

```
                  Estimate of Lag Distribution
Variable          -0.116                                      0

PMG_PGNP(0)   |                              **********|
PMG_PGNP(1)   |                  *************************|
PMG_PGNP(2)   |         *************************************|
PMG_PGNP(3)   |*******************************************|
PMG_PGNP(4)   |  *************************************|
```

ALMON LAG(S=8,P=2)

PDLREG Procedure
Dependent Variable = QMG_CAR

Ordinary Least Squares Estimates

SSE	0.020741	DFE	24
MSE	0.000864	Root MSE	0.029398
SBC	-112.761	AIC	-121.168
Reg Rsq	0.9438	Total Rsq	0.9438
Durbin-Watson	0.9531		

Variable	DF	B Value	Std Error	t Ratio	Approx Prob
Intercept	1	-7.71363805	2.3053	-3.346	0.0027
RGNP_POP	1	-0.53016065	0.2041	-2.597	0.0158
CAR_POP	1	0.17117375	0.2099	0.815	0.4229
PMG_PGNP**0	1	-0.28572518	0.0267	-10.698	0.0001
PMG_PGNP**1	1	-0.09282151	0.0417	-2.225	0.0358
PMG_PGNP**2	1	-0.12948786	0.0512	-2.527	0.0185

Variable	Parameter Value	Std Error	t Ratio	Approx Prob
PMG_PGNP(0)	-0.11617	0.028	-4.09	0.0004
PMG_PGNP(1)	-0.07651	0.016	-4.73	0.0001
PMG_PGNP(2)	-0.05160	0.015	-3.34	0.0027
PMG_PGNP(3)	-0.04145	0.018	-2.30	0.0301
PMG_PGNP(4)	-0.04605	0.017	-2.63	0.0146
PMG_PGNP(5)	-0.06541	0.013	-4.85	0.0001
PMG_PGNP(6)	-0.09953	0.012	-8.10	0.0001
PMG_PGNP(7)	-0.14841	0.025	-5.97	0.0001
PMG_PGNP(8)	-0.21204	0.047	-4.53	0.0001

Estimate of Lag Distribution

Variable   -0.212   0

```
PMG_PGNP(0)  |                     **********************|
PMG_PGNP(1)  |                        ***************   |
PMG_PGNP(2)  |                             **********   |
PMG_PGNP(3)  |                              ********    |
PMG_PGNP(4)  |                             *********    |
PMG_PGNP(5)  |                        *************     |
PMG_PGNP(6)  |                   *******************    |
PMG_PGNP(7)  |           ****************************   |
PMG_PGNP(8)  |******************************************|
```

d. Third Degree Polynomial Almon Lag(S=6,P=3)

```
PDLREG Procedure
Dependent Variable = QMG_CAR
```

Ordinary Least Squares Estimates

SSE	0.034308	DFE		25
MSE	0.001372	Root MSE	0.037045
SBC	-103.747	AIC		-114.007
Reg Rsq	0.9144	Total Rsq	0.9144
Durbin-Watson	0.6763			

Variable	DF	B Value	Std Error	t Ratio	Approx Prob
Intercept	1	-7.31542415	3.0240	-2.419	0.0232
RGNP_POP	1	-0.57343614	0.2771	-2.069	0.0490
CAR_POP	1	0.23462358	0.2790	0.841	0.4084
PMG_PGNP**0	1	-0.24397597	0.0320	-7.613	0.0001
PMG_PGNP**1	1	-0.07041380	0.0417	-1.690	0.1035
PMG_PGNP**2	1	-0.11318734	0.0637	-1.778	0.0876
PMG_PGNP**3	1	-0.19730731	0.0955	-2.067	0.0493

Variable	Parameter Value	Std Error	t Ratio	Approx Prob
PMG_PGNP(0)	-0.03349	0.059	-0.57	0.5725
PMG_PGNP(1)	-0.14615	0.041	-3.54	0.0016
PMG_PGNP(2)	-0.12241	0.043	-2.87	0.0082
PMG_PGNP(3)	-0.04282	0.026	-1.64	0.1130
PMG_PGNP(4)	0.01208	0.045	0.27	0.7890
PMG_PGNP(5)	-0.03828	0.043	-0.90	0.3788
PMG_PGNP(6)	-0.27443	0.067	-4.10	0.0004

Estimate of Lag Distribution

Variable	-0.274		0.0121
PMG_PGNP(0)		| ****|	
PMG_PGNP(1)		| ********************|	
PMG_PGNP(2)		| ****************|	
PMG_PGNP(3)		| *****|	
PMG_PGNP(4)		| |**	
PMG_PGNP(5)		| ****|	
PMG_PGNP(6)	|**************************************|		

e. Almon Lag(S=6,P=2) with Near End-Point Restriction

```
PDLREG Procedure
Dependent Variable = QMG_CAR
```

Ordinary Least Squares Estimates

SSE	0.046362	DFE	27
MSE	0.001717	Root MSE	0.041438
SBC	-101.043	AIC	-108.372
Reg Rsq	0.8843	Total Rsq	0.8843
Durbin-Watson	0.5360		

Variable	DF	B Value	Std Error	t Ratio	Approx Prob
Intercept	1	-3.81408793	3.0859	-1.236	0.2271
RGNP_POP	1	-0.28069982	0.2843	-0.988	0.3322
CAR_POP	1	-0.05768233	0.2873	-0.201	0.8424
PMG_PGNP**0	1	-0.21562744	0.0317	-6.799	0.0001
PMG_PGNP**1	1	-0.07238330	0.0458	-1.581	0.1255
PMG_PGNP**2	1	0.02045576	0.0268	0.763	0.4519

Restriction	DF	L Value	Std Error	t Ratio	Approx Prob
PMG_PGNP(-1)	-1	0.03346081	0.0176	1.899	0.0683

Variable	Parameter Value	Std Error	t Ratio	Approx Prob
PMG_PGNP(0)	-0.02930	0.013	-2.31	0.0286
PMG_PGNP(1)	-0.05414	0.020	-2.75	0.0105
PMG_PGNP(2)	-0.07452	0.021	-3.49	0.0017
PMG_PGNP(3)	-0.09043	0.018	-4.91	0.0001
PMG_PGNP(4)	-0.10187	0.015	-6.80	0.0001
PMG_PGNP(5)	-0.10886	0.022	-4.88	0.0001
PMG_PGNP(6)	-0.11138	0.042	-2.66	0.0130

Estimate of Lag Distribution

Variable	-0.111	0
PMG_PGNP(0)		| **********|
PMG_PGNP(1)		| ******************|
PMG_PGNP(2)		| *************************|
PMG_PGNP(3)		|******************************|
PMG_PGNP(4)		|**********************************|
PMG_PGNP(5)		|************************************|
PMG_PGNP(6)		|*************************************|

Chapter 6: Distributed Lags and Dynamic Models

ALMON LAG(S=6,P=2) with FAR END-POINT RESTRICTION

PDLREG Procedure
Dependent Variable = QMG_CAR

Ordinary Least Squares Estimates

SSE	0.050648	DFE		27
MSE	0.001876	Root MSE	0.043311	
SBC	-98.2144	AIC	-105.543	
Reg Rsq	0.8736	Total Rsq	0.8736	
Durbin-Watson	0.5690			

Variable	DF	B Value	Std Error	t Ratio	Approx Prob
Intercept	1	-6.02848892	3.2722	-1.842	0.0764
RGNP_POP	1	-0.49021381	0.2948	-1.663	0.1079
CAR_POP	1	0.15878127	0.2982	0.532	0.5988
PMG_PGNP**0	1	-0.20851840	0.0337	-6.195	0.0001
PMG_PGNP**1	1	-0.00242944	0.0418	-0.058	0.9541
PMG_PGNP**2	1	0.06159671	0.0240	2.568	0.0161

Restriction	DF	L Value	Std Error	t Ratio	Approx Prob
PMG_PGNP(7)	-1	0.03803694	0.0161	2.363	0.0256

Variable	Parameter Value	Std Error	t Ratio	Approx Prob
PMG_PGNP(0)	-0.04383	0.039	-1.12	0.2730
PMG_PGNP(1)	-0.07789	0.022	-3.49	0.0017
PMG_PGNP(2)	-0.09852	0.016	-6.20	0.0001
PMG_PGNP(3)	-0.10570	0.018	-5.90	0.0001
PMG_PGNP(4)	-0.09943	0.020	-5.01	0.0001
PMG_PGNP(5)	-0.07973	0.018	-4.43	0.0001
PMG_PGNP(6)	-0.04659	0.011	-4.05	0.0004

Estimate of Lag Distribution

Variable	-0.106	0
PMG_PGNP(0)		| ****************|
PMG_PGNP(1)		| *****************************|
PMG_PGNP(2)		| ***************************************|
PMG_PGNP(3)	|***|
PMG_PGNP(4)		| ***|
PMG_PGNP(5)		| ******************************|
PMG_PGNP(6)		| ******************|

SAS PROGRAM

```
Data RAWDATA;
Input Year CAR QMG PMG POP RGNP PGNP;
Cards;
Data USGAS; set RAWDATA;
LNQMG=LOG(QMG);
LNCAR=LOG(CAR);
LNPOP=LOG(POP);
LNRGNP=LOG(RGNP);
LNPGNP=LOG(PGNP);
LNPMG=LOG(PMG);
QMG_CAR=LOG(QMG/CAR);
RGNP_POP=LOG(RGNP/POP);
CAR_POP=LOG(CAR/POP);
PMG_PGNP=LOG(PMG/PGNP);
PMPG_1=LAG1(PMG_PGNP);
PMPG_2=LAG2(PMG_PGNP);
PMPG_3=LAG3(PMG_PGNP);
PMPG_4=LAG4(PMG_PGNP);
PMPG_5=LAG5(PMG_PGNP);
PMPG_6=LAG6(PMG_PGNP);

Proc autoreg data=USGAS;
    Model QMG_CAR=RGNP_POP CAR_POP PMG_PGNP PMPG_1 PMPG_2 PMPG_3
                PMPG_4 PMPG_5 PMPG_6;
  TITLE 'UNRESTRICTED MODEL';

PROC PDLREG DATA=USGAS;
    MODEL QMG_CAR=RGNP_POP CAR_POP PMG_PGNP(6,2);
TITLE 'ALMON LAG(S=6,P=2)';

PROC PDLREG DATA=USGAS;
    MODEL QMG_CAR=RGNP_POP CAR_POP PMG_PGNP(4,2);
TITLE 'ALMON LAG(S=4,P=2)';

PROC PDLREG DATA=USGAS;
    MODEL QMG_CAR=RGNP_POP CAR_POP PMG_PGNP(8,2);
TITLE 'ALMON LAG(S=8,P=2)';

PROC PDLREG DATA=USGAS;
    MODEL QMG_CAR=RGNP_POP CAR_POP PMG_PGNP(6,3);
TITLE 'Third Degree Polynomial ALMON LAG(S=6,P=3)';

PROC PDLREG DATA=USGAS;
    MODEL QMG_CAR=RGNP_POP CAR_POP PMG_PGNP(6,2,,FIRST);
TITLE 'ALMON LAG(S=6,P=2) with NEAR END-POINT RESTRICTION';

PROC PDLREG DATA=USGAS;
    MODEL QMG_CAR=RGNP_POP CAR_POP PMG_PGNP(6,2,,LAST);
TITLE 'ALMON LAG(S=6,P=2) with FAR END-POINT RESTRICTION';
RUN;
```

CHAPTER 7
The General Linear Model: The Basics

7.1 *Invariance of the fitted values and residuals to non-singular transformations of the independent variables.*

The regression model in (7.1) can be written as $y = XCC^{-1}\beta + u$ where C is a non-singular matrix. Let $X^* = XC$, then $y = X^*\beta^* + u$ where $\beta^* = C^{-1}\beta$.

a. $P_{X^*} = X^*(X^{*'}X^*)^{-1}X^{*'} = XC[C'X'XC]^{-1}C'X' = XCC^{-1}(X'X)^{-1}C'^{-1}C'X' = P_X$.

Hence, the regression of y on X^* yields

$$\hat{y} = X^*\hat{\beta}^*_{ols} = P_{X^*}\cdot y = P_X y = X\hat{\beta}_{ols}$$

which is the same fitted values as those from the regression of y on X. Since the dependent variable y is the same, the residuals from both regressions will be the same.

b. Multiplying each X by a constant is equivalent to post-multiplying the matrix X by a diagonal matrix C with a typical k-th element c_k. Each X_k will be multiplied by the constant c_k for $k = 1,2,...,K$. This diagonal matrix C is non-singular. Therefore, using the results in part (a), the fitted values and the residuals will remain the same.

c. In this case, $X = [X_1, X_2]$ is of dimension nx2 and

$$X^* = [X_1-X_2, X_1+X_2] = [X_1,X_2]\begin{bmatrix} 1 & 1 \\ -1 & 1 \end{bmatrix} = XC$$

where $C = \begin{bmatrix} 1 & 1 \\ -1 & 1 \end{bmatrix}$ is non-singular with $C^{-1} = \frac{1}{2}\begin{bmatrix} 1 & -1 \\ 1 & 1 \end{bmatrix}$. Hence, the results of part (a) apply, and we get the same fitted values and residuals when we regress y on (X_1-X_2) and (X_1+X_2) as in the regression of y on X_1 and X_2.

7.2 *The FWL Theorem.*

a. The inverse of a partitioned matrix

$$A = \begin{bmatrix} A_{11} & A_{12} \\ A_{21} & A_{22} \end{bmatrix}$$

is given by

$$A^{-1} = \begin{bmatrix} A_{11}^{-1} + A_{11}^{-1}A_{12}B_{22}A_{21}A_{11}^{-1} & -A_{11}^{-1}A_{12}B_{22} \\ -B_{22}A_{21}A_{11}^{-1} & B_{22} \end{bmatrix}$$

where $B_{22} = (A_{22} - A_{21}A_{11}^{-1}A_{12})^{-1}$. From (7.9), we get

$B_{22} = (X_2'X_2 - X_2'X_1(X_1'X_1)^{-1}X_1'X_2)^{-1} = (X_2'X_2 - X_2'P_{X_1}X_2)^{-1} = (X_2'\bar{P}_{X_1}X_2)^{-1}$.

Also, $-B_{22}A_{21}A_{11}^{-1} = -(X_2'\bar{P}_{X_1}X_2)^{-1}X_2'X_1(X_1'X_1)^{-1}$. Hence, from (7.9), we solve for $\hat{\beta}_{2,ols}$ to get

$$\hat{\beta}_{2,ols} = -B_{22}A_{21}A_{11}^{-1}X_1'y + B_{22}X_2'y$$

which yields

$\hat{\beta}_{2,ols} = -(X_2'\bar{P}_{X_1}X_2)^{-1}X_2'P_{X_1}y + (X_2'\bar{P}_{X_1}X_2)^{-1}X_2'y = (X_2'\bar{P}_{X_1}X_2)^{-1}X_2'\bar{P}_{X_1}y$

as required in (7.10).

b. Alternatively, one can write (7.9) as

$$(X_1'X_1)\hat{\beta}_{1,ols} + (X_1'X_2)\hat{\beta}_{2,ols} = X_1'y$$
$$(X_2'X_1)\hat{\beta}_{1,ols} + (X_2'X_2)\hat{\beta}_{2,ols} = X_2'y.$$

Solving for $\hat{\beta}_{1,ols}$ in terms of $\hat{\beta}_{2,ols}$ by multiplying the first equation by $(X_1'X_1)^{-1}$ we get

$$\hat{\beta}_{1,ols} = (X_1'X_1)^{-1}X_1'y - (X_1'X_1)^{-1}X_1'X_2\hat{\beta}_{2,ols} = (X_1'X_1)^{-1}X_1'(y - X_2\hat{\beta}_{2,ols}).$$

Substituting $\hat{\beta}_{1,ols}$ in the second equation, we get

$$X_2'X_1(X_1'X_1)^{-1}X_1'y - X_2'P_{X_1}X_2\hat{\beta}_{2,ols} + (X_2'X_2)\hat{\beta}_{2,ols} = X_2'y.$$

Collecting terms, we get $(X_2'\bar{P}_{X_1}X_2)\hat{\beta}_{2,ols} = X_2'\bar{P}_{X_1}y$.

Hence, $\hat{\beta}_{2,ols} = (X_2'\bar{P}_{X_1}X_2)^{-1}X_2'\bar{P}_{X_1}y$ as given in (7.10).

c. In this case, $X = [\iota_n, X_2]$ where ι_n is a vector of ones of dimension n.

$P_{X_1} = \iota_n(\iota_n'\iota_n)^{-1}\iota_n' = \iota_n\iota_n'/n = J_n/n$ where $J_n = \iota_n\iota_n'$ is a matrix of ones of dimension n. But $J_n y = \sum_{i=1}^{n} y_i$ and $J_n y/n = \bar{y}$. Hence, $\bar{P}_{X_1} = I_n - P_{X_1} = I_n - J_n/n$ and $\bar{P}_{X_1}y = (I_n - J_n/n)y$ has a typical element $(y_i - \bar{y})$. From the FWL Theorem,

$\hat{\beta}_{2,ols}$ can be obtained from the regression of $(y_i - \bar{y})$ on the set of variables in X_2 expressed as deviations from their respective means, i.e.,

$$\bar{P}_{X_1} X_2 = (I_n - J_n/n) X_2.$$

From part (b),

$$\hat{\beta}_{1,ols} = (X_1' X_1)^{-1} X_1' (y - X_2 \hat{\beta}_{2,ols}) = (\iota_n' \iota_n)^{-1} \iota_n' (y - X_2 \hat{\beta}_{2,ols})$$

$$= \frac{\iota_n'}{n} (y - X_2 \hat{\beta}_{2,ols}) = \bar{y} - \bar{X}_2' \hat{\beta}_{2,ols}$$

where $\bar{X}_2' = \iota_n' X_2 / n$ is the vector of sample means of the independent variables in X_2.

7.3 $D_i = (0,0,..,1,0,..,0)'$ where all the elements of this n×1 vector are zeroes except for the i-th element which takes the value 1. In this case, $P_{D_i} = D_i (D_i' D_i)^{-1} D_i'$ = $D_i D_i'$ which is a matrix of zeroes except for the i-th diagonal element which takes the value 1. Hence, $I_n - P_{D_i}$ is an identity matrix except for the i-th diagonal element which takes the value zero. Therefore, $(I_n - P_{D_i}) y$ returns the vector y except for the i-th element which is zero. Using the FWL Theorem, the OLS regression

$$y = X\beta + D_i \gamma + u$$

yields the same estimates as $(I_n - P_{D_i}) y = (I_n - P_{D_i}) X\beta + (I_n - P_{D_i}) u$ which can be rewritten as $\tilde{y} = \tilde{X}\beta + \tilde{u}$ with $\tilde{y} = (I_n - P_{D_i}) y$, $\tilde{X} = (I_n - P_{D_i}) X$. The OLS normal equations yield $(\tilde{X}' \tilde{X}) \hat{\beta}_{ols} = \tilde{X}' \tilde{y}$ and the i-th OLS normal equation can be ignored since it gives $0' \hat{\beta}_{ols} = 0$. Ignoring the i-th observation equation yields $(X^{*'} X^*) \hat{\beta}_{ols} = X^{*'} y^*$ where X^* is the matrix X without the i-th observation and y^* is the vector y without the i-th observation. The FWL

Theorem also states that the residuals from $\tilde{y}$ on $\tilde{X}$ are the same as those from y on X and D_i. For the i-th observation, $\tilde{y}_i = 0$ and $\tilde{x}_i = 0$. Hence the i-th residual must be zero. This also means that the i-th residual in the original regression with the dummy variable D_i is zero, i.e., $y_i - x_i' \hat{\beta}_{ols} - \hat{\gamma}_{ols} = 0$. Rearranging terms, we get $\hat{\gamma}_{ols} = y_i - x_i' \hat{\beta}_{ols}$. In other words, $\hat{\gamma}_{ols}$ is the forecasted OLS residual for the i-th observation from the regression of y^* on X^*. The i-th observation was excluded from the estimation of $\hat{\beta}_{ols}$ by the inclusion of the dummy variable D_i.

7.5 If $u \sim N(0, \sigma^2 I_n)$ then $(n-K)s^2/\sigma^2 \sim \chi^2_{n-K}$. In this case,

a. $E[(n-K)s^2/\sigma^2] = E(\chi^2_{n-K}) = n-K$ since the expected value of a χ^2 random variable with (n-K) degrees of freedom is (n-K). Hence,

$[(n-K)/\sigma^2]E(s^2) = (n-K)$ or $E(s^2) = \sigma^2$.

b. $\text{var}[(n-K)s^2/\sigma^2] = \text{var}(\chi^2_{n-K}) = 2(n-K)$ since the variance of a χ^2 random variable with (n-K) degrees of freedom is 2(n-K). Hence,

$[(n-K)^2/\sigma^4] \text{var}(s^2) = 2(n-K)$ or $\text{var}(s^2) = 2\sigma^4/(n-K)$.

7.6 a. Using the results in problem 4, we know that $\hat{\sigma}^2_{mle} = e'e/n = (n-K)s^2/n$. Hence,
$E(\hat{\sigma}^2_{mle}) = (n-K) E(s^2)/n = (n-K)\sigma^2/n$.

This means that $\hat{\sigma}^2_{mle}$ is biased for σ^2, but asymptotically unbiased. The bias is equal to $-K\sigma^2/n$ which goes to zero as $n \to \infty$.

b. $\text{var}(\hat{\sigma}^2_{mle}) = (n-K)^2 \text{var}(s^2)/n^2 = (n-K)^2 2\sigma^4/n^2(n-K) = 2(n-K)\sigma^4/n^2$ and
$\text{MSE}(\hat{\sigma}^2_{mle}) = \text{Bias}^2(\hat{\sigma}^2_{mle}) + \text{var}(\hat{\sigma}^2_{mle}) = K^2\sigma^4/n^2 + 2(n-K)\sigma^4/n^2$
$= (K^2 + 2n - 2K)\sigma^4/n^2$.

Chapter 7: The General Linear Model: The Basics   141

c. Similarly, $\tilde{\sigma}^2 = e'e/r = (n-K)s^2/r$ with $E(\tilde{\sigma}^2) = (n-K)\sigma^2/r$ and

$$\text{var}(\tilde{\sigma}^2) = (n-K)^2 \text{var}(s^2)/r^2 = (n-K)^2 2\sigma^4/r^2(n-K) = 2(n-K)\sigma^4/r^2$$

$$\text{MSE}(\tilde{\sigma}^2) = \text{Bias}^2(\tilde{\sigma}^2) + \text{var}(\tilde{\sigma}^2) = (n-K-r)^2\sigma^4/r^2 + 2(n-K)\sigma^4/r^2.$$

Minimizing $\text{MSE}(\tilde{\sigma}^2)$ with respect to r yields the first-order condition

$$\frac{\partial \text{MSE}(\tilde{\sigma}^2)}{\partial r} = \frac{-2(n-K-r)\sigma^4 r^2 - 2r(n-K-r)^2\sigma^4}{r^4} - \frac{4r(n-K)\sigma^4}{r^4} = 0$$

which yields

$$(n-K-r)r + (n-K-r)^2 + 2(n-K) = 0$$

$$(n-K-r)(r+n-K-r) + 2(n-K) = 0$$

$$(n-K)(n-K-r+2) = 0$$

since $n > K$, this is zero for $r = n-K+2$. Hence, the Minimum MSE is obtained at $\tilde{\sigma}_*^2 = e'e/(n-K+2)$ with

$$\text{MSE}(\tilde{\sigma}_*^2) = 4\sigma^4/(n-K+2)^2 + 2(n-K)\sigma^4/(n-K+2)^2$$

$$= 2\sigma^4(n-K+2)/(n-K+2)^2 = 2\sigma^4/(n-K+2).$$

Note that $s^2 = e'e/(n-K)$ with $\text{MSE}(s^2) = \text{var}(s^2) = 2\sigma^4/(n-K) > \text{MSE}(\tilde{\sigma}_*^2)$. Also, it can be easily verified that $\text{MSE}(\hat{\sigma}_{mle}^2) = (K^2-2K+2n)\sigma^4/n^2 \geq \text{MSE}(\tilde{\sigma}_*^2)$ for $2 \leq K < n$, with the equality holding for $K = 2$.

7.7 *Computing Forecasts and Forecast Standard Errors Using a Regression Package.* This is based on Salkever (1976). From (7.23) one gets

$$X^{*'}X^* = \begin{bmatrix} X' & X_o' \\ 0 & I_{T_o} \end{bmatrix} \begin{bmatrix} X & 0 \\ X_o & I_{T_o} \end{bmatrix} = \begin{bmatrix} X'X+X_o'X_o & X_o' \\ X_o & I_{T_o} \end{bmatrix}$$

and

$$X^{*'}y^* = \begin{bmatrix} X'y+X_o'y_o \\ y_o \end{bmatrix}.$$

The OLS normal equations yield

$$X^{*\prime}X^* \begin{bmatrix} \hat{\beta}_{ols} \\ \hat{\gamma}_{ols} \end{bmatrix} = X^{*\prime}y^*$$

or $(X'X)\hat{\beta}_{ols} + (X'_o X_o)\hat{\beta}_{ols} + X'_o \hat{\gamma}_{ols} = X'y + X'_o y_o$

and $X_o \hat{\beta}_{ols} + \hat{\gamma}_{ols} = y_o$

From the second equation, it is obvious that $\hat{\gamma}_{ols} = y_o - X_o \hat{\beta}_{ols}$. Substituting this in the first equation yields

$$(X'X)\hat{\beta}_{ols} + (X'_o X_o)\hat{\beta}_{ols} + X'_o y_o - X'_o X_o \hat{\beta}_{ols} = X'y + X'_o y_o$$

which upon cancellations gives $\hat{\beta}_{ols} = (X'X)^{-1}X'y$.

Alternatively, one could apply the FWL Theorem using $X_1 = \begin{bmatrix} X \\ X_o \end{bmatrix}$ and $X_2 = \begin{bmatrix} 0 \\ I_{T_o} \end{bmatrix}$.

In this case, $X'_2 X_2 = I_{T_o}$ and

$$P_{X_2} = X_2 (X'_2 X_2)^{-1} X'_2 = X_2 X'_2 = \begin{bmatrix} 0 & 0 \\ 0 & I_{T_o} \end{bmatrix}.$$

This means that

$$\bar{P}_{X_2} = I_{n+T_o} - P_{X_2} = \begin{bmatrix} I_n & 0 \\ 0 & 0 \end{bmatrix}.$$

Premultiplying (7.23) by $\bar{P}_{X_2}$ is equivalent to omitting the last T_o observations. The resulting regression is that of y on X which yields $\hat{\beta}_{ols} = (X'X)^{-1}X'y$ as obtained above. Also, premultiplying by $\bar{P}_{X_2}$, the last T_o observations yield zero residuals because the observations on both the dependent and independent variables are zero. For this to be true in the original regression, we must have $y_o - X_o \hat{\beta}_{ols} - \hat{\gamma}_{ols} = 0$. This means that $\hat{\gamma}_{ols} = y_o - X_o \hat{\beta}_{ols}$ as required.

b. The OLS residuals of (7.23) yield the usual least squares residuals

$$e_{ols} = y - X\hat{\beta}_{ols}$$

for the first n observations and zero residuals for the next T_o observations. This means that $e^{*\prime} = (e'_{ols}, 0')$ and $e^{*\prime}e^* = e'_{ols} e_{ols}$ with the same residual sum of squares. The number of observations in (7.23) is $n+T_o$ and the number of parameters estimated is $K+T_o$. Hence the new degrees of freedom in (7.23) is $(n+T_o) - (K+T_o) = (n-K) =$ the old degrees of freedom in the regression of y on X. Hence, $s^{*2} = e^{*'}e^*/(n-K) = e'_{ols}e_{ols}/(n-K) = s^2$.

c. Using partitioned inverse formulas on $(X^{*'}X^*)$ one gets

$$(X^{*'}X^*)^{-1} = \begin{bmatrix} (X'X)^{-1} & -(X'X)^{-1}X'_o \\ -X_o(X'X)^{-1} & I_{T_o} + X_o(X'X)^{-1}X'_o \end{bmatrix}.$$

This uses the fact that the inverse of

$$A = \begin{bmatrix} A_{11} & A_{12} \\ A_{21} & A_{22} \end{bmatrix} \quad \text{is} \quad A^{-1} = \begin{bmatrix} B_{11} & -B_{11}A_{12}A_{22}^{-1} \\ -A_{22}^{-1}A_{21}B_{11} & A_{22}^{-1} + A_{22}^{-1}A_{21}B_{11}A_{12}A_{22}^{-1} \end{bmatrix}$$

where $B_{11} = (A_{11} - A_{12}A_{22}^{-1}A_{21})^{-1}$. Hence, $s^{*2}(X^{*'}X^*)^{-1} = s^2(X^{*'}X^*)^{-1}$ and is given by (7.25).

d. If we replace y_o by 0 and I_{T_o} by $-I_{T_o}$ in (7.23), we get

$$\begin{bmatrix} y \\ 0 \end{bmatrix} = \begin{bmatrix} X & 0 \\ X_o & -I_{T_o} \end{bmatrix} \begin{bmatrix} \beta \\ \gamma \end{bmatrix} + \begin{bmatrix} u \\ u_o \end{bmatrix}$$

or $y^* = X^*\delta + u^*$. Now $X^{*'}X^* = \begin{bmatrix} X' & X'_o \\ 0 & -I_{T_o} \end{bmatrix} \begin{bmatrix} X & 0 \\ X_o & -I_{T_o} \end{bmatrix} = \begin{bmatrix} X'X + X'_oX_o & -X'_o \\ -X_o & I_{T_o} \end{bmatrix}$

and $X^{*'}y^* = \begin{bmatrix} X'y \\ 0 \end{bmatrix}$. The OLS normal equations yield

$(X'X)\hat{\beta}_{ols} + (X'_oX_o)\hat{\beta}_{ols} - X'_o\hat{\gamma}_{ols} = X'y$ and $-X_o\hat{\beta}_{ols} + \hat{\gamma}_{ols} = 0$.

From the second equation, it immediately follows that $\hat{\gamma}_{ols} = X_o\hat{\beta}_{ols} = \hat{y}_o$ the forecast of the T_o observations using the estimates from the first n observations. Substituting this in the first equation yields

$(X'X)\hat{\beta}_{ols} + (X'_oX_o)\hat{\beta}_{ols} - X'_oX_o\hat{\beta}_{ols} = X'y$

which gives $\hat{\beta}_{ols} = (X'X)^{-1}X'y$.

Alternatively, one could apply the FWL Theorem using $X_1 = \begin{bmatrix} X \\ X_o \end{bmatrix}$ and

$X_2 = \begin{bmatrix} 0 \\ -I_{T_o} \end{bmatrix}$. In this case, $X_2'X_2 = I_{T_o}$ and $P_{X_2} = X_2X_2' = \begin{bmatrix} 0 & 0 \\ 0 & I_{T_o} \end{bmatrix}$ as before.

This means that $\bar{P}_{X_2} = I_{n+T_o} - P_{X_2} = \begin{bmatrix} I_n & 0 \\ 0 & 0 \end{bmatrix}$.

As in part (a), premultiplying by $\bar{P}_{X_2}$ omits the last T_o observations and yields $\hat{\beta}_{ols}$ based on the regression of y on X from the first n observations only. The last T_o observations yield zero residuals because the dependent and independent variables for these T_o observations have zero values. For this to be true in the original regression, it must be true that $0 - X_o\hat{\beta}_{ols} + \hat{\gamma}_{ols} = 0$ which yields $\hat{\gamma}_{ols} = X_o\hat{\beta}_{ols} = \hat{y}_o$ as expected. The residuals are still $(e'_{ols}, 0')$ and $s^{*2} = s^2$ for the same reasons given in part (b). Also, using partitioned inverse as in part (c) above, we get

$$(X^{*\prime}X^*)^{-1} = \begin{bmatrix} (X'X)^{-1} & (X'X)^{-1}X_o' \\ X_o(X'X)^{-1} & I_{T_o} + X_o(X'X)^{-1}X_o' \end{bmatrix}.$$

Hence, $s^{*2}(X^{*\prime}X^*)^{-1} = s^2(X^{*\prime}X^*)^{-1}$ and the diagonal elements are as given in (7.25).

7.8 a. $\text{cov}(\hat{\beta}_{ols}, e) = E(\hat{\beta}_{ols} - \beta)e' = E[(X'X)^{-1}X'uu'\bar{P}_X] = \sigma^2(X'X)^{-1}X'\bar{P}_X = 0$ where the second equality uses the fact that $e = \bar{P}_X u$ and $\hat{\beta}_{ols} = \beta + (X'X)^{-1}X'u$. The third equality uses the fact that $E(uu') = \sigma^2I_n$ and the last equality uses the fact that $\bar{P}_X X = 0$. But $e \sim N(0, \sigma^2\bar{P}_X)$ and $\hat{\beta}_{ols} \sim N(\beta, \sigma^2(X'X)^{-1})$, therefore zero covariance and normality imply independence of $\hat{\beta}_{ols}$ and e.

b. $\hat{\beta}_{ols} - \beta = (X'X)^{-1}X'u$ is linear in u, and $(n-K)s^2 = e'e = u'\bar{P}_X u$ is quadratic in u. A linear and quadratic forms in normal random variables $u \sim N(0, \sigma^2I_n)$ are independent if $(X'X)^{-1}X'\bar{P}_X = 0$, see Graybill (1961), Theorem 4.17. This is true since $\bar{P}_X X = 0$.

Chapter 7: The General Linear Model: The Basics   145

7.9 Replacing R by c' in (7.29) one gets $(c'\hat{\beta}_{ols}-c'\beta)'[c'(X'X)^{-1}c]^{-1}(c'\hat{\beta}_{ols}-c'\beta)/\sigma^2$.

Since $c'(X'X)^{-1}c$ is a scalar, this can be rewritten as

$(c'\hat{\beta}_{ols}-c'\beta)^2/\sigma^2 \, c'(X'X)^{-1}c$

which is exactly the square of z_{obs} in (7.26). Since $z_{obs} \sim N(0,1)$ under the null hypothesis, its square is χ_1^2 under the null hypothesis.

b. Dividing the statistic given in part (a) by $(n-K)s^2/\sigma^2 \sim \chi_{n-K}^2$ divided by its degrees of freedom $(n-K)$ results in replacing σ^2 by s^2, i.e.,

$(c'\hat{\beta}_{ols}-c'\beta)^2/s^2c'(X'X)^{-1}c$.

This is the square of the t-statistic given in (7.27). But, the numerator is $z_{obs}^2 \sim \chi_1^2$ and the denominator is $\chi_{n-K}^2 / (n-K)$. Hence, if the numerator and denominator are independent, the resulting statistic is distributed as $F(1,n-K)$ under the null hypothesis.

7.10

a. The quadratic form $u'Au/\sigma^2$ in (7.30) has

$A = X(X'X)^{-1}R'[R(X'X)^{-1}R']^{-1}R(X'X)^{-1}X'$.

This is symmetric, and idempotent since

$A^2 = X(X'X)^{-1}R'[R(X'X)^{-1}R']^{-1}R(X'X)^{-1}X'X(X'X)^{-1}$
$\quad R'[R(X'X)^{-1}R']^{-1}R(X'X')^{-1}X'$
$= X(X'X)^{-1}R'[R(X'X)^{-1}R']^{-1}R(X'X)^{-1}X' = A$

and rank $(A) = tr(A) = tr(R(X'X)^{-1}(X'X)(X'X)^{-1}R'[R(X'X)^{-1}R']^{-1})$

$\quad = tr(I_g) = g$ since R is gxK.

b. From lemma 1, $u'Au/\sigma^2 \sim \chi_g^2$ since A is symmetric and idempotent of rank g and $u \sim N(0,\sigma^2 I_n)$.

7.11

a. The two quadratic forms $s^2 = u'\bar{P}_X u/(n-K)$ and $u'Au/\sigma^2$ given in (7.30) are independent if and only if $\bar{P}_X A = 0$, see Graybill (1961), Theorem 4.10. This is true since $\bar{P}_X X = 0$.

b. $(n-K)s^2/\sigma^2$ is χ^2_{n-K} and $u'Au/\sigma^2 \sim \chi^2_g$ and both quadratic forms are independent of each other. Hence, dividing χ^2_g by g we get $u'Au/g\sigma^2$. Also, χ^2_{n-K} by (n-K) we get s^2/σ^2. Dividing $u'Au/g\sigma^2$ by s^2/σ^2 we get $u'Au/gs^2$ which is another way of writing (7.31). This is distributed as F(g,n-K) under the null hypothesis.

7.12 Restricted Least Squares

a. From (7.36), taking expected value we get

$$E(\hat{\beta}_{rls}) = E(\hat{\beta}_{ols}) + (X'X)^{-1}R'[R(X'X)^{-1}R']^{-1}(r-RE(\hat{\beta}_{ols}))$$
$$= \beta + (X'X)^{-1}R'[R(X'X)^{-1}R']^{-1}(r-R\beta)$$

since $E(\hat{\beta}_{ols}) = \beta$. It is clear that $\hat{\beta}_{rls}$ is in general biased for β unless $r = R\beta$ is satisfied, in which case the second term above is zero.

b. $\text{var}(\hat{\beta}_{rls}) = E[\hat{\beta}_{rls} - E(\hat{\beta}_{rls})][\hat{\beta}_{rls} - E(\hat{\beta}_{rls})]'$

But from (7.36) and part (a), we have

$$\hat{\beta}_{rls} - E(\hat{\beta}_{rls}) = (\hat{\beta}_{ols} - \beta) + (X'X)^{-1}R'[R(X'X)^{-1}R']^{-1}R(\beta - \hat{\beta}_{ols})$$

using $\hat{\beta}_{ols} - \beta = (X'X)^{-1}X'u$, one gets $\hat{\beta}_{rls} - E(\hat{\beta}_{rls}) = A(X'X)^{-1}X'u$ where $A = I_K - (X'X)^{-1}R'[R(X'X)^{-1}R']^{-1}R$. It is obvious that A is not symmetric, i.e., $A \neq A'$. However, $A^2 = A$, since

$$A^2 = I_K - (X'X)^{-1}R'[R(X'X)^{-1}R']^{-1}R - (X'X)^{-1}R'[R(X'X)^{-1}R']^{-1}R$$
$$+ (X'X)^{-1}R'[R(X'X)^{-1}R']^{-1}R(X'X)^{-1}R'[R(X'X)^{-1}R']^{-1}R$$
$$= I_K - (X'X)^{-1}R'[R(X'X)^{-1}R']^{-1}R = A.$$

Therefore,

$$\text{var}(\hat{\beta}_{rls}) = E[A(X'X)^{-1}X'uu'X(X'X)^{-1}A'] = \sigma^2 A(X'X)^{-1}A'$$
$$= \sigma^2[(X'X)^{-1} - (X'X)^{-1}R'[R(X'X)^{-1}R']^{-1}R(X'X)^{-1}$$
$$- (X'X)^{-1}R'[R(X'X)^{-1}R']^{-1}R(X'X)^{-1}$$
$$+ (X'X)^{-1}R'[R(X'X)^{-1}R']^{-1}R(X'X)^{-1}R'[R(X'X)^{-1}R']^{-1}R(X'X)^{-1}$$

$$= \sigma^2[(X'X)^{-1}-(X'X)^{-1}R'[R(X'X)^{-1}R']^{-1}R(X'X)^{-1}]$$

c. Using part (b) and the fact that $\text{var}(\hat{\beta}_{ols}) = \sigma^2(X'X)^{-1}$ gives

$$\text{var}(\hat{\beta}_{ols}) - \text{var}(\hat{\beta}_{rls}) = \sigma^2(X'X)^{-1}R'[R(X'X)^{-1}R']^{-1}R(X'X)^{-1}$$

and this is positive semi-definite, since $R(X'X)^{-1}R'$ is positive definite.

7.13 The Chow Test

a. OLS on (7.47) yields

$$\begin{pmatrix} \hat{\beta}_{1,ols} \\ \hat{\beta}_{2,ols} \end{pmatrix} = \begin{bmatrix} X_1'X_1 & 0 \\ 0 & X_2'X_2 \end{bmatrix}^{-1} \begin{bmatrix} X_1'y_1 \\ X_2'y_2 \end{bmatrix}$$

$$= \begin{bmatrix} (X_1'X_1)^{-1} & 0 \\ 0 & (X_2'X_2)^{-1} \end{bmatrix} \begin{pmatrix} X_1'y_1 \\ X_2'y_2 \end{pmatrix} = \begin{pmatrix} (X_1'X_1)^{-1}X_1'y_1 \\ (X_2'X_2)^{-1}X_2'y_2 \end{pmatrix}$$

which is OLS on each equation in (7.46) separately.

b. The vector of OLS residuals for (7.47) can be written as $e' = (e_1', e_2')$ where $e_1 = y_1 - X_1\hat{\beta}_{1,ols}$ and $e_2 = y_2 - X_2\hat{\beta}_{2,ols}$ are the vectors of OLS residuals from the two equations in (7.46) separately. Hence, the residual sum of squares $= e'e = e_1'e_1 + e_2'e_2 = $ sum of the residual sum of squares from running y_i on X_i for $i = 1,2$.

c. From (7.47), one can write

$$\begin{pmatrix} y_1 \\ y_2 \end{pmatrix} = \begin{bmatrix} X_1 \\ 0 \end{bmatrix} \beta_1 + \begin{bmatrix} 0 \\ X_2 \end{bmatrix} \beta_2 + \begin{pmatrix} u_1 \\ u_2 \end{pmatrix}.$$

This has the same OLS fitted values and OLS residuals as

$$\begin{pmatrix} y_1 \\ y_2 \end{pmatrix} = \begin{bmatrix} X_1 \\ X_2 \end{bmatrix} \gamma_1 + \begin{bmatrix} 0 \\ X_2 \end{bmatrix} \gamma_2 + \begin{pmatrix} u_1 \\ u_2 \end{pmatrix} \text{ where } \begin{bmatrix} X_1 \\ X_2 \end{bmatrix} = \begin{bmatrix} X_1 \\ 0 \end{bmatrix} + \begin{bmatrix} 0 \\ X_2 \end{bmatrix}. \text{ This follows}$$

from problem 1 with $X = \begin{bmatrix} X_1 & 0 \\ 0 & X_2 \end{bmatrix}$ and $C = \begin{bmatrix} I_K & 0 \\ I_K & I_K \end{bmatrix} = \begin{bmatrix} 1 & 0 \\ 1 & 1 \end{bmatrix} \otimes I_K$

where C is non-singular since $C^{-1} = \begin{bmatrix} 1 & 0 \\ -1 & 1 \end{bmatrix} \otimes I_K = \begin{bmatrix} I_K & 0 \\ -I_K & I_K \end{bmatrix}$. Hence,

the X matrix in (7.47) is related to that in (7.48) as follows:

$$\begin{bmatrix} X_1 & 0 \\ X_2 & X_2 \end{bmatrix} = \begin{bmatrix} X_1 & 0 \\ 0 & X_2 \end{bmatrix} C$$

and the coefficients are therefore related as follows:

$$\begin{pmatrix} \gamma_1 \\ \gamma_2 \end{pmatrix} = C^{-1} \begin{pmatrix} \beta_1 \\ \beta_2 \end{pmatrix} = \begin{bmatrix} I_K & 0 \\ -I_K & I_K \end{bmatrix} \begin{pmatrix} \beta_1 \\ \beta_2 \end{pmatrix} = \begin{pmatrix} \beta_1 \\ \beta_2 - \beta_1 \end{pmatrix}$$

as required in (7.49). Hence, (7.49) yields the same URSS as (7.47). The RRSS sets $(\beta_2 - \beta_1) = 0$ which yields the same regression as (7.48).

7.14 The FWL Theorem states that $\hat{\beta}_{2,ols}$ from $\bar{P}_{X_1} y = \bar{P}_{X_1} X_2 \beta_2 + \bar{P}_{X_1} u$ will be identical to the estimate of β_2 obtained from (7.8). Also, the residuals from both regressions will be identical. This means that the RSS from (7.8) given by $y' \bar{P}_X y = y'y - y' P_X y$ is identical to that from the above regression. The latter is similarly obtained as

$$y' \bar{P}_{X_1} y - y' \bar{P}_{X_1} \bar{P}_{X_1} X_2 (X_2' \bar{P}_{X_1} X_2)^{-1} X_2' \bar{P}_{X_1} \bar{P}_{X_1} y = y' \bar{P}_{X_1} y - y' \bar{P}_{X_1} X_2 (X_2' \bar{P}_{X_1} X_2)^{-1} X_2' \bar{P}_{X_1} y$$

b. For testing H_o; $\beta_2 = 0$, the RRSS $= y' \bar{P}_{X_1} y$ and the URSS is given in part (a). Hence, the numerator of the Chow F-statistics given in (7.45) is given by

$$(\text{RRSS-URSS}) / k_2 = y' \bar{P}_{X_1} X_2 (X_2' \bar{P}_{X_1} X_2)^{-1} X_2' \bar{P}_{X_1} y / k_2$$

Substituting $y = X_1 \beta_1 + u$ under the null hypothesis, yields

$$u' \bar{P}_{X_1} X_2 (X_2' \bar{P}_{X_1} X_2)^{-1} X_2' \bar{P}_{X_1} u / k_2 \text{ since } \bar{P}_{X_1} X_1 = 0.$$

c. Let $v = X_2' \bar{P}_{X_1} u$. Given that $u \sim N(0, \sigma^2 I_n)$, then v is Normal with mean zero and $\text{var}(v) = X_2' \bar{P}_{X_1} \text{var}(u) \bar{P}_{X_1} X_2 = \sigma^2 X_2' \bar{P}_{X_1} X_2$ since $\bar{P}_{X_1}$ is idempotent. Hence, $v \sim N(0, \sigma^2 X_2' \bar{P}_{X_1} X_2)$. Therefore, the numerator of the Chow F-statistic given in part (b) when divided by σ^2 can be written as $v'[\text{var}(v)]^{-1} v / k_2$. This is distributed as $\chi^2_{k_2}$ divided by its degrees of freedom k_2 under the null hypothesis. In fact, from part (b), $A = \bar{P}_{X_1} X_2 (X_2' \bar{P}_{X_1} X_2)^{-1} X_2' \bar{P}_{X_1}$ is symmetric and idempotent and of rank equal to its trace equal to k_2. Hence, by lemma 1, $u'Au/\sigma^2$ is $\chi^2_{k_2}$ under the null hypothesis.

d. The numerator $u'Au/k_2$ is independent of the denominator
$(n-k)s^2/(n-k) = u'\bar{P}_X u/(n-k)$

provided $\bar{P}_X A = 0$ as seen in problem 11. This is true because
$\bar{P}_X \bar{P}_{X_1} = \bar{P}_X(I_n - \bar{P}_{X_1}) = \bar{P}_X - \bar{P}_X X_1 (X_1' X_1)^{-1} X_1' = \bar{P}_X$ since $\bar{P}_X X_1 = 0$. Hence,
$\bar{P}_X A = \bar{P}_X \bar{P}_{X_1} X_2 (X_2' \bar{P}_{X_1} X_2)^{-1} X_2' \bar{P}_{X_1} = \bar{P}_X X_2 (X_2' \bar{P}_{X_1} X_2)^{-1} X_2' \bar{P}_{X_1} = 0$
since $\bar{P}_X X_2 = 0$. Recall, $\bar{P}_X X = \bar{P}_X [X_1, X_2] = [\bar{P}_X X_1, \bar{P}_X X_2] = 0$.

e. The Wald statistic for H_o; $\beta_2 = 0$ given in (7.41) boils down to replacing R by $[0, I_{k_2}]$ and r by 0. Also, $\hat{\beta}_{mle}$ by $\hat{\beta}_{ols}$ from the unrestricted model given in (7.8) and σ^2 is replaced by its estimate $s^2 = URSS/(n-k)$ to make the Wald statistic feasible. This yields $W = \hat{\beta}_2' [R(X'X)^{-1}R']^{-1} \hat{\beta}_2/s^2$. From problem 2, we showed that the partitioned inverse of $X'X$ yields $B_{22} = (X_2' \bar{P}_{X_1} X_2)^{-1}$ for its second diagonal $(k_2 \times k_2)$ block. Hence,

$$R(X'X)^{-1}R' = [0, I_{k_2}](X'X)^{-1}\begin{bmatrix} 0 \\ I_{k_2} \end{bmatrix} = B_{22} = (X_2' \bar{P}_{X_1} X_2)^{-1}.$$

Also, from problem 2, $\hat{\beta}_{2,ols} = (X_2' \bar{P}_{X_1} X_2)^{-1} X_2' \bar{P}_{X_1} y = (X_2' \bar{P}_{X_1} X_2)^{-1} X_2' \bar{P}_{X_1} u$ after substituting $y = X_1 \beta_1 + u$ under the null hypothesis and using $\bar{P}_{X_1} X_1 = 0$. Hence,
$s^2 W = u' \bar{P}_{X_1} X_2 (X_2' \bar{P}_{X_1} X_2)^{-1} (X_2' \bar{P}_{X_1} X_2)(X_2' \bar{P}_{X_1} X_2)^{-1} X_2' \bar{P}_{X_1} u$
$= u' \bar{P}_{X_1} X_2 (X_2' \bar{P}_{X_1} X_2)^{-1} X_2' \bar{P}_{X_1} u$

which is exactly k_2 times the expression in part (b), i.e., the numerator of the Chow F-statistic.

f. The restricted MLE of β is $(\hat{\beta}_{1,rls}', 0')$ since $\beta_2 = 0$ under the null hypothesis. Hence the score form of the LM test given in (7.44) yields
$(y - X_1 \hat{\beta}_{1,rls})' X(X'X)^{-1} X' (y - X_1 \hat{\beta}_{1,rls}) / \sigma^2$.

In order to make this feasible, we replace σ^2 by $\tilde{s}^2 = RRSS/(n-k_1)$ where RRSS is the restricted residual sum of squares from running y on X_1. But this expression is exactly the regression sum of squares from running $(y-X_1\hat{\beta}_{1,rls})/\tilde{s}$ on the matrix X. In order to see this, the regression sum of squares of y on X is usually $y'P_X y$. Here, y is replaced by $(y-X_1\hat{\beta}_{1,rls})/\tilde{s}$.

7.15 *Iterative Estimation in Partitioned Regression Models.* This is based on Baltagi (1996).

a. The least squares residuals of y on X_1 are given by $\bar{P}_{X_1} y$, where $\bar{P}_{X_1} = I - P_{X_1}$ and $P_{X_1} = X_1(X_1'X_1)^{-1}X_1'$. Regressing these residuals on x_2 yields $b_2^{(1)} = (x_2'x_2)^{-1}x_2'\bar{P}_{X_1} y$. Substituting for y from (7.8) and using $\bar{P}_{X_1} X_1 = 0$ yields $b_2^{(1)} = (x_2'x_2)^{-1}x_2'\bar{P}_{X_1}(x_2\beta_2+u)$ with

$$E(b_2^{(1)}) = (x_2'x_2)^{-1}x_2'\bar{P}_{X_1}x_2\beta_2 = \beta_2 - (x_2'x_2)^{-1}x_2'P_{X_1}x_2\beta_2 = (1-a)\beta_2$$

where $a = (x_2'P_{X_1}x_2)/(x_2'x_2)$ is a scalar, with $0 \leq a < 1$. $a \neq 1$ as long as x_2 is linearly independent of X_1. Therefore, the bias$(b_2^{(1)}) = -a\beta_2$.

b. $b_1^{(1)} = (X_1'X_1)^{-1}X_1'(y-x_2 b_2^{(1)}) = (X_1'X_1)^{-1}X_1'(I-P_{x_2}\bar{P}_{X_1})y$ and

$$b_2^{(2)} = (x_2'x_2)^{-1}x_2'(y-X_1 b_1^{(1)}) = (x_2'x_2)^{-1}x_2'[I-P_{X_1}(I-P_{x_2}\bar{P}_{X_1})]y$$

$$= (x_2'x_2)^{-1}x_2'[\bar{P}_{X_1}+P_{X_1}P_{x_2}\bar{P}_{X_1}]y = (1+a)b_2^{(1)}.$$

Similarly,

$$b_1^{(2)} = (X_1'X_1)^{-1}X_1'(y-x_2 b_2^{(2)}) = (X_1'X_1)^{-1}X_1'(y-(1+a)x_2 b_2^{(1)})$$

$$b_2^{(3)} = (x_2'x_2)^{-1}x_2'(y-X_1 b_1^{(2)}) = (x_2'x_2)^{-1}x_2'[y-P_{X_1}(y-(1+a)x_2 b_2^{(1)})]$$

$$= b_2^{(1)} + (1+a)(x_2'x_2)^{-1}x_2'P_{X_1}x_2 b_2^{(1)} = (1+a+a^2)b_2^{(1)}.$$

By induction, one can infer that

$$b_2^{(j+1)} = (1+a+a^2+..+a^j)b_2^{(1)} \quad \text{for } j=0,1,2,...$$

Therefore,

$E(b_2^{(j+1)}) = (1+a+a^2+..+a^j) E(b_2^{(1)}) = (1+a+a^2+..a^j)(1-a)\beta_2 = (1-a^{j+1})\beta_2$

and the bias($b_2^{(j+1)}$) = $-a^{j+1}\beta_2$ tends to zero as $j\to\infty$, since $|a| < 1$.

c. Using the Frisch-Waugh-Lovell Theorem, least squares on the original model yields

$$\hat{\beta}_2 = (x_2'\bar{P}_{X_1}x_2)^{-1}x_2'\bar{P}_{X_1}y = (x_2'x_2 - x_2'P_{X_1}x_2)^{-1}x_2'\bar{P}_{X_1}y$$

$$= (1-a)^{-1}(x_2'x_2)^{-1}x_2'\bar{P}_{X_1}y = b_2^{(1)}/(1-a).$$

As $j\to\infty$, $\lim b_2^{(j+1)} \to (\sum_{j=0}^{\infty} a^j)b_2^{(1)} = b_2^{(1)}/(1-a) = \hat{\beta}_2$.

7.16 Maddala (1992, pp.120-127).

a. For H_o: $\beta = 0$, the RRSS is based on a regression of y_i on a constant. This yields $\hat{\alpha} = \bar{y}$ and the RRSS = $\sum_{i=1}^{n}(y_i-\bar{y})^2$ = usual TSS. The URSS is the usual least squares residual sum of squares based on estimating α and β. The log-likelihood in this case is given by

$$\log L(\alpha,\beta,\sigma^2) = -\frac{n}{2}\log 2\pi - \frac{n}{2}\log\sigma^2 - \frac{1}{2\sigma^2}\sum_{i=1}^{n}(y_i-\alpha-\beta X_i)^2$$

with the unrestricted MLE of α and β yielding $\hat{\alpha}_{ols} = \bar{y} - \hat{\beta}_{ols}\bar{X}$ and

$$\hat{\beta}_{ols} = \sum_{i=1}^{n}(X_i-\bar{X})y_i / \sum_{i=1}^{n}(X_i-\bar{X})^2$$

and $\hat{\sigma}^2_{mle}$ = URSS/n. In this case, the unrestricted log-likelihood yields

$$\log L(\hat{\alpha}_{mle}, \hat{\beta}_{mle}, \hat{\sigma}^2_{mle}) = -\frac{n}{2}\log 2\pi - \frac{n}{2}\log\hat{\sigma}^2_{mle} - \frac{URSS}{(2\,URSS/n)}$$

$$= -\frac{n}{2}\log 2\pi - \frac{n}{2} - \frac{n}{2}\log(URSS/n).$$

Similarly, the restricted MLE yields $\hat{\alpha}_{rmle} = \bar{y}$ and $\hat{\beta}_{rmle} = 0$ and

$$\hat{\sigma}^2_{rmle} = RRSS/n = \sum_{i=1}^{n}(y_i-\bar{y})^2/n.$$

The restricted log-likelihood yields

$$\log L(\hat{\alpha}_{rmle}, \hat{\beta}_{rmle}, \hat{\sigma}^2_{rmle}) = -\frac{n}{2}\log 2\pi - \frac{n}{2}\log\hat{\sigma}^2_{rmle} - \frac{RRSS}{(2\,RRSS/n)}$$

$$= -\frac{n}{2} \log 2\pi - \frac{n}{2} - \frac{n}{2} \log(\text{RRSS}/n).$$

Hence, the LR test is given by

$$\text{LR} = n(\log \text{RRSS} - \log \text{URSS}) = n \log(\text{RRSS}/\text{URSS})$$
$$= n\log(\text{TSS}/\text{RSS}) = n \log(1/1 - r^2)$$

where TSS and RSS are the total and residual sum of squares from the unrestricted regression. By definition $R^2 = 1 - (\text{RSS}/\text{TSS})$ and for the simple regression $r_{XY}^2 = R^2$ of that regression, see Chapter 3.

b. The Wald statistic for H_o; $\beta = 0$ is based upon $r(\hat{\beta}_{mle}) = (\hat{\beta}_{mle} - 0)$ and $R(\hat{\beta}_{mle}) = 1$ and from (7.40), we get $W = \hat{\beta}_{mle}^2/\text{var}(\hat{\beta}_{mle}) = \hat{\beta}_{ols}^2/\text{var}(\hat{\beta}_{ols})$. This is the square of the usual t-statistic for $\beta = 0$ with $\hat{\sigma}_{mle}^2$ used instead of s^2 in estimating σ^2. Using the results in Chapter 3, we get

$$W = \frac{\left(\sum_{i=1}^{n} x_i y_i\right)^2 / \left(\sum_{i=1}^{n} x_i^2\right)^2}{\hat{\sigma}_{mle}^2 / \left(\sum_{i=1}^{n} x_i^2\right)} = \frac{\left(\sum_{i=1}^{n} x_i y_i\right)^2}{\hat{\sigma}_{mle}^2 \sum_{i=1}^{n} x_i^2}$$

with $\hat{\sigma}_{mle}^2 = \text{URSS}/n = \text{TSS}(1-R^2)/n$ from the definition of R^2. Hence,

$$W = \frac{n\left(\sum_{i=1}^{n} x_i y_i\right)^2}{\left(\sum_{i=1}^{n} y_i^2\right)\left(\sum_{i=1}^{n} x_i^2\right)(1-R^2)} = \frac{nr^2}{1-r^2}$$

using the definition of $r_{XY}^2 = r^2 = R^2$ for the simple regression.

c. The LM statistic given in (7.43) is based upon $\text{LM} = \hat{\beta}_{ols}^2 / \hat{\sigma}_{rmle}^2 \left(\sum_{i=1}^{n} x_i^2\right)$. This is the t-statistic on $\beta = 0$ using $\hat{\sigma}_{rmle}^2$ as an estimate for σ^2. In this case, $\hat{\sigma}_{rmle}^2 = \text{RRSS}/n = \sum_{i=1}^{n} y_i^2 / n$. Hence, $\text{LM} = n\left(\sum_{i=1}^{n} x_i y_i\right)^2 / \sum_{i=1}^{n} y_i^2 \sum_{i=1}^{n} x_i^2 = nr^2$ from the definition of $r_{XY}^2 = r^2$.

d. Note that from part (b), we get $W/n = r^2/(1-r^2)$ and $1 + (W/n) = 1/(1-r^2)$.

Hence, from part (a), we have $(LR/n) = \log(1/1-r^2) = \log[1+(W/n)]$.

From part (c), we get $(LM/n) = (W/n)/[1+(W/n)]$. Using the inequality $x \geq \log(1+x) \geq x/(1+x)$ with $x = W/n$ we get $W \geq LR \geq LM$.

e. From Chapter 3, R^2 of the regression of $\log C$ on $\log P$ is 0.2913 and $n = 46$.

Hence, $W = nr^2/1-r^2 = (46)(0.2913)/(0.7087) = 18.91$,

$LM = nr^2 = (46)(0.2913) = 13.399$

and $LR = n\log(1/1-r^2) = 46 \log(1/0.7087) = 46 \log(1.4110342) = 15.838$.

It is clear that $W > LR > LM$ in this case.

7.17 *Engle (1984, pp.785-786)*.

a. For the Bernoulli random variable y_t with probability of success θ, the log-likelihood function is given by $\log L(\theta) = \sum_{t=1}^{T} [y_t \log\theta + (1-y_t)\log(1-\theta)]$.

The score is given by

$$S(\theta) = \partial \log L(\theta)/\partial\theta = \sum_{t=1}^{T} y_t/\theta - \sum_{t=1}^{T}(1-y_t)/(1-\theta)$$

$$= \left[\sum_{t=1}^{T} y_t - \theta \sum_{t=1}^{T} y_t - \theta T + \theta \sum_{t=1}^{T} y_t\right]/\theta(1-\theta) = \sum_{t=1}^{T}(y_t-\theta)/\theta(1-\theta)$$

The MLE is given by setting $S(\theta) = 0$ giving $\hat{\theta}_{mle} = \sum_{t=1}^{T} y_t/T = \bar{y}$.

Now $\partial^2 \log L(\theta)/\partial\theta^2 = [-T\theta(1-\theta) - \sum_{t=1}^{T}(y_t-\theta)(1-2\theta)]/\theta^2(1-\theta)^2$.

Therefore, the information matrix is

$$I(\theta) = -E\left[\frac{\partial^2 \log L(\theta)}{\partial\theta^2}\right] = (T\theta(1-\theta) + (1-2\theta)\sum_{t=1}^{T}[E(y_t)-\theta])/\theta^2(1-\theta^2) = T/\theta(1-\theta).$$

b. For testing $H_o; \theta = \theta_o$ versus $H_A; \theta \neq \theta_o$, the Wald statistic given in (7.40) has $r(\theta) = \theta - \theta_o$ and $R(\theta) = 1$ with $I^{-1}(\hat{\theta}_{mle}) = \hat{\theta}_{mle}(1-\hat{\theta}_{mle})/T$. Hence,

$$W = T(\hat{\theta}_{mle}-\theta_o)^2/\hat{\theta}_{mle}(1-\hat{\theta}_{mle}) = T(\bar{y}-\theta_o)^2/\bar{y}(1-\bar{y}).$$

The LM statistic given in (7.42) has $S(\theta_o) = T(\bar{y}-\theta_o)/\theta_o(1-\theta_o)$ and $I^{-1}(\theta_o) = \theta_o(1-\theta_o)/T$. Hence,

$$LM = \frac{T^2(\bar{y}-\theta_o)^2}{[\theta_o(1-\theta_o)]^2} \cdot \frac{\theta_o(1-\theta_o)}{T} = \frac{T(\bar{y}-\theta_o)^2}{\theta_o(1-\theta_o)}.$$

The unrestricted log-likelihood is given by

$$\log L(\hat{\theta}_{mle}) = \log L(\bar{y}) = \sum_{t=1}^{T} [y_t \log\bar{y}+(1-y_t)\log(1-\bar{y})]$$

$$= T\bar{y} \log\bar{y} + T(1-\bar{y})\log(1-\bar{y}).$$

The restricted log-likelihood is given by

$$\log L(\theta_o) = \sum_{t=1}^{T} [y_t \log\theta_o+(1-y_t)\log(1-\theta_o)]$$

$$= T\bar{y} \log\theta_o + T(1-\bar{y})\log(1-\theta_o).$$

Hence, the likelihood ratio test gives

$$LR = 2T\bar{y}(\log\bar{y}-\log\theta_o) + 2T(1-\bar{y})[\log(1-\bar{y})-\log(1-\theta_o)]$$

$$= 2T\bar{y} \log(\bar{y}/\theta_o) + 2T(1-\bar{y})\log[(1-\bar{y})/(1-\theta_o)].$$

All three statistics have a limiting χ_1^2 distribution under H_o. Each statistic will reject when $(\bar{y}-\theta_o)^2$ is large. Hence, for finite sample exact results one can refer to the binomial distribution and compute exact critical values.

7.18 For the regression model

$$y = X\beta + u \qquad \text{with} \quad u \sim N(0,\sigma^2 I_T).$$

a. $L(\beta,\sigma^2) = (1/2\pi\sigma^2)^{T/2}\exp\{-(y-X\beta)'(y-X\beta)/2\sigma^2\}$

$\log L = -(T/2)(\log 2\pi + \log \sigma^2) - (y-X\beta)'(y-X\beta)/2\sigma^2$

$\partial \log L/\partial \beta = -(-2X'y+2X'X\beta)/2\sigma^2 = 0$

so that $\hat{\beta}_{mle} = (X'X)^{-1}X'y$ and $\partial \log L/\partial \sigma^2 = -T/2\sigma^2 + \hat{u}'\hat{u}/2\sigma^4 = 0$. Hence,

Chapter 7: The General Linear Model: The Basics

$$\hat{\sigma}^2_{mle} = \hat{u}'\hat{u}/T \qquad \text{(where } \hat{u} = y - X\hat{\beta}_{mle}\text{)}.$$

b. The score for β is $S(\beta) = \partial \log L(\beta)/\partial \beta = (X'y - X'X\beta)/\sigma^2$.

The Information matrix is given by

$$I(\beta,\sigma^2) = -E\left[\frac{\partial^2 \log L}{\partial(\beta,\sigma^2)\partial(\beta,\sigma^2)'}\right] = -E\begin{bmatrix} -\dfrac{X'X}{\sigma^2} & -\dfrac{X'y - X'X\beta}{\sigma^4} \\ -\dfrac{X'y - X'X\beta}{\sigma^4} & \dfrac{T}{2\sigma^4} - \dfrac{(y-X\beta)'(y-X\beta)}{\sigma^6} \end{bmatrix}$$

since $E(X'y - X'X\beta) = X'E(y - X\beta) = 0$ and $E(y-X\beta)'(y-X\beta) = T\sigma^2$

$$I(\beta,\sigma^2) = \begin{bmatrix} \dfrac{X'X}{\sigma^2} & 0 \\ 0 & \dfrac{T}{2\sigma^4} \end{bmatrix}$$

which is block-diagonal and also given in (7.19).

c. The Wald statistic given in (7.41) needs

$r(\beta) = \beta_1 - \beta_1^0$

$R(\beta) = [I_{k_1}, 0]$

$W = (\hat{\beta}_1 - \beta_1^0)'[(I_{k_1}, 0)\hat{\sigma}^2(X'X)^{-1}\begin{pmatrix} I_{k_1} \\ 0 \end{pmatrix}]^{-1}(\hat{\beta}_1 - \beta_1^0)$

with

$$(X'X)^{-1} = \begin{bmatrix} X_1'X_1 & X_1'X_2 \\ X_2'X_1 & X_2'X_2 \end{bmatrix}^{-1} = \begin{bmatrix} a & b \\ c & d \end{bmatrix}$$

and

$a = (X_1'X_1 - X_1'X_2(X_2'X_2)^{-1}X_2'X_1)^{-1}$

$\quad = [X_1'[I - X_2(X_2'X_2)^{-1}X_2']X_1]^{-1} = [X_1'\bar{P}_{X_2}X_1]^{-1}$

by partitioned inverse. Therefore,

$W = (\beta_1^0 - \hat{\beta}_1)'[X_1'\bar{P}_{X_2}X_1](\beta_1^0 - \hat{\beta}_1)/\hat{\sigma}^2$

as required. For the LR statistic $LR = -2(\log L_r^* - \log L_u^*)$ where L_r^* is the restricted likelihood and L_u^* is the unrestricted likelihood.

$$LR = -2\left[\left(-\frac{T}{2} - \frac{T}{2}\log 2\pi - \frac{T}{2}\log\left(\frac{RRSS}{T}\right)\right)\right.$$
$$\left. - \left(-\frac{T}{2} - \frac{T}{2}\log 2\pi - \frac{T}{2}\log\left(\frac{URSS}{T}\right)\right)\right]$$
$$= T\log(RRSS/URSS) = T\log\left[\frac{\tilde{u}'\tilde{u}}{\hat{u}'\hat{u}}\right]$$

For the LM statistic, the score version is given in (7.42) as

$$LM = S(\tilde{\beta})'I^{-1}(\tilde{\beta})S(\tilde{\beta})$$

where

$$S(\tilde{\beta}) = \begin{bmatrix} S_1(\tilde{\beta}) \\ S_2(\tilde{\beta}) \end{bmatrix} = \begin{bmatrix} X_1'(y-X\tilde{\beta}) \\ X_2'(y-X\tilde{\beta}) \end{bmatrix}.$$

The restriction on β_1 is $(\beta_1 = \beta_1^0)$, but there are no restrictions on β_2. Note that $\tilde{\beta}_2$ can be obtained from the regression of $(y - X_1\beta_1^0)$ on X_2. This yields

$$\tilde{\beta}_2 = (X_2'X_2)^{-1}X_2'(y - X_1\beta_1^0).$$

Therefore,

$$S_2(\tilde{\beta}) = X_2'(y - X\tilde{\beta}) = X_2'y - X_2'X_1\beta_1^0 - X_2'X_2\,\tilde{\beta}_2$$
$$= X_2'y - X_2'X_1\beta_1^0 - X_2'y - X_2'X_1\beta_1^0 = 0.$$

Hence, $LM = S_1(\tilde{\beta})'I^{11}(\tilde{\beta})S_1(\tilde{\beta})$ where $I^{11}(\beta)$ is obtained from the partitioned inverse of $I^{-1}(\beta)$. Since $S_1(\tilde{\beta}) = X_1'(y-X\tilde{\beta})/\tilde{\sigma}^2 = X_1'\tilde{u}/\tilde{\sigma}^2$ and $I^{-1}(\tilde{\beta}) = \tilde{\sigma}^2(X'X)^{-1}$ with $I^{11}(\tilde{\beta}) = \tilde{\sigma}^2[X_1'\bar{P}_{X_2}X_1]^{-1}$ we get

$$LM = \tilde{u}'X_1\,[X_1'\bar{P}_{X_2}X_1]^{-1}X_1'\tilde{u}/\tilde{\sigma}^2$$

d. $W = (\beta_1^0 - \hat{\beta}_1)'[X_1'\bar{P}_{X_2}X_1](\beta_1^0 - \hat{\beta}_1)/\hat{\sigma}^2$
$= (\beta_1^0 - \hat{\beta}_1)'[R(X'X)^{-1}R']^{-1}(\beta_1^0 - \hat{\beta}_1)/\hat{\sigma}^2$

$$= (r-R\hat{\beta})'[R(X'X)^{-1}R']^{-1}(r-R\hat{\beta})/\hat{\sigma}^2.$$

From (7.39) we know that $\tilde{u}'\tilde{u} - \hat{u}'\hat{u} = (r-R\hat{\beta})'[R(X'X)^{-1}R']^{-1}(r-R\hat{\beta})$.

Also, $\hat{\sigma}^2 = \hat{u}'\hat{u}/T$. Therefore, $W = T(\tilde{u}'\tilde{u} - \hat{u}'\hat{u})/\hat{u}'\hat{u}$ as required.

$$LM = \tilde{u}'X_1[X_1'\bar{P}_{X_2}X_1]^{-1}X_1'\tilde{u}/\tilde{\sigma}^2$$

$$= S_1(\tilde{\beta})'I^{11}(\tilde{\beta})S_1(\tilde{\beta}) = S_1(\tilde{\beta})'I^{-1}(\tilde{\beta})S_1(\tilde{\beta}).$$

From (7.43) we know that $LM = (r-R\tilde{\beta})'[R(X'X)^{-1}R']^{-1}(r-R\tilde{\beta})/\tilde{\sigma}^2$

and $\tilde{\sigma}^2 = \tilde{u}'\tilde{u}/T$. Using (7.39) we can rewrite this as $LM = T(\tilde{u}'\tilde{u} - \hat{u}'\hat{u})/\tilde{u}'\tilde{u}$

as required. Finally,

$$LR = T\log(\tilde{u}'\tilde{u}/\hat{u}'\hat{u}) = T\log(1 + \frac{(\tilde{u}'\tilde{u} - \hat{u}'\hat{u})}{\hat{u}'\hat{u}}) = T\log(1+W/T).$$

Also, $W/LM = \dfrac{\tilde{u}'\tilde{u}}{\hat{u}'\hat{u}} = 1 + W/T$. Hence, $LM = W/(1+W/T)$ and from (7.45)

$$((T-k)/T) \cdot W/k_1 = \frac{(\tilde{u}'\tilde{u} - \hat{u}'\hat{u})/k_1}{\hat{u}'\hat{u}/T-k} \sim F_{k_1,T-k}.$$

Using the inequality $x \geq \log(1+x) \geq x/(1+x)$ with $x = W/T$ we get

$(W/T) \geq \log(1+W/T) \geq (W/T)/(1+W/T)$ or $(W/T) \geq (LR/T) \geq (LM/T)$ or

$W \geq LR \geq LM$. However, it is important to note that all the statistics are monotonic functions of the F-statistic and exact tests for each would produce identical critical regions.

e. For the cigarette consumption data given in Table 3.2 the following test statistics were computed for $H_o; \beta = -1$

Wald $= 1.16 \geq LR = 1.15 \geq LM = 1.13$

and the SAS program that produces these results is given below.

f. The Wald statistic for $H_0^A; \beta = 1$ yields 1.16, for $H_0^B; \beta^5 = -1$ yields 0.43 and for $H_0^C; \beta^{-5} = -1$ yields 7.89. The SAS program that produces these results is given below.

SAS PROGRAM

```
Data CIGARETT;
Input OBS STATE $ LNC LNP LNY;
Cards;

Proc IML; Use CIGARETT;
Read all into Temp;

N=NROW(TEMP);           ONE=Repeat(1,N,1);
Y=Temp[,2];    X=ONE||Temp[,3]||Temp[,4];
BETA_U=INV(X`*X)*X`*Y;
R={0 1 0};
Ho=BETA_U[2,]+1;
BETA_R=BETA_U+INV(X`*X)*R`*INV(R*INV(X`*X)*R`)*Ho;
ET_U=Y-X*BETA_U;
ET_R=Y-X*BETA_R;
SIG_U=(ET_U`*ET_U)/N;
SIG_R=(ET_R`*ET_R)/N;
X1=X[,2];
X2=X[,1]||X[,3];
Q_X2=I(N)-X2*INV(X2`*X2)*X2`;
VAR_D=SIG_U*INV(X1`*Q_X2*X1);
WALD=Ho`*INV(VAR_D)*Ho;
LR=N*LOG(1+(WALD/N));
LM=(ET_R`*X1*INV(X1`*Q_X2*X1)*X1`*ET_R)/SIG_R;

  *WALD=N*(ET_R`*ET_R-ET_U`*ET_U)/(ET_U`*ET_U);
  *LR=N*Log(ET_R`*ET_R/(ET_U`*ET_U));
  *LM=N*(ET_R`*ET_R-ET_U`*ET_U)/(ET_R`*ET_R);

PRINT 'Chapter7 Problem18.(e)',, WALD;
PRINT LR;
PRINT LM;

BETA=BETA_U[2,];
H1=BETA+1;
H2=BETA**5+1;
H3=BETA**(-5)+1;
```

```
VAR_D1=SIG_U*INV(X1`*Q_X2*X1);
VAR_D2=(5*BETA**4)*VAR_D1*(5*BETA**4);
VAR_D3=(-5*BETA**(-6))*VAR_D1*(-5*BETA**(-6));

WALD1=H1`*INV(VAR_D1)*H1;
WALD2=H2`*INV(VAR_D2)*H2;
WALD3=H3`*INV(VAR_D3)*H3;

PRINT 'Chapter7 Problem18.(f)',, WALD1;
PRINT WALD2;
PRINT WALD3;
```

7.19 *Gregory and Veall (1985).*

a. For H^A: $\beta_1 - 1/\beta_2 = 0$, we have $r^A(\beta) = \beta_1 - 1/\beta_2$ and $\beta' = (\beta_0, \beta_1, \beta_2)$. In this case, $R^A(\beta) = (0, 1, 1/\beta_2^2)$ and the unrestricted MLE is OLS on (7.50) with variance-covariance matrix $\hat{V}(\hat{\beta}_{ols}) = \hat{\sigma}^2 (X'X)^{-1}$ where $\hat{\sigma}^2_{mle} = URSS/n$. Let v_{ij} denote the corresponding elements of $\hat{V}(\hat{\beta}_{ols})$ for $i,j = 0,1,2$. Therefore,

$$W^A = (\hat{\beta}_1 - 1/\hat{\beta}_2)[(0,1,1/\hat{\beta}_2^2) \; \hat{V}(\hat{\beta}_{ols})(0,1,1/\hat{\beta}_2^2)']^{-1}(\hat{\beta}_1 - 1/\hat{\beta}_2)$$

$$= (\hat{\beta}_1 \hat{\beta}_2 - 1)^2/(\hat{\beta}_2^2 \; v_{11} + 2v_{12} + v_{22}/\hat{\beta}_2^2)$$

as required in (7.52). Similarly, for H^B; $\beta_1\beta_2 - 1 = 0$, we have $r^B(\beta) = \beta_1\beta_2 - 1$.

In this case, $R^B(\beta) = (0, \beta_2, \beta_1)$ and

$$W^B = (\hat{\beta}_1\hat{\beta}_2 - 1)[(0, \hat{\beta}_2, \hat{\beta}_1) \; \hat{V}(\hat{\beta}_{ols})(0, \hat{\beta}_2, \hat{\beta}_1)']^{-1}(\hat{\beta}_1\hat{\beta}_2 - 1)$$

$$= (\hat{\beta}_1\hat{\beta}_2 - 1)^2/(\hat{\beta}_2^2 v_{11} + 2\hat{\beta}_1\hat{\beta}_2 \; v_{12} + \hat{\beta}_1^2 \; v_{22})$$

as required in (7.53).

7.20 *Gregory and Veall (1986).*

a. From (7.51), we get $W = r(\hat{\beta}_{ols})'[R(\hat{\beta}_{ols})\hat{\sigma}^2 (X'X)^{-1} R(\hat{\beta}_{ols})']^{-1} r(\hat{\beta}_{ols})$.

For H^A; $\beta_1\rho + \beta_2 = 0$, we have $r(\beta) = \beta_1\rho + \beta_2$ and $R(\beta) = (\beta_1, \rho, 1)$ where $\beta' = (\rho, \beta_1, \beta_2)$. Hence,

$$W^A = (\hat{\beta}_1\hat{\rho}+\hat{\beta}_2)[(\hat{\beta}_1,\hat{\rho},1)\hat{\sigma}^2(X'X)^{-1}(\hat{\beta}_1,\hat{\rho},1)']^{-1}(\hat{\beta}_1\hat{\rho}+\hat{\beta}_2)$$

where the typical element of the matrix are $[y_{t-1}, x_t, x_{t-1}]$. For H^B;
$\beta_1 + (\beta_2/\rho) = 0$, we have $r(\beta) = \beta_1 + (\beta_2/\rho)$ and

$$R(\beta) = \left(-\frac{\beta_2}{\rho^2}, 1, \frac{1}{\rho}\right).$$ Hence,

$$W^B = (\hat{\beta}_1+\hat{\beta}_2/\hat{\rho})[(-\frac{\hat{\beta}_2}{\hat{\rho}^2}, 1, \frac{1}{\hat{\rho}})\hat{\sigma}^2(X'X)^{-1}(-\frac{\hat{\beta}_2}{\hat{\rho}^2}, 1, \frac{1}{\hat{\rho}})']^{-1}(\hat{\beta}_1+\hat{\beta}_2/\hat{\rho}).$$

For H^C; $\rho + (\beta_2/\beta_1) = 0$, we have

$$r(\beta) = \rho + \beta_2/\beta_1 \quad \text{and} \quad R(\beta) = (1, -\frac{\beta_2}{\beta_1^2}, \frac{1}{\beta_1}).$$

Hence, $W^C = (\hat{\rho}+\hat{\beta}_2/\hat{\beta}_1)[(1,-\frac{\hat{\beta}_2}{\hat{\beta}_1^2}, \frac{1}{\hat{\beta}_1})\hat{\sigma}^2(X'X)^{-1}(1,-\frac{\hat{\beta}_2}{\hat{\beta}_1^2}, \frac{1}{\hat{\beta}_1})']^{-1}(\hat{\rho}+\frac{\hat{\beta}_2}{\hat{\beta}_1}).$

for H^D, $(\beta_1\rho/\beta_2) + 1 = 0$, we have

$$r(\beta) = (\beta_1\rho/\beta_2) + 1 \quad \text{and} \quad R(\beta) = \left(\frac{\beta_1}{\beta_2}, \frac{\rho}{\beta_2}, -\frac{\beta_1\rho}{\beta_2^2}\right).$$

Hence,

$$W^D = (\frac{\hat{\beta}_1\hat{\rho}}{\hat{\beta}_2}+1)[(\frac{\hat{\beta}_1}{\hat{\beta}_2}, \frac{\hat{\rho}}{\hat{\beta}_2}, -\frac{\hat{\beta}_1\hat{\rho}}{\hat{\beta}_2^2})\hat{\sigma}^2(X'X)^{-1}(\frac{\hat{\beta}_1}{\hat{\beta}_2}, \frac{\hat{\rho}}{\hat{\beta}_2}, -\frac{\hat{\beta}_1\hat{\rho}}{\hat{\beta}_2^2})']^{-1}(\frac{\hat{\beta}_1\hat{\rho}}{\hat{\beta}_2}+1).$$

b. Applying these four Wald statistics to the equation relating real per-capita consumption to real per-capita disposable income in the U.S. over the post World War II period 1950-1993, we get:

$W^A = 2.792$, $W^B = 2.063$, $W^C = 2.166$ and $W^D = 1.441$.

These are distributed as χ_1^2 under the null hypothesis. All of them being insignificant at the 5% level, ($\chi_{1,.05}^2 = 3.84$). The SAS program that generated these Wald statistics is given below

SAS PROGRAM

```
Data CONSUMP;
Input YEAR Y C;
cards;

PROC IML; USE CONSUMP;
READ ALL VAR {Y C};

Yt=Y[2:NROW(Y)];
YLAG=Y[1:NROW(Y)-1];
Ct=C[2:NROW(C)];
CLAG=C[1:NROW(C)-1];

X=CLAG||Yt||YLAG;
BETA=INV(X`*X)*X`*Ct;
RHO=BETA[1];
BT1=BETA[2];
BT2=BETA[3];

Px=X*INV(X`*X)*X`;
Qx=I(NROW(X))-Px;
et_U=Qx*Ct;
SIG_U=SSQ(et_U)/NROW(X);

Ha=BT1*RHO+BT2;
Hb=BT1+BT2/RHO;
Hc=RHO+BT2/BT1;
Hd=BT1*RHO/BT2+1;
Ra=BT1||RHO||{1};
Rb=(-BT2/RHO**2)||{1}||(1/RHO);
Rc={1}||(-BT2/BT1**2)||(1/BT1);
Rd=(BT1/BT2)||(RHO/BT2)||(-BT1*RHO/BT2**2);

VAR_a=Ra*SIG_U*INV(X`*X)*Ra`;
VAR_b=Rb*SIG_U*INV(X`*X)*Rb`;
VAR_c=Rc*SIG_U*INV(X`*X)*Rc`;
VAR_d=Rd*SIG_U*INV(X`*X)*Rd`;
```

```
WALD_a=Ha`*INV(VAR_a)*Ha;
WALD_b=Hb`*INV(VAR_b)*Hb;
WALD_c=Hc`*INV(VAR_c)*Hc;
WALD_d=Hd`*INV(VAR_d)*Hd;

PRINT 'Chapter7 Problem20.(b)',,WALD_a;
PRINT WALD_b;
PRINT WALD_c;
PRINT WALD_d;
```

7.21 *Effect of Additional Regressors on R^2.* For the regression equation $y = X\beta + u$ the OLS residuals are given by $e = y - X\hat{\beta}_{ols} = \bar{P}_X y$ where $\bar{P}_X = I_n - P_X$, and $P_X = X(X'X)^{-1}X'$ is the projection matrix. Therefore, the SSE for this regression is $e'e = y'\bar{P}_X y$. In particular, $SSE_1 = y'\bar{P}_{X_1} y$, for $X = X_1$ and $SSE_2 = y'\bar{P}_X y$ for $X = (X_1, X_2)$. Therefore,

$$SSE_1 - SSE_2 = y'(\bar{P}_{X_1} - \bar{P}_X)y = y'(P_X - P_{X_1})y = y'Ay$$

where $A = P_X - P_{X_1}$. This difference in the residual sums of squares is non-negative for any vector y because $y'Ay$ is positive semi-definite. The latter result holds because A is symmetric and idempotent. In fact, A is the difference between two idempotent matrices that also satisfy the following property: $P_X P_{X_1} = P_{X_1} P_X = P_{X_1}$. Hence, $A^2 = P_X^2 - P_{X_1} - P_{X_1} + P_{X_1}^2 = P_X - P_{X_1} = A$. $R^2 = 1 - (SSE/SST)$ where SST is the total sum of squares to be explained by the regression and this depends only on the y's. SST is fixed for both regressions. Hence $R_2^2 \geq R_1^2$, since $SSE_1 \geq SSE_2$.

Reference

Baltagi, B. H. (1996), "Iterative Estimation in Partitioned Regression Models," *Econometric Theory*, Solutions 95.5.1, 12:869-870.

CHAPTER 8
Regression Diagnostics and Specification Tests

8.1 Since $H = P_X$ is idempotent, it is positive semi-definite with $b'H b \geq 0$ for any arbitrary vector b. Specifically, for $b' = (1,0,..,0)$ we get $h_{11} \geq 0$. Also, $H^2 = H$. Hence,

$$h_{11} = \sum_{j=1}^{n} h_{1j}^2 \geq h_{11}^2 > 0.$$

From this inequality, we deduce that $h_{11}^2 - h_{11} \leq 0$ or that $h_{11}(h_{11}-1) \leq 0$. But $h_{11} \geq 0$, hence $0 \leq h_{11} \leq 1$. There is nothing particular about our choice of h_{11}. The same proof holds for h_{22} or h_{33} or in general h_{ii}. Hence, $0 \leq h_{ii} \leq 1$ for $i = 1,2,..,n$.

8.2 *A Simple Regression With No Intercept.* Consider

$$y_i = x_i\beta + u_i \quad \text{for} \quad i = 1,2,..,n$$

a. $H = P_x = x(x'x)^{-1}x' = xx'/x'x$ since $x'x$ is a scalar. Therefore, $h_{ii} = x_i^2 / \sum_{i=1}^{n} x_i^2$ for $i = 1,2,..,n$. Note that the x_i's are not in deviation form as in the case of a simple regression with an intercept. In this case $tr(H) = tr(P_x) = tr(xx')/x'x = tr(x'x)/x'x = x'x/x'x = 1$. Hence, $\sum_{i=1}^{n} h_{ii} = 1$.

b. From (8.13), $\hat{\beta} - \hat{\beta}_{(i)} = \dfrac{(x'x)^{-1}x_i e_i}{1 - h_{ii}} = \dfrac{x_i e_i}{\sum_{j=1}^{n} x_j^2 - x_i^2} = \dfrac{x_i e_i}{\sum_{j \neq i} x_j^2}$.

From (8.18), $(n-2) s_{(i)}^2 = (n-1)s^2 - \dfrac{e_i^2}{1 - h_{ii}} = (n-1)s^2 - e_i^2 (\sum_{i=1}^{n} x_i^2 / \sum_{j \neq i} x_j^2)$.

From (8.19), $DFBETAS_i = (\hat{\beta} - \hat{\beta}_{(i)}) / s_{(i)} \sqrt{(x'x)^{-1}} = \dfrac{x_i e_i}{\sum_{j \neq i} x_j^2} \cdot \left(\sum_{i=1}^{n} x_i^2 \right)^{1/2} / s_{(i)}$.

c. From (8.21), $DFFIT_i = \hat{y}_i - \hat{y}_{(i)} = x_i'[\hat{\beta} - \hat{\beta}_{(i)}] = \dfrac{h_{ii} e_i}{(1 - h_{ii})} = \dfrac{x_i^2 e_i}{\sum_{j \neq i} x_j^2}$.

From (8.22),

$$\text{DFFITS}_i = \left(\frac{h_{ii}}{1-h_{ii}}\right)^{1/2} \frac{e_i}{s_{(i)}\sqrt{1-h_{ii}}} = \left(\frac{x_i^2}{\sum_{j\neq i} x_j^2}\right)^{1/2} \frac{e_i}{s_{(i)}\left(\sum_{j\neq i} x_j^2 / \sum_{i=1}^{n} x_i^2\right)^{1/2}}$$

$$= \frac{x_i e_i \left(\sum_{i=1}^{n} x_i^2\right)^{1/2}}{s_{(i)} \sum_{j\neq i} x_j^2}$$

d. From (8.24), $D_i^2(s) = \dfrac{(\hat{\beta}-\hat{\beta}_{(i)})^2 \sum_{i=1}^{n} x_i^2}{ks^2} = \dfrac{x_i^2 e_i^2}{(\sum_{j\neq i} x_j^2)^2} \cdot \dfrac{\sum_{i=1}^{n} x_i^2}{ks^2}$ from part (b).

From (8.25), $D_i^2(s) = \dfrac{e_i^2}{ks^2} \left[\dfrac{x_i^2 \sum_{j\neq i}^{n} x_j^2}{(\sum_{j\neq i} x_j^2)^2}\right]$.

e. $\det[X'_{(i)} X_{(i)}] = x'_{(i)} x_{(i)} = \sum_{j\neq i} x_j^2$, while

$\det[X'X] = x'x = \sum_{i=1}^{n} x_i^2$ and $(1-h_{ii}) = 1 - (x_i^2/\sum_{i=1}^{n} x_i^2) = \sum_{j\neq i} x_j^2 / \sum_{i=1}^{n} x_i^2$.

The last term is the ratio of the two determinants. Rearranging terms, one can easily verify (8.27). From (8.26),

$$\text{COVRATIO}_i = \frac{s_{(i)}^2}{s^2} \left[\frac{\sum_{i=1}^{n} x_i^2}{\sum_{j\neq i} x_j^2}\right].$$

8.3 From (8.17) $s_{(i)}^2 = \dfrac{1}{n-k-1} \sum_{t\neq i} (y_t - x'_t \hat{\beta}_{(i)})^2$ substituting (8.13), one gets

$$s_{(i)}^2 = \frac{1}{n-k-1} \sum_{t\neq i} \left(y_t - x'_t\hat{\beta} + \frac{x'_t(X'X)^{-1}x_i e_i}{1-h_{ii}}\right)^2 \quad \text{or} \quad (n-k-1)s_{(i)}^2 = \sum_{t\neq i}\left(e_t + \frac{h_{it} e_i}{1-h_{ii}}\right)^2$$

where $h_{it} = x'_t(X'X)^{-1} x_i$. Adding and subtracting the i-th term of this summation, yields

Chapter 8: Regression Diagnostics and Specification Tests   165

$$(n-k-1)s_{(i)}^2 = \sum_{t=1}^{n}\left(e_t + \frac{h_{it}e_i}{1-h_{ii}}\right)^2 - \left(e_i + \frac{h_{ii}e_i}{1-h_{ii}}\right)^2$$

$$= \sum_{t=1}^{n}\left(e_t + \frac{h_{it}e_i}{1-h_{ii}}\right)^2 - \frac{e_i^2}{(1-h_{ii})^2}$$

which is the first equation in (8.18). The next two equations in (8.18) simply expand this square and substitute $\sum_{t=1}^{n}e_t h_{it} = 0$ and $\sum_{t=1}^{n}h_{it}^2 = h_{ii}$ which follow from the fact that $He = 0$ and $H^2 = H$.

8.4 Obtaining e_i^* from an Augmented Regression

a. In order to get $\hat{\beta}^*$ from the augmented regression given in (8.5), one can premultiply by $\bar{P}_{d_i}$ as described in (8.14) and perform OLS. The Frisch-Waugh Lovell Theorem guarantees that the resulting estimate of β^* is the same as that from (8.5). The effect of $\bar{P}_{d_i}$ is to wipe out the i-th observation from the regression and hence $\hat{\beta}^* = \hat{\beta}_{(i)}$ as required.

b. In order to get $\hat{\phi}$ from (8.5), one can premultiply by $\bar{P}_x$ and perform OLS on the transformed equation $\bar{P}_x y = \bar{P}_x d_i \phi + \bar{P}_x u$. The Frisch-Waugh Lovell Theorem guarantees that the resulting estimate of ϕ is the same as that from (8.5). OLS yields $\hat{\phi} = (d_i'\bar{P}_x d_i)^{-1} d_i'\bar{P}_x y$. Using the fact that $e = \bar{P}_x y$, $d_i'e = e_i$ and $d_i'\bar{P}_x d_i = d_i'Hd_i = h_{ii}$ one gets $\hat{\phi} = e_i/(1-h_{ii})$ as required.

c. The Frisch-Waugh Lovell Theorem also states that the residuals from (8.5) are the same as those obtained from (8.14). This means that the i-th observation residual is zero. This also gives us the fact that $\hat{\phi} = y_i - x_i'\hat{\beta}_{(i)}$ = the forecasted residual for the i-th observation, see below (8.14). Hence the RSS from (8.5) $= \sum_{t \neq i}(y_t - x_t'\hat{\beta}_{(i)})^2$ since the i-th observation contributes a zero residual. As in (8.17) and (8.18) and problem 3 one can substitute (8.13) to get

$$\sum_{t \neq i} (y_t - x_t'\hat{\beta}_{(i)})^2 = \sum_{t \neq i} \left(y_t - x_t'\hat{\beta} + \frac{x_i'(X'X)^{-1}x_i e_i}{1-h_{ii}} \right)^2 = \sum_{t \neq i} \left(e_t + \frac{h_{it} e_i}{1-h_{ii}} \right)^2$$

Using the same derivation as in the solution to problem 3, one gets

$$\sum_{t \neq i} (y_t - x_t'\hat{\beta}_{(i)})^2 = \sum_{t=1}^{n} e_t^2 - \frac{e_i^2}{1-h_{ii}}$$ which is an alternative way of writing (8.18).

d. Under Normality of the disturbances, $u \sim N(0, \sigma^2 I_n)$, we get the standard t-statistic on $\hat{\phi}$ as $t = \hat{\phi}/s.e.(\hat{\phi}) \sim t_{n-k-1}$ under the null hypothesis that $\phi = 0$. But $\hat{\phi} = (d_i' \bar{P}_X d_i)^{-1} d_i' \bar{P}_X y = \phi + (d_i' \bar{P}_X d_i)^{-1} d_i' \bar{P}_X u$ with $E(\hat{\phi}) = \phi$ and $var(\hat{\phi}) = \sigma^2 (d_i' \bar{P}_X d_i)^{-1} = \sigma^2/(1-h_{ii})$. This is estimated by $\hat{var}(\hat{\phi}) = s_{(i)}^2/(1-h_{ii})$.

Hence, $t = \hat{\phi}/s.e.(\hat{\phi}) = \frac{e_i}{1-h_{ii}} \cdot \left(\frac{1-h_{ii}}{s_{(i)}^2} \right)^{1/2} = \frac{e_i}{s_{(i)} \sqrt{1-h_{ii}}} = e_i^*$ as in (8.3).

8.5 a. Applying OLS on this regression equation yields

$$\begin{pmatrix} \hat{\beta}^* \\ \hat{\phi}^* \end{pmatrix} = \begin{bmatrix} X'X & 0 \\ 0 & D_p' \bar{P}_X D_p \end{bmatrix}^{-1} \begin{bmatrix} X'y \\ D_p' \bar{P}_X y \end{bmatrix}$$

using the fact that $\bar{P}_X X = 0$. Hence $\hat{\beta}^* = (X'X)^{-1} X'y = \hat{\beta}_{ols}$ and

b. $\hat{\phi}^* = (D_p' \bar{P}_X D_p)^{-1} D_p' \bar{P}_X y = (D_p' \bar{P}_X D_p)^{-1} D_p' e = (D_p' \bar{P}_X D_p)^{-1} e_p$

since $e = \bar{P}_X y$ and $D_p' e = e_p$

c. The residuals are given by

$$y - X\hat{\beta}_{ols} - \bar{P}_X D_p (D_p' \bar{P}_X D_p)^{-1} e_p = e - \bar{P}_X D_p (D_p' \bar{P}_X D_p)^{-1} D_p' e$$

so that the residuals sum of squares is

$$e'e + e_p' (D_p' \bar{P}_X D_p)^{-1} D_p' \bar{P}_X D_p (D_p' \bar{P}_X D_p)^{-1} e_p - 2e' \bar{P}_X D_p (D_p' \bar{P}_X D_p)^{-1} e_p$$
$$= e'e - e_p' (D_p' \bar{P}_X D_p)^{-1} e_p$$

since $\bar{P}_X e = e$ and $e' D_p = e_p'$. From the Frisch-Waugh Lovell Theorem, (8.6) has the same residuals sum of squares as (8.6) premultiplied by $\bar{P}_X$, i.e.,

$\bar{P}_X y = \bar{P}_X D_p \phi^* + \bar{P}_X u$. This is also the same residuals sum of squares as the

Chapter 8: Regression Diagnostics and Specification Tests 167

augmented regression in problem 5, since this also yields the above regression when premultiplied by $\bar{P}_X$.

d. From (8.7), the denominator residual sum of squares is given in part (c). The numerator residual sum of squares is $e'e$ - the RSS obtained in part (c). This yields $e_p' (D_p' \bar{P}_X D_p)^{-1} e_p$ as required.

e. For problem 4, consider the augmented regression $y = X\beta^* + \bar{P}_X d_i \phi + u$. This has the same residual sum of squares by the Frisch-Waugh Lovell Theorem as the following regression:

$$\bar{P}_X y = \bar{P}_X d_i \phi + \bar{P}_X u$$

This last regression also have the same residual sum of squares as

$$y = X\beta^* + d_i \phi + u$$

Hence, using the first regression and the results in part (c) we get

Residual Sum of Squares = (Residual Sum of Squares with d_i deleted)

$$- e' d_i (d_i' \bar{P}_X d_i)^{-1} d_i' e$$

when D_p is replaced by d_i. But $d_i' e = e_i$ and $d_i' \bar{P}_X d_i = 1 - h_{ii}$, hence this last term is $e_i^2/(1-h_{ii})$ which is exactly what we proved in problem 4, part (c). The F-statistic for $\phi = 0$, would be exactly the square of the t-statistic in problem 4, part (d).

8.6 Let $A = X'X$ and $a = b = x_i'$. Using the updated formula

$$(A - a'b)^{-1} = A^{-1} + A^{-1} a' (I - bA^{-1} a')^{-1} b A^{-1}$$

$$(X'X - x_i x_i')^{-1} = (X'X)^{-1} + (X'X)^{-1} x_i (1 - x_i'(X'X)^{-1} x_i)^{-1} x_i' (X'X)^{-1}$$

Note that $X' = [x_1, \ldots, x_n]$ where x_i is $k \times 1$, so that

$$X'X = \sum_{j=1}^{n} x_j x_j' \qquad \text{and} \qquad X_{(i)}' X_{(i)} = \sum_{j \neq i} x_j x_j' = X'X - x_i x_i'.$$

Therefore,

$$(X'_{(i)}X_{(i)})^{-1} = (X'X)^{-1} + \frac{(X'X)^{-1}x_ix_i'(X'X)^{-1}}{1-h_{ii}}.$$

where $h_{ii} = x_i'(X'X)^{-1}x_i$. This verifies (8.12). Note that $X'y = \sum_{j=1}^{n} x_jy_j$ and

$$X'_{(i)}y_{(i)} = \sum_{j \neq i} x_jy_j = X'y - x_iy_i.$$

Therefore,

$$\hat{\beta}_{(i)} = (X'_{(i)}X_{(i)})^{-1}X'_{(i)}y_{(i)} = \hat{\beta} - (X'X)^{-1}x_iy_i + \frac{(X'X)^{-1}x_ix_i'\hat{\beta}}{1-h_{ii}}$$

$$- \frac{(X'X)^{-1}x_ix_i'(X'X)^{-1}x_iy_i}{1-h_{ii}}$$

and

$$\hat{\beta} - \hat{\beta}_{(i)} = -\frac{(X'X)^{-1}x_ix_i'\hat{\beta}}{1-h_{ii}} + \left(\frac{1-h_{ii}+h_{ii}}{1-h_{ii}}\right)(X'X)^{-1}x_iy_i$$

$$= \frac{(X'X)^{-1}x_i(y-x_i'\hat{\beta})}{1-h_{ii}} = \frac{(X'X)^{-1}x_ie_i}{1-h_{ii}}$$

which verifies (8.13).

8.7 From (8.22),

$$DFFITS_i(\sigma) = \left(\frac{h_{ii}}{1-h_{ii}}\right)^{1/2} \frac{e_i}{\sigma\sqrt{1-h_{ii}}} = \frac{\sqrt{h_{ii}}\, e_i}{\sigma(1-h_{ii})}$$

while from (8.25)

$$\sqrt{k}\, D_i(\sigma) = \frac{e_i}{\sigma}\left(\frac{\sqrt{h_{ii}}}{1-h_{ii}}\right)$$

which is identical.

8.8 $\det[X'_{(i)}X_{(i)}] = \det[X'X-x_ix_i'] = \det[\{I_k - x_ix_i'(X'X)^{-1}\}X'X]$. Let $a = x_i$ and $b' = x_i'(X'X)^{-1}$. Using $\det[I_k-ab'] = 1 - b'a$ one gets $\det[I_k-x_ix_i'(X'X)^{-1}] = 1 - x_i'(X'X)^{-1}x_i = 1 - h_{ii}$. Using $\det(AB) = \det(A)\det(B)$, one gets $\det[X'_{(i)}X_{(i)}] = (1-h_{ii})\det(X'X)$ which verifies (8.27).

8.9 The cigarette data example given in Table 3.2.

a. Tables 8.1 and 8.2 were generated by SAS with PROC REG asking for all the diagnostic options allowed by this procedure.

b. For the New Hampshire (NH) observation number 27 in Table 8.2, the leverage $h_{NH} = 0.13081$ which is larger than $2\bar{h} = 2k/n = 0.13043$. The internally studentized residual $\tilde{e}_{NH}$ is computed from (8.1) as follows:

$$\tilde{e}_{NH} = \frac{e_{NH}}{s\sqrt{1-h_{NH}}} = \frac{0.15991}{0.16343\sqrt{1-0.13081}} = 1.0495$$

The externally studentized residual e^*_{NH} is computed from (8.3) as follows:

$$e^*_{(NH)} = \frac{e_{NH}}{s_{(NH)}\sqrt{1-h_{NH}}} = \frac{0.15991}{0.163235\sqrt{1-0.13081}} = 1.0508$$

where $s^2_{(NH)}$ is obtained from (8.2) as follows:

$$s^2_{(NH)} = \frac{(n-k)s^2 - e^2_{NH}/(1-h_{NH})}{(n-k-1)} = \frac{(46-3)(0.16343)^2 - (0.15991)^2/(1-0.13081)}{(46-3-1)}$$

$$= \frac{1.14854 - 0.02941945}{42} = 0.0266457$$

both $\tilde{e}_{NH}$ and $e^*_{(NH)}$ are less than 2 in absolute value. From (8.13), the change in the regression coefficients due to the omission of the NH observation is given by $\hat{\beta} - \hat{\beta}_{NH} = (X'X)^{-1}x_{NH}e_{NH}/(1-h_{NH})$ where $(X'X)^{-1}$ is given in the empirical example and $x'_{NH} = (1, 0.15852, 5.00319)$ with $e_{NH} = 0.15991$ and $h_{NH} = 0.13081$. This gives $(\hat{\beta} - \hat{\beta}_{NH})' = (-0.3174, -0.0834, 0.0709)$. In order to assess whether this change is large or small, we compute DFBETAS given in (8.19). For the NH observation, these are given by

$$DFBETAS_{NH,1} = \frac{\hat{\beta}_1 - \hat{\beta}_{1,NH}}{s_{(NH)}\sqrt{(X'X)^{-1}_{11}}} = \frac{-0.3174}{0.163235\sqrt{30.9298169}} = -0.34967$$

Similarly, $DFBETAS_{NH,2} = -0.2573$ and $DFBETAS_{NH,3} = 0.3608$. These are not larger than 2 in absolute value. However, $DFBETAS_{NH,1}$ and

DFBETAS$_{NH,3}$ are both larger than $2/\sqrt{n} = 2/\sqrt{46} = 0.2949$ in absolute value.

The change in the fit due to omission of the NH observation is given by (8.21). In fact,

$$\text{DFFIT}_{NH} = \hat{y}_{NH} - \hat{y}_{(NH)} = x'_{NH}[\hat{\beta} - \hat{\beta}_{NH}]$$

$$= (1, 0.15852, 5.00319)\begin{pmatrix} -0.3174 \\ -0.0834 \\ 0.0709 \end{pmatrix} = 0.02407.$$

or simply

$$\text{DFFIT}_{NH} = \frac{h_{NH}e_{NH}}{(1-h_{NH})} = \frac{(0.13081)(0.15991)}{(1-0.13081)} = 0.02407.$$

Scaling it by the variance of $\hat{y}_{(NH)}$ we get from (8.22)

$$\text{DFFITS}_{NH} = \left(\frac{h_{NH}}{(1-h_{NH})}\right)^{1/2} e^*_{NH}$$

$$= \left(\frac{0.13081}{1-0.13081}\right)^{1/2} (1.0508) = 0.40764.$$

This is not larger than the size adjusted cutoff of $2/\sqrt{k/n} = 0.511$. Cook's distance measure is given by (8.25) and for NH can be computed as follows:

$$D^2_{NH}(s) = \frac{e^2_{NH}}{ks^2}\left(\frac{h_{NH}}{(1-h_{NH})^2}\right) = \left(\frac{(0.15991)^2}{3(0.16343)^2}\right)\left(\frac{0.13081}{(1-0.13081)^2}\right)$$

$$= 0.05526$$

COVRATIO omitting the NH observation can be computed from (8.28) as follows:

$$\text{COVRATIO}_{NH} = \left(\frac{s^2_{(NH)}}{s^2}\right)^k \frac{1}{1-h_{NH}} = \left(\frac{0.0266457}{(0.16343)^2}\right)^3 \left(\frac{1}{(1-0.13081)}\right)$$

$$= 1.1422$$

which means that $|\text{COVRATIO}_{NH} - 1| = 0.1422$ is less than $3k/n = 0.1956$. Finally, FVARATIO omitting the NH observation can be computed from (8.29) as

Chapter 8: Regression Diagnostics and Specification Tests   171

$$\text{FVARATIO}_{NH} = \frac{s^2_{(NH)}}{s^2(1-h_{NH})} = \frac{0.0266457}{(0.16343)^2(1-0.13081)} = 1.14775$$

c. Similarly, the same calculations can be obtained for the observations of the states of AR, CT, NJ and UT.

d. Also, the states of NV, ME, NM and ND.

8.10 The Consumption - Income data given in Table 5.1.

The following SAS output gives the predicted values along with the 95% confidence interval for this prediction.

Obs	Dep Var C	Predict Value	Std Err Predict	Lower95% Mean	Upper95% Mean	Lower95% Predict	Upper95% Predict
1	5820.0	5688.0	40.847	5605.5	5770.4	5367.2	6008.7
2	5843.0	5785.0	40.095	5704.1	5866.0	5464.7	6105.4
3	5917.0	5863.8	39.490	5784.1	5943.5	5543.7	6183.8
4	6054.0	6013.9	38.351	5936.5	6091.3	5694.4	6333.4
5	6099.0	6003.0	38.433	5925.4	6080.5	5683.4	6322.5
6	6365.0	6232.8	36.724	6158.7	6306.9	5914.1	6551.5
7	6440.0	6416.8	35.392	6345.4	6488.2	6098.7	6734.9
8	6465.0	6447.9	35.170	6377.0	6518.9	6129.9	6765.9
9	6449.0	6447.0	35.177	6376.0	6518.0	6129.0	6765.0
10	6658.0	6578.0	34.255	6508.8	6647.1	6260.4	6895.6
11	6698.0	6585.3	34.204	6516.3	6654.3	6267.7	6902.9
12	6740.0	6693.3	33.460	6625.8	6760.9	6376.1	7010.6
13	6931.0	6877.4	32.228	6812.3	6942.4	6560.6	7194.1
14	7089.0	7001.0	31.427	6937.6	7064.4	6684.6	7317.4
15	7384.0	7387.4	29.084	7328.7	7446.1	7071.9	7702.9
16	7703.0	7724.3	27.277	7669.3	7779.4	7409.5	8039.2
17	8005.0	8011.8	25.944	7959.5	8064.2	7697.5	8326.2
18	8163.0	8279.2	24.908	8228.9	8329.5	7965.2	8593.2
19	8506.0	8540.1	24.109	8491.5	8588.8	8226.4	8853.9
20	8737.0	8729.7	23.673	8681.9	8777.5	8416.0	9043.3
21	8842.0	8976.0	23.301	8929.0	9023.0	8662.5	9289.5
22	9022.0	9192.1	23.163	9145.3	9238.8	8878.6	9505.6
23	9425.0	9469.5	23.248	9422.6	9516.4	9156.0	9783.0
24	9752.0	10018.0	24.264	9969.0	10066.9	9704.1	10331.8
25	9602.0	9852.2	23.843	9804.1	9900.4	9538.5	10165.9
26	9711.0	9920.0	24.004	9871.5	9968.4	9606.2	10233.7
27	10121.0	10181.9	24.771	10131.9	10231.8	9867.9	10495.8
28	10425.0	10377.8	25.487	10326.4	10429.2	10063.6	10692.0
29	10744.0	10785.3	27.319	10730.1	10840.4	10470.4	11100.1
30	10876.0	10957.4	28.216	10900.5	11014.3	10642.2	11272.6
31	10746.0	10926.3	28.049	10869.7	10982.9	10611.2	11241.4
32	10770.0	11064.5	28.806	11006.4	11122.7	10749.1	11379.9
33	10782.0	11055.4	28.754	10997.3	11113.4	10740.0	11370.7
34	11179.0	11241.2	29.829	11181.0	11301.4	10925.5	11557.0
35	11617.0	11863.9	33.845	11795.6	11932.2	11546.4	12181.3
36	12015.0	12073.5	35.315	12002.3	12144.8	11755.5	12391.6
37	12336.0	12342.7	37.272	12267.5	12417.9	12023.8	12661.7
38	12568.0	12336.3	37.225	12261.2	12411.4	12017.4	12655.3
39	12903.0	12652.2	39.605	12572.3	12732.1	12332.1	12972.3
40	13029.0	12757.5	40.416	12675.9	12839.1	12437.0	13078.0
41	13093.0	12845.4	41.099	12762.5	12928.3	12524.5	13166.3
42	12899.0	12755.7	40.401	12674.1	12837.2	12435.2	13076.2

| 43 | 13110.0 | 13008.4 | 42.379 | 12922.9 | 13093.9 | 12686.8 | 13330.0 |
| 44 | 13391.0 | 13065.2 | 42.829 | 12978.7 | 13151.6 | 12743.4 | 13387.0 |

Next, we compute the residuals, internally studentized residuals, Cook's D statistic and externally studentized residuals.

Obs	Residual	Std Err Residual	Student Residual	-2-1-0 1 2	Cook's D	Rstudent
1	132.0	148.071	0.892	*	0.030	0.8894
2	57.9635	148.276	0.391		0.006	0.3869
3	53.2199	148.438	0.359		0.005	0.3548
4	40.0577	148.737	0.269		0.002	0.2663
5	96.0452	148.715	0.646	*	0.014	0.6413
6	132.2	149.147	0.887	*	0.024	0.8842
7	23.1835	149.468	0.155		0.001	0.1533
8	17.0523	149.521	0.114		0.000	0.1127
9	1.9679	149.519	0.013		0.000	0.0130
10	80.0338	149.733	0.535	*	0.007	0.5299
11	112.7	149.745	0.753	*	0.015	0.7487
12	46.6653	149.913	0.311		0.002	0.3079
13	53.6250	150.182	0.357		0.003	0.3533
14	88.0159	150.352	0.585	*	0.007	0.5808
15	-3.3771	150.823	-0.022		0.000	-0.0221
16	-21.3264	151.160	-0.141		0.000	-0.1394
17	-6.8321	151.395	-0.045		0.000	-0.0446
18	-116.2	151.569	-0.767	*	0.008	-0.7628
19	-34.1467	151.698	-0.225		0.001	-0.2225
20	7.3193	151.766	0.048		0.000	0.0477
21	-134.0	151.824	-0.882	*	0.009	-0.8801
22	-170.1	151.845	-1.120	**	0.015	-1.1235
23	-44.5043	151.832	-0.293		0.001	-0.2899
24	-266.0	151.673	-1.754	***	0.039	-1.7997
25	-250.2	151.740	-1.649	***	0.034	-1.6848
26	-209.0	151.714	-1.378	**	0.024	-1.3929
27	-60.8591	151.591	-0.401		0.002	-0.3974
28	47.1975	151.472	0.312		0.001	0.3082
29	-41.2548	151.152	-0.273		0.001	-0.2699
30	-81.3920	150.988	-0.539	*	0.005	-0.5345
31	-180.3	151.019	-1.194	**	0.025	-1.1999
32	-294.5	150.876	-1.952	***	0.069	-2.0226
33	-273.4	150.886	-1.812	***	0.060	-1.8644
34	-62.2352	150.677	-0.413		0.003	-0.4089
35	-246.9	149.826	-1.648	***	0.069	-1.6832
36	-58.5367	149.487	-0.392		0.004	-0.3876
37	-6.7299	149.011	-0.045		0.000	-0.0446
38	231.7	149.023	1.555	***	0.075	1.5822
39	250.8	148.408	1.690	***	0.102	1.7295
40	271.5	148.189	1.832	***	0.125	1.8871
41	247.6	148.001	1.673	***	0.108	1.7109
42	143.3	148.193	0.967	*	0.035	0.9664
43	101.6	147.639	0.688	*	0.020	0.6839
44	325.8	147.509	2.209	****	0.206	2.3215

Finally, the leverage, covariance-ratio, Dffits and DFBETAS are computed.

Obs	Hat Diag H	Cov Ratio	Dffits	INTERCEP Dfbetas	Y Dfbetas
1	0.0707	1.0869	0.2453	0.2309	-0.2021
2	0.0681	1.1179	0.1046	0.0980	-0.0854
3	0.0661	1.1168	0.0944	0.0880	-0.0765
4	0.0623	1.1153	0.0687	0.0635	-0.0547
5	0.0626	1.0973	0.1657	0.1533	-0.1323
6	0.0572	1.0717	0.2177	0.1984	-0.1690
7	0.0531	1.1069	0.0363	0.0326	-0.0275
8	0.0524	1.1067	0.0265	0.0237	-0.0200
9	0.0524	1.1074	0.0031	0.0027	-0.0023
10	0.0497	1.0893	0.1212	0.1073	-0.0893
11	0.0496	1.0745	0.1710	0.1512	-0.1259
12	0.0475	1.0966	0.0687	0.0601	-0.0496

Chapter 8: Regression Diagnostics and Specification Tests 173

```
13    0.0440   1.0910    0.0758   0.0649  -0.0527
14    0.0419   1.0774    0.1214   0.1021  -0.0821
15    0.0359   1.0884   -0.0043  -0.0034   0.0026
16    0.0315   1.0825   -0.0252  -0.0183   0.0133
17    0.0285   1.0801   -0.0076  -0.0051   0.0034
18    0.0263   1.0478   -0.1254  -0.0743   0.0462
19    0.0246   1.0733   -0.0354  -0.0182   0.0098
20    0.0238   1.0748    0.0074   0.0033  -0.0015
21    0.0230   1.0346   -0.1351  -0.0487   0.0150
22    0.0227   1.0106   -0.1714  -0.0476   0.0041
23    0.0229   1.0696   -0.0444  -0.0074  -0.0039
24    0.0250   0.9244   -0.2879   0.0133  -0.0860
25    0.0241   0.9405   -0.2647  -0.0044  -0.0631
26    0.0244   0.9806   -0.2204   0.0020  -0.0581
27    0.0260   1.0691   -0.0649   0.0069  -0.0231
28    0.0275   1.0741    0.0519  -0.0090   0.0217
29    0.0316   1.0798   -0.0488   0.0145  -0.0259
30    0.0337   1.0710   -0.0999   0.0343  -0.0571
31    0.0333   1.0132   -0.2228   0.0748  -0.1258
32    0.0352   0.8992   -0.3862   0.1431  -0.2297
33    0.0350   0.9242   -0.3553   0.1309  -0.2106
34    0.0377   1.0817   -0.0810   0.0334  -0.0510
35    0.0486   0.9650   -0.3802   0.2020  -0.2773
36    0.0529   1.0999   -0.0916   0.0516  -0.0691
37    0.0589   1.1149   -0.0112   0.0067  -0.0087
38    0.0587   0.9902    0.3952  -0.2367   0.3095
39    0.0665   0.9764    0.4615  -0.2934   0.3744
40    0.0692   0.9544    0.5147  -0.3329   0.4218
41    0.0716   0.9847    0.4751  -0.3115   0.3925
42    0.0692   1.0777    0.2635  -0.1704   0.2159
43    0.0761   1.1104    0.1963  -0.1317   0.1644
44    0.0777   0.8888    0.6741  -0.4556   0.5670
```

Sum of Residuals   0
Sum of Squared Residuals   990923.1310
Predicted Resid SS (Press)   1096981.5476

SAS PROGRAM

```
data CONSUM;
input YEAR Y C;
cards;
Proc reg data=CONSUM;
    Model C=Y / influence  p r cli clm;
run;
```

8.12 The Gasoline data used in Chapter 10. The following SAS output gives the diagnostics for two countries: Austria and Belgium.

a. AUSTRIA:

Obs	Dep Var Y	Predict Value	Std Err Predict	Lower95% Mean	Upper95% Mean	Lower95% Predict	Upper95% Predict
1	4.1732	4.1443	0.026	4.0891	4.1996	4.0442	4.2445
2	4.1010	4.1121	0.021	4.0680	4.1563	4.0176	4.2066
3	4.0732	4.0700	0.016	4.0359	4.1041	3.9798	4.1602
4	4.0595	4.0661	0.014	4.0362	4.0960	3.9774	4.1548
5	4.0377	4.0728	0.012	4.0469	4.0987	3.9854	4.1603
6	4.0340	4.0756	0.013	4.0480	4.1032	3.9877	4.1636
7	4.0475	4.0296	0.014	3.9991	4.0601	3.9407	4.1185
8	4.0529	4.0298	0.016	3.9947	4.0650	3.9392	4.1205
9	4.0455	4.0191	0.019	3.9794	4.0587	3.9266	4.1115
10	4.0464	4.0583	0.015	4.0267	4.0898	3.9690	4.1475

11	4.0809	4.1108	0.016	4.0767	4.1450	4.0206	4.2011
12	4.1067	4.1378	0.022	4.0909	4.1847	4.0420	4.2336
13	4.1280	4.0885	0.015	4.0561	4.1210	3.9989	4.1782
14	4.1994	4.1258	0.022	4.0796	4.1721	4.0303	4.2214
15	4.0185	4.0270	0.017	3.9907	4.0632	3.9359	4.1180
16	4.0290	3.9799	0.016	3.9449	4.0149	3.8893	4.0704
17	3.9854	4.0287	0.017	3.9931	4.0642	3.9379	4.1195
18	3.9317	3.9125	0.024	3.8606	3.9643	3.8141	4.0108
19	3.9227	3.9845	0.019	3.9439	4.0250	3.8916	4.0773

Obs	Residual	Std Err Residual	Student Residual	-2-1-0 1 2	Cook's D	Rstudent
1	0.0289	0.029	0.983	*	0.188	0.9821
2	-0.0111	0.033	-0.335		0.011	-0.3246
3	0.00316	0.036	0.088		0.000	0.0853
4	-0.00661	0.037	-0.180		0.001	-0.1746
5	-0.0351	0.037	-0.942	*	0.024	-0.9387
6	-0.0417	0.037	-1.126	**	0.039	-1.1370
7	0.0179	0.036	0.491		0.009	0.4787
8	0.0231	0.036	0.649	*	0.023	0.6360
9	0.0264	0.034	0.766	*	0.043	0.7554
10	-0.0119	0.036	-0.328		0.004	-0.3178
11	-0.0300	0.036	-0.838	*	0.035	-0.8287
12	-0.0310	0.032	-0.957	*	0.105	-0.9541
13	0.0395	0.036	1.093	**	0.053	1.1008
14	0.0735	0.033	2.254	****	0.563	2.6772
15	-0.00846	0.035	-0.239		0.003	-0.2318
16	0.0492	0.036	1.381	**	0.101	1.4281
17	-0.0433	0.035	-1.220	**	0.082	-1.2415
18	0.0192	0.031	0.625	*	0.061	0.6118
19	-0.0617	0.034	-1.802	***	0.250	-1.9663

Obs	Hat Diag H	Cov Ratio	Dffits	INTERCEP Dfbetas	X1 Dfbetas	X2 Dfbetas	X3 Dfbetas
1	0.4377	1.7954	0.8666	0.2112	0.4940	-0.0549	-0.5587
2	0.2795	1.7750	-0.2022	-0.0321	-0.0859	-0.0141	0.1009
3	0.1665	1.5779	0.0381	-0.0095	-0.0042	0.0134	0.0008
4	0.1279	1.4980	-0.0669	0.0284	0.0188	-0.0202	-0.0118
5	0.0959	1.1418	-0.3057	0.1381	0.0735	-0.0382	-0.0387
6	0.1091	1.0390	-0.3980	0.2743	0.1728	0.0437	-0.1219
7	0.1331	1.4246	0.1876	-0.1200	-0.1390	0.0742	0.1286
8	0.1767	1.4284	0.2947	-0.2174	-0.2359	0.0665	0.2176
9	0.2254	1.4500	0.4075	-0.2985	-0.3416	0.0910	0.3220
10	0.1424	1.4930	-0.1295	0.0812	0.0693	0.0280	-0.0607
11	0.1669	1.3061	-0.3710	0.0513	-0.0795	0.2901	0.1027
12	0.3150	1.4954	-0.6470	0.0762	-0.1710	0.5599	0.2111
13	0.1508	1.1134	0.4639	0.0284	0.1127	-0.3194	-0.1143
14	0.3070	0.3639	1.7819	0.9050	1.3184	-1.3258	-1.2772
15	0.1884	1.5990	-0.1117	-0.0945	-0.0574	-0.0153	0.0407
16	0.1754	0.9276	0.6587	0.3450	-0.0416	0.3722	0.1521
17	0.1810	1.0595	-0.5836	-0.4710	-0.2819	-0.0069	0.1923
18	0.3854	1.9295	0.4845	0.1093	-0.2126	0.3851	0.2859
19	0.2357	0.6501	-1.0918	-0.8004	-0.2221	-0.4001	0.0226

Sum of Residuals   0
Sum of Squared Residuals   0.0230
Predicted Resid SS (Press)   0.0399

b. BELGIUM:

Obs	Dep Var Y	Predict Value	Std Err Predict	Lower95% Mean	Upper95% Mean	Lower95% Predict	Upper95% Predict
1	4.1640	4.1311	0.019	4.0907	4.1715	4.0477	4.2144
2	4.1244	4.0947	0.017	4.0593	4.1301	4.0137	4.1757
3	4.0760	4.0794	0.015	4.0471	4.1117	3.9997	4.1591
4	4.0013	4.0412	0.013	4.0136	4.0688	3.9632	4.1192
5	3.9944	4.0172	0.012	3.9924	4.0420	3.9402	4.0942
6	3.9515	3.9485	0.015	3.9156	3.9814	3.8685	4.0285
7	3.8205	3.8823	0.017	3.8458	3.9189	3.8008	3.9639

Chapter 8: Regression Diagnostics and Specification Tests   175

8	3.9069	3.9045	0.012	3.8782	3.9309	3.8270	3.9820
9	3.8287	3.8242	0.019	3.7842	3.8643	3.7410	3.9074
10	3.8546	3.8457	0.014	3.8166	3.8748	3.7672	3.9242
11	3.8704	3.8516	0.011	3.8273	3.8759	3.7747	3.9284
12	3.8722	3.8537	0.011	3.8297	3.8776	3.7769	3.9304
13	3.9054	3.8614	0.014	3.8305	3.8922	3.7822	3.9405
14	3.8960	3.8874	0.013	3.8606	3.9142	3.8097	3.9651
15	3.8182	3.8941	0.016	3.8592	3.9289	3.8133	3.9749
16	3.8778	3.8472	0.013	3.8185	3.8760	3.7689	3.9256
17	3.8641	3.8649	0.016	3.8310	3.8988	3.7845	3.9453
18	3.8543	3.8452	0.014	3.8163	3.8742	3.7668	3.9237
19	3.8427	3.8492	0.028	3.7897	3.9086	3.7551	3.9433

Obs	Residual	Std Err Residual	Student Residual	-2-1-0 1 2	Cook's D	Rstudent
1	0.0329	0.028	1.157	|**	0.148	1.1715
2	0.0297	0.030	0.992	|*	0.076	0.9911
3	-0.00343	0.031	-0.112	|	0.001	-0.1080
4	-0.0399	0.032	-1.261	**|	0.067	-1.2890
5	-0.0228	0.032	-0.710	*|	0.016	-0.6973
6	0.00302	0.031	0.099	|	0.001	0.0956
7	-0.0618	0.030	-2.088	****|	0.366	-2.3952
8	0.00236	0.032	0.074	|	0.000	0.0715
9	0.00445	0.029	0.156	|	0.003	0.1506
10	0.00890	0.031	0.284	|	0.004	0.2748
11	0.0188	0.032	0.583	|*	0.011	0.5700
12	0.0186	0.032	0.575	|*	0.010	0.5620
13	0.0440	0.031	1.421	|**	0.110	1.4762
14	0.00862	0.032	0.271	|	0.003	0.2624
15	-0.0759	0.030	-2.525	*****|	0.472	-3.2161
16	0.0305	0.031	0.972	|*	0.043	0.9698
17	-0.00074	0.030	-0.025	|	0.000	-0.0237
18	0.00907	0.031	0.289	|	0.004	0.2798
19	-0.00645	0.020	-0.326	|	0.053	-0.3160

Obs	Hat Diag H	Cov Ratio	Dffits	INTERCEP Dfbetas	X1 Dfbetas	X2 Dfbetas	X3 Dfbetas
1	0.3070	1.3082	0.7797	-0.0068	0.3518	0.0665	-0.4465
2	0.2352	1.3138	0.5497	0.0273	0.2039	0.1169	-0.2469
3	0.1959	1.6334	-0.0533	0.0091	-0.0199	0.0058	0.0296
4	0.1433	0.9822	-0.5272	0.1909	-0.0889	0.1344	0.2118
5	0.1158	1.3002	-0.2524	0.0948	-0.0430	0.0826	0.1029
6	0.2039	1.6509	0.0484	-0.0411	-0.0219	-0.0356	0.0061
7	0.2516	0.4458	-1.3888	0.0166	0.9139	-0.6530	-1.0734
8	0.1307	1.5138	0.0277	-0.0028	-0.0169	0.0093	0.0186
9	0.3019	1.8754	0.0990	-0.0335	-0.0873	0.0070	0.0884
10	0.1596	1.5347	0.1198	-0.0551	-0.0959	-0.0243	0.0886
11	0.1110	1.3524	0.2014	-0.0807	-0.1254	-0.0633	0.1126
12	0.1080	1.3513	0.1956	-0.0788	-0.0905	-0.0902	0.0721
13	0.1792	0.9001	0.6897	0.5169	0.0787	0.5245	0.1559
14	0.1353	1.4943	0.1038	0.0707	0.0596	0.0313	-0.0366
15	0.2284	0.1868	-1.7498	-1.4346	-1.1919	-0.7987	0.7277
16	0.1552	1.2027	0.4157	0.3104	0.1147	0.2246	0.0158
17	0.2158	1.6802	-0.0124	-0.0101	-0.0071	-0.0057	0.0035
18	0.1573	1.5292	0.1209	0.0564	0.0499	0.0005	-0.0308
19	0.6649	3.8223	-0.4451	0.1947	-0.0442	0.3807	0.1410

Sum of Residuals   0
Sum of Squared Residuals   0.0175
Predicted Resid SS (Press)   0.0289

SAS PROGRAM

```
Data GASOLINE;
Input COUNTRY $ YEAR Y X1 X2 X3;
CARDS;
```

```
DATA AUSTRIA; SET GASOLINE;
IF COUNTRY='AUSTRIA';
Proc reg data=AUSTRIA;
      Model Y=X1 X2 X3 / influence  p r cli clm;
RUN;

DATA BELGIUM; SET GASOLINE;
IF COUNTRY='BELGIUM';
Proc reg data=BELGIUM;
      Model Y=X1 X2 X3 / influence  p r cli clm;
RUN;
```

8.13 *Independence of Recursive Residual.* This is based on Johnston (1984, p 386).

    a. Using the updating formula given in (8.11) with $A = (X_t'X_t)$ and $a = -b = x_{t+1}'$, we get

$$(X_t'X_t + x_{t+1}x_{t+1}')^{-1} = (X_t'X_t)^{-1}$$

$$- (X_t'X_t)^{-1} x_{t+1} (1 + x_{t+1}'(X_t'X_t)^{-1} x_{t+1}')^{-1} x_{t+1}'(X_t'X_t)^{-1}$$

But $X_{t+1} = \begin{pmatrix} X_t \\ x_{t+1}' \end{pmatrix}$, hence $X_{t+1}'X_{t+1} = X_t'X_t + x_{t+1}x_{t+1}'$. Substituting this expression we get

$$(X_{t+1}'X_{t+1})^{-1} = (X_t'X_t)^{-1} - \frac{(X_t'X_t)^{-1} x_{t+1} x_{t+1}' (X_t'X_t)^{-1}}{1 + x_{t+1}'(X_t'X_t)^{-1} x_{t+1}}$$

which is exactly (8.31).

    b. $\hat{\beta}_{t+1} = (X_{t+1}'X_{t+1})^{-1} X_{t+1}' Y_{t+1}$

Replacing X_{t+1}' by (X_t', x_{t+1}) and Y_{t+1} by $\begin{pmatrix} Y_t \\ y_{t+1} \end{pmatrix}$ yields

$$\hat{\beta}_{t+1} = (X_{t+1}'X_{t+1})^{-1}(X_t'Y_t + x_{t+1}y_{t+1})$$

Replacing $(X_{t+1}'X_{t+1})^{-1}$ by its expression in (8.31) we get

$$\hat{\beta}_{t+1} = \hat{\beta}_t + (X_t'X_t)^{-1} x_{t+1} y_{t+1} - \frac{(X_t'X_t)^{-1} x_{t+1} x_{t+1}' \hat{\beta}_t}{f_{t+1}}$$

$$- \frac{(X_t'X_t)^{-1} x_{t+1} x_{t+1}' (X_t'X_t)^{-1} x_{t+1} y_{t+1}}{f_{t+1}} = \hat{\beta}_t - \frac{(X_t'X_t)^{-1} x_{t+1} x_{t+1}' \hat{\beta}_t}{f_{t+1}}$$

$$+ \left(\frac{f_{t+1} - f_{t+1} + 1}{f_{t+1}}\right)(X_t'X_t)^{-1} x_{t+1} y_{t+1} = \hat{\beta}_t - (X_t'X_t)^{-1} x_{t+1} (y_{t+1} - x_{t+1}' \hat{\beta}_t)/f_{t+1}$$

where we used the fact that $x'_{t+1}(X'_tX_t)^{-1}x_{t+1} = f_{t+1} - 1$.

c. Using (8.30), $w_{t+1} = (y_{t+1} - x'_{t+1}\hat{\beta}_t)/\sqrt{f_{t+1}}$ where $f_{t+1} = 1 + x'_{t+1}(X'_tX_t)^{-1}x_{t+1}$. Defining $v_{t+1} = \sqrt{f_{t+1}}\, w_{t+1}$ we get $v_{t+1} = y_{t+1} - x'_{t+1}\hat{\beta}_t = x'_{t+1}(\beta - \hat{\beta}_t) + u_{t+1}$ for $t = k,..,T-1$. Since $u_t \sim IIN(0,\sigma^2)$, then w_{t+1} has zero mean and $var(w_{t+1}) = \sigma^2$. Furthermore, w_{t+1} are linear in the y's. Therefore, they are themselves normally distributed. Given normality of the w_{t+1}'s it is sufficient to show that $cov(w_{t+1}, w_{s+1}) = 0$ for $t \neq s$; $t = s = k,..,T-1$. But f_{t+1} is fixed. Therefore, it suffices to show that

$cov(v_{t+1}, v_{s+1}) = 0$ for $t \neq s$.

Using the fact that $\hat{\beta}_t = (X'_tX_t)^{-1}X'_tY_t = \beta + (X'_tX_t)^{-1}X'_tu_t$ where $u'_t = (u_1,..,u_t)$, we get $v_{t+1} = u_{t+1} - x'_{t+1}(X'_tX_t)^{-1}X'_tu_t$. Therefore,

$E(v_{t+1}v_{s+1}) = E\{[u_{t+1} - x'_{t+1}(X'_tX_t)^{-1}X'_tu_t][u_{s+1} - x'_{s+1}(X'_sX_s)^{-1}X'_su_s]\}$

$= E(u_{t+1}u_{s+1}) - x'_{t+1}(X'_tX_t)^{-1}X'_t E(u_tu_{s+1}) - E(u_{t+1}u'_s)X_s(X'_sX_s)^{-1}x_{s+1}$

$+ x'_{t+1}(X'_tX_t)X'_tE(u_tu'_s)X_s(X'_sX_s)^{-1}x_{s+1}$.

Assuming, without loss of generality, that $t < s$, we get $E(u_{t+1}u_{s+1}) = 0$,

$E(u_tu_{s+1}) = 0$ since $t < s < s+1$,

$E(u_tu'_s) = (0,..,\sigma^2,..0)$ where the σ^2 is in the (t+1)-th position, and

$E(u_tu'_s) = E\begin{pmatrix} u_1 \\ u_2 \\ \vdots \\ u_t \end{pmatrix}(u_1,..,u_t,..,u_{t+1},..,u_s) = \sigma^2(I_t, 0)$

Substituting these covariances in $E(v_{t+1}v_{s+1})$ yields immediately zero for the first two terms and what remains is

$E(v_{t+1}v_{s+1}) = -\sigma^2 x'_{t+1}(X'_sX_s)^{-1}x_{s+1} + \sigma^2 x'_{t+1}(X'_tX_t)X'_t[I_t, 0]X_s(X'_sX_s)^{-1}x_{s+1}$

But $[I_t, 0]X_s = X_t$ for $t < s$. Hence

$E(v_{t+1}v_{s+1}) = -\sigma^2 x'_{t+1}(X'_sX_s)^{-1}x_{s+1} + \sigma^2 x'_{t+1}(X'_sX_s)^{-1}x_{s+1} = 0$ for $t \neq s$.

A simpler proof using the C matrix defined in (8.34) is given in problem 14.

8.14 *Recursive Residuals are Linear Unbiased With Scalar Covariance Matrix (LUS).*

a. The first element of $w = Cy$ where C is defined in (8.34) is

$$w_{k+1} = -x'_{k+1}(X'_k X_k)^{-1} X'_k Y_k / \sqrt{f_{k+1}} + y_{k+1}/\sqrt{f_{k+1}}$$

where $Y'_k = (y_1,...,y_k)$. Hence, using the fact that $\hat{\beta}_k = (X'_k X_k)^{-1} X'_k Y_k$, we get $w_{k+1} = (y_{k+1} - x'_{k+1}\hat{\beta}_k)/\sqrt{f_{k+1}}$ which is (8.30) for $t = k$. Similarly, the t-th element of $w = Cy$ from (8.34) is $w_t = -x'_t (X'_{t-1} X_{t-1})^{-1} X'_{t-1} Y_{t-1}/\sqrt{f_t} + y_t/\sqrt{f_t}$
where $Y'_{t-1} = (y_1,...,y_{t-1})$. Hence, using the fact that $\hat{\beta}_{t-1} = (X'_{t-1} X_{t-1})^{-1} X'_{t-1} Y_{t-1}$,
we get $w_t = (y_t - x'_t \hat{\beta}_{t-1})/\sqrt{f_t}$ which is (8.30) for $t = t - 1$. Also, the last term can be expressed in the same way.

$$w_T = -x'_T (X'_{T-1} X_{T-1})^{-1} X'_{T-1} Y_{T-1}/\sqrt{f_T} + y_T/\sqrt{f_T} = (y_T - x'_T \hat{\beta}_{T-1})/\sqrt{f_T}$$

which is (8.30) for $t = T - 1$.

b. The first row of CX is obtained by multiplying the first row of C by

$$X = \begin{bmatrix} X_k \\ x'_{k+1} \\ \vdots \\ x'_T \end{bmatrix}.$$

This yields $-x'_{k+1}(X'_k X_k)^{-1} X'_k X_k/\sqrt{f_{k+1}} + x'_{k+1}/\sqrt{f_{k+1}} = 0$.
Similarly, the t-th row of CX is obtained by multiplying the t-th row of C by

$$X = \begin{bmatrix} X_{t-1} \\ x'_t \\ \vdots \\ x'_T \end{bmatrix}.$$

This yields $-x'_t (X'_{t-1} X_{t-1})^{-1} X'_{t-1} X_{t-1}/\sqrt{f_t} + x'_t/\sqrt{f_t} = 0$.
Also, the last row of CX is obtained by multiplying the T-th row of C by

$$X = \begin{bmatrix} X_{T-1} \\ x'_{T+1} \end{bmatrix}.$$

This yields $-x'_T (X'_{T-1} X_{T-1})^{-1} X'_{T-1}/\sqrt{f_T} + x'_{T+1}/\sqrt{f_T} = 0$.

This proves that $CX = 0$. This means that

$w = Cy = CX\beta + Cu = Cu$ since $CX = 0$.

Chapter 8: Regression Diagnostics and Specification Tests

Hence, $E(w) = CE(u) = 0$. Note that

$$C = \begin{bmatrix} \dfrac{-X_k(X_k'X_k)^{-1}x_{k+1}}{\sqrt{f_{k+1}}} & \dfrac{-X_{t-1}(X_{t-1}'X_{t-1})^{-1}x_t}{\sqrt{f_t}} & \cdots & \dfrac{-X_{T-1}(X_{T-1}'X_{T-1})^{-1}x_T}{\sqrt{f_T}} \\ \dfrac{1}{\sqrt{f_{k+1}}} & \dfrac{1}{\sqrt{f_t}} & \cdots & \cdot \\ 0 & 0 & & \cdot \\ \vdots & \vdots & & \vdots \\ 0 & 0 & \cdots & \dfrac{1}{\sqrt{f_T}} \end{bmatrix}$$

so that, the first diagonal element of CC' is

$$\frac{x_{k+1}'(X_k'X_k)^{-1}X_k'X_k(X_k'X_k)^{-1}x_{k+1}}{f_{k+1}} + \frac{1}{f_{k+1}} = \frac{1 + x_{k+1}'(X_k'X_k)^{-1}x_{k+1}}{f_{k+1}} = \frac{f_{k+1}}{f_{k+1}} = 1$$

similarly, the t-th diagonal element of CC' is

$$\frac{x_t'(X_{t-1}'X_{t-1})^{-1}X_{t-1}'X_{t-1}(X_{t-1}'X_{t-1})^{-1}x_t}{f_t} + \frac{1}{f_t} = \frac{f_t}{f_t} = 1$$

and the T-th diagonal element of CC' is

$$\frac{x_T'(X_{T-1}'X_{T-1})^{-1}X_{T-1}'X_{T-1}(X_{T-1}'X_{T-1})^{-1}x_T}{f_T} + \frac{1}{f_T} = \frac{f_T}{f_T} = 1$$

Using the fact that

$$X_{t-1} = \begin{bmatrix} X_k \\ x_{k+1}' \\ \vdots \\ x_{t-1}' \end{bmatrix},$$

one gets the (1,t) element of CC' by multiplying the first row of C by the t-th column of C'. This yields

$$\frac{x_{k+1}'(X_k'X_k)^{-1}X_k'X_k(X_{t-1}'X_{t-1})^{-1}x_t}{\sqrt{f_{k+1}}\sqrt{f_t}} - \frac{x_{k+1}'(X_{t-1}'X_{t-1})^{-1}x_t}{\sqrt{f_{k+1}}\sqrt{f_t}} = 0.$$

Similarly, the (1,T) element of CC' is obtained by multiplying the first row of C by the T-th column of C'. This yields

$$\frac{x'_{k+1}(X'_kX_k)^{-1}X'_kX_k(X'_{T-1}X_{T-1})^{-1}x_T}{\sqrt{f_{k+1}}\sqrt{f_T}} - \frac{x'_{k+1}(X'_{T-1}X_{T-1})^{-1}x_T}{\sqrt{f_{k+1}}\sqrt{f_T}} = 0.$$

Similarly, one can show that the (t,1)-th element, the (t,T)-th element, the (T,1)-th element and the (T,t)-th element of CC' are all zero. This proves that $CC' = I_{T-k}$. Hence, w is linear in y, unbiased with mean zero, and

$$\text{var}(w) = \text{var}(Cu) = C\,E(uu')C' = \sigma^2 CC' = \sigma^2 I_{T-k},$$

i.e., the variance-covariance matrix is a scalar times an identity matrix. Using Theil's (1971) terminology, these recursive residuals are (LUS) linear unbiased with a scalar covariance matrix. A further result holds for all LUS residuals. In fact, $C'C = \bar{P}_x$ see Theil (1971, p. 208). Hence,

$$\sum_{t=k+1}^{T} w_t^2 = w'w = y'C'Cy = y'\bar{P}_x y = e'e = \sum_{t=1}^{T} e_t^2$$

Therefore, the sum of squares of the (T-k) recursive residuals is equal to the sum of squares of the T least squares residuals.

c. If $u \sim N(0, \sigma^2 I_T)$ then $w = Cu$ is also Normal since it is a linear function of normal random variables. From part (b) we proved that w has mean zero and variance $\sigma^2 I_{T-k}$. Hence, $w \sim N(0, \sigma^2 I_{T-k})$ as required.

d. Let us express the (t+1) residuals as follows:

$$Y_{t+1} - X_{t+1}\hat{\beta}_{t+1} = Y_{t+1} - X_{t+1}\hat{\beta}_t - X_{t+1}(\hat{\beta}_{t+1} - \hat{\beta}_t)$$

so that

$$\begin{aligned}
RSS_{t+1} &= (Y_{t+1} - X_{t+1}\hat{\beta}_{t+1})'(Y_{t+1} - X_{t+1}\hat{\beta}_{t+1}) \\
&= (Y_{t+1} - X_{t+1}\hat{\beta}_t)'(Y_{t+1} - X_{t+1}\hat{\beta}_t) + (\hat{\beta}_{t+1} - \hat{\beta}_t)'X'_{t+1}X_{t+1}(\hat{\beta}_{t+1} - \hat{\beta}_t) \\
&\quad - 2(\hat{\beta}_{t+1} - \hat{\beta}_t)'X'_{t+1}(Y_{t+1} - X_{t+1}\hat{\beta}_t)
\end{aligned}$$

Partition $Y_{t+1} = \begin{pmatrix} Y_t \\ y_{t+1} \end{pmatrix}$ and $X_{t+1} = \begin{pmatrix} X_t \\ x'_{t+1} \end{pmatrix}$ so that the first term of RSS_{t+1}

becomes $(Y_{t+1} - X_{t+1}\hat{\beta}_t)'(Y_{t+1} - X_{t+1}\hat{\beta}_t)$

$$= [(Y_t - X_t\hat{\beta}_t)', (y_{t+1} - x'_{t+1}\hat{\beta}_t)'] \begin{pmatrix} Y_t - X_t\hat{\beta}_t \\ y_{t+1} - x'_{t+1}\hat{\beta}_t \end{pmatrix}$$

$$= RSS_t + f_{t+1}\, w_{t+1}^2$$

where $f_{t+1} = 1 + x'_{t+1}(X'_t X_t)^{-1} x_{t+1}$ is defined in (8.32) and w_{t+1} is the recursive residual defined in (8.30). Next, we make the substitution for $(\hat{\beta}_{t+1} - \hat{\beta}_t)$ from (8.32) into the second term of RSS_{t+1} to get $(\hat{\beta}_{t+1} - \hat{\beta}_t)' X'_{t+1} X_{t+1} (\hat{\beta}_{t+1} - \hat{\beta}_t)$

$$= (y_{t+1} - x'_{t+1} \hat{\beta}_t)' x'_{t+1} (X'_t X_t)^{-1} (X'_{t+1} X_{t+1})(X'_t X_t)^{-1} x_{t+1}(y_{t+1} - x'_{t+1}\hat{\beta}_t)$$

Postmultiplying (8.31) by x'_{t+1} one gets

$$(X'_{t+1} X_{t+1})^{-1} x_{t+1} = (X'_t X_t)^{-1} x_{t+1} - \frac{(X'_t X_t)^{-1} x_{t+1}\, c}{1 + c}$$

where $c = x'_{t+1}(X'_t X_t)^{-1} x_{t+1}$. Collecting terms one gets

$$(X'_{t+1} X_{t+1})^{-1} x_{t+1} = (X'_t X_t)^{-1} x_{t+1} / f_{t+1}$$

where $f_{t+1} = 1 + c$.

Substituting this above, the second term of RSS_{t+1} reduces to

$$x'_{t+1}(X'_t X_t)^{-1} x_{t+1}\, w_{t+1}^2.$$

For the third term of RSS_{t+1}, we observe that $X'_{t+1}(Y_{t+1} - X'_{t+1}\hat{\beta}_t)$

$$= [X'_t, x_{t+1}] \begin{bmatrix} Y_t - X_t \hat{\beta}_t \\ y_{t+1} - x'_{t+1}\hat{\beta}_t \end{bmatrix} = X'_t(Y_t - X_t \hat{\beta}_t) + x_{t+1}(y_{t+1} - x'_{t+1}\hat{\beta}_t).$$

The first term is zero since X_t and its least squares residuals are orthogonal. Hence, $X'_{t+1}(Y_{t+1} - X_{t+1}\hat{\beta}_t) = x_{t+1} w_{t+1} \sqrt{f_{t+1}}$ and the third term of RSS_{t+1} becomes $-2(\hat{\beta}_{t+1} - \hat{\beta}_t)' x_{t+1} w_{t+1} \sqrt{f_{t+1}}$ using (8.32), this reduces to $-2 x'_{t+1}(X'_t X_t)^{-1} x_{t+1} w_{t+1}^2$. Adding all three terms yields

$$RSS_{t+1} = RSS_t + f_{t+1} w_{t+1}^2 + x'_{t+1}(X'_t X_t)^{-1} x_{t+1} w_{t+1}^2 - 2 x'_{t+1}(X'_t X_t)^{-1} x_{t+1} w_{t+1}^2$$

$$= RSS_t + f_{t+1} w_{t+1}^2 - (f_{t+1} - 1) w_{t+1}^2 = RSS_t - w_{t+1}^2 \text{ as required.}$$

8.15 *The Harvey and Collier (1977) Misspecification t-test as a Variable Additions Test.* This is based on Wu (1993).

a. The Chow F-test for H_o: $\gamma = 0$ in (8.44) is given by

$$F = \frac{\text{RRSS} - \text{URSS}}{\text{URSS}/(T-k-1)} = \frac{y'\bar{P}_X y - y'\bar{P}_{[X,z]} y}{y'\bar{P}_{[X,z]} y/(T-k-1)}.$$

Using the fact that $z = C'\iota_{T-k}$ where C is defined in (8.34) and ι_{T-k} is a vector of ones of dimension (T-k). From (8.35) we know that $CX = 0$, $CC' = I_{T-k}$ and $C'C = \bar{P}_X$. Hence, $z'X = \iota'_{T-k} CX = 0$ and

$$P_{[X,z]} = [X,z] \begin{bmatrix} X'X & 0 \\ 0 & z'z \end{bmatrix}^{-1} \begin{pmatrix} X' \\ z' \end{pmatrix} = P_X + P_z$$

Therefore, URSS $= y'\bar{P}_{[X,z]}y = y'y - y'P_X y - y'P_z y$ and RRSS $= y'\bar{P}_X y = y'y - y'P_X y$ and the F-statistic reduces to

$$F = \frac{y'P_z y}{y'(\bar{P}_X - P_z)y/(T-k-1)}$$

which is distributed as $F(1, T-k-1)$ under the null hypothesis $H_o; \gamma = 0$.

b. The numerator of the F-statistic is

$$y'P_z y = y'z(z'z)^{-1} z'y = y'C'\iota_{T-k}(\iota'_{T-k} CC'\iota_{T-k})^{-1} \iota'_{T-k} C'y$$

But $CC' = I_{T-k}$ and $\iota'_{T-k} CC'\iota_{T-k} = \iota'_{T-k} \iota_{T-k} = (T-k)$. Therefore $y'P_z y = y'C'\iota_{T-k}\iota'_{T-k}C'y/(T-k)$. The recursive residuals are constructed as $w = Cy$, see below (8.34). Hence

$$y'P_z y = w'\iota_{T-k}\iota'_{T-k}w/(T-k) = (\sum_{t=k+1}^{T} w_t)^2/(T-k) = (T-k)\bar{w}^2$$

where $\bar{w} = \sum_{t=k+1}^{T} w_t/(T-k)$. Using $\bar{P}_X = C'C$, we can write

$$y'\bar{P}_X y = y'C'Cy = w'w = \sum_{t=k+1}^{T} w_t^2$$

Hence, the denominator of the F-statistic is $y'(\bar{P}_X - P_z)y/(T-k-1)$

$$= [\sum_{t=k+1}^{T} w_t^2 - (T-k)\bar{w}^2]/(T-k-1) = \sum_{t=k+1}^{T} (w_t - \bar{w})^2/(T-k-1) = s_w^2$$

where $s_w^2 = \sum_{t=k+1}^{T} (w_t - \bar{w})^2/(T-k-1)$ was given below (8.43). Therefore, the F-statistic in part (a) is $F = (T-k)\bar{w}^2/s_w^2$ which is the square of the Harvey and Collier (1977) t_{T-k-1}-statistic given in (8.43).

8.16 The Gasoline data model given in Chapter 10.

a. Using Eviews, and the data for AUSTRIA one can generate the regression of y on X_1, X_2 and X_3 and the corresponding recursive residuals.

```
LS // Dependent Variable is Y
Date: 08/15/97   Time: 00:42
Sample: 1960 1978
Included observations: 19

Variable  Coefficient   Std. Error    t-Statistic    Prob.

C          3.726605     0.373018      9.990422       0.0000
X1         0.760721     0.211471      3.597289       0.0026
X2        -0.793199     0.150086     -5.284956       0.0001
X3        -0.519871     0.113130     -4.595336       0.0004

R-squared             0.733440    Mean dependent var      4.056487
Adjusted R-squared    0.680128    S.D. dependent var      0.069299
S.E. of regression    0.039194    Akaike info criterion  -6.293809
Sum squared resid     0.023042    Schwarz criterion      -6.094980
Log likelihood       36.83136     F-statistic            13.75753
Durbin-Watson stat    1.879634    Prob(F-statistic)       0.000140
```

```
                RECURSIVE RESIDUALS:

                       0.001047
                      -0.010798
                       0.042785
                       0.035901
                       0.024071
                      -0.009944
                      -0.007583
                       0.018521
                       0.006881
                       0.003415
                      -0.092971
                       0.013540
                      -0.069668
                       0.000706
                      -0.070614
```

b. EViews also gives as an option a plot of the recursive residuals plus or minus twice their standard error. Next, the CUSUM plot is given along with 5% upper and lower lines.

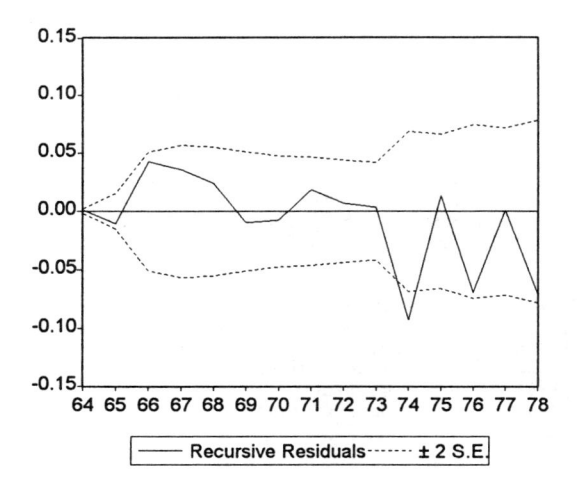

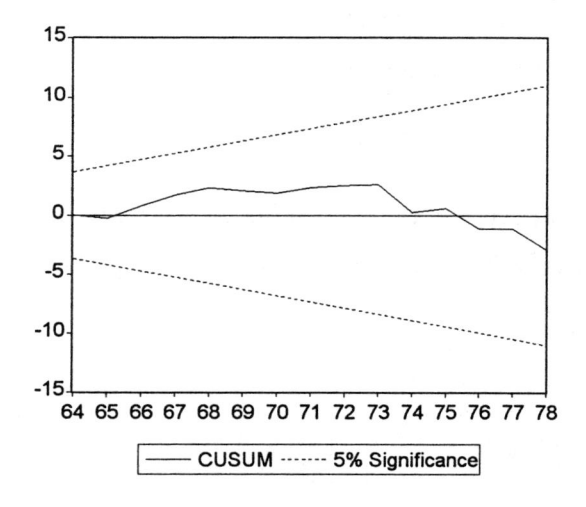

Chapter 8: Regression Diagnostics and Specification Tests

d. Chow Forecast Test: Forecast from 1978 to 1978

```
F-statistic                  3.866193    Probability        0.069418
Log likelihood ratio         4.633206    Probability        0.031359


Test Equation:
LS // Dependent Variable is Y
Sample: 1960 1977
Included observations: 18

Variable  Coefficient    Std. Error    t-Statistic    Prob.

C          4.000172      0.369022      10.83991      0.0000
X1         0.803752      0.194999       4.121830     0.0010
X2        -0.738174      0.140340      -5.259896     0.0001
X3        -0.522216      0.103666      -5.037483     0.0002

R-squared              0.732757    Mean dependent var        4.063917
Adjusted R-squared     0.675490    S.D. dependent var        0.063042
S.E. of regression     0.035913    Akaike info criterion    -6.460203
Sum squared resid      0.018056    Schwarz criterion        -6.262343
Log likelihood        36.60094     F-statistic              12.79557
Durbin-Watson stat     2.028488    Prob(F-statistic)         0.000268
```

8.17 *The Differencing Test in a Regression with Equicorrelated Disturbances.* This is based on Baltagi (1990).

a. For the equicorrelated case Ω can be written as $\Omega = \sigma^2(1-\rho)[E_T + \theta \bar{J}_T]$ where $E_T = I_T - \bar{J}_T$, $\bar{J}_T = J_T/T$ and $\theta = [1+(T-1)\rho]/(1-\rho)$.

$\Omega^{-1} = [E_T + (1/\theta)\bar{J}_T]/\sigma^2(1-\rho)$

$\iota'\Omega^{-1} = \iota'/\theta\sigma^2(1-\rho)$

and $L = E_T/\sigma^2(1-\rho)$. Hence $\hat{\beta} = (X'E_T X)^{-1} X'E_T Y$, which is the OLS estimator of β, since E_T is a matrix that transforms each variable into deviations from its mean. That GLS is equivalent to OLS for the equicorrelated case, is a standard result in the literature.

Also, $D\Omega = \sigma^2(1-\rho)D$ since $D\iota = 0$ and $D\Omega D' = \sigma^2(1-\rho)DD'$. Therefore, $M = P_{D'}/\sigma^2(1-\rho)$ where $P_{D'} = D'(DD')^{-1}D$. In order to show that $M = L$, it remains to show that $P_{D'} = E_T$ or equivalently that $P_{D'} + \bar{J}_T = I_T$ from the definition of E_T. Note that both $P_{D'}$ and $\bar{J}_T$ are symmetric idempotent matrices which are orthogonal to each other. ($D\bar{J}_T = 0$ since $D\iota = 0$). Using Graybill's

(1961) Theorem 1.69, these two properties of $P_{D'}$ and $\bar{J}_T$ imply a third: Their sum is idempotent with rank equal to the sum of the ranks. But rank of $P_{D'}$ is (T-1) and rank of $\bar{J}_T$ is 1, hence their sum is idempotent of rank T, which could only be I_T. This proves the Maeshiro and Wichers (1989) result, i.e., $\hat{\beta} = \tilde{\beta}$, which happens to be the OLS estimator from (1) in this particular case.

b. The Plosser, Schwert and White differencing test is based on the difference between the OLS estimator from (1) and the OLS estimator from (2). But OLS on (1) is equivalent to GLS on (1). Also, part (a) proved that GLS on (1) is in fact GLS from (2). Hence, the differencing test can be based upon the difference between the OLS and GLS estimators from the differenced equation. An alternative solution is given by Koning (1992).

8.20 For the Cigarette data used in Chapter 3, the following SAS output gives the OLS regression of logC on an intercept, logP and logY to obtain the predicted logC.

```
Dependent Variable: LNC
                         Analysis of Variance
                          Sum of         Mean
        Source    DF    Squares       Square      F Value    Prob>F
        Model      2    0.50098       0.25049       9.378    0.0004
        Error     43    1.14854       0.02671
        C Total   45    1.64953

            Root MSE       0.16343    R-square      0.3037
            Dep Mean       4.84784    Adj R-sq      0.2713
            C.V.           3.37125

                         Parameter Estimates
                    Parameter    Standard     T for H0:
    Variable  DF    Estimate        Error   Parameter=0    Prob > |T|
    INTERCEP   1    4.299662    0.90892571        4.730        0.0001
    LNP        1   -1.338335    0.32460147       -4.123        0.0002
    LNY        1    0.172386    0.19675440        0.876        0.3858

                         Covariance of Estimates
        COVB           INTERCEP             LNP                  LNY
        INTERCEP     0.8261459377    0.128504021         -0.178406868
        LNP          0.128504021     0.1053661138        -0.031443637
        LNY         -0.178406868    -0.031443637          0.0387122922
```

The second regression runs Ramsey's (1969) RESET including only LNCHAT2 = (predicted logC)^2. The coefficient of this last variable is 2.84 with a t-statistic of 1.79 which is statistically insignificant at the 5 percent level.

Ramsey(1969) Reset Test

Dependent Variable: LNC

Analysis of Variance

Source	DF	Sum of Squares	Mean Square	F Value	Prob>F
Model	3	0.58211	0.19404	7.635	0.0003
Error	42	1.06742	0.02541		
C Total	45	1.64953			

Root MSE	0.15942	R-square	0.3529	
Dep Mean	4.84784	Adj R-sq	0.3067	
C.V.	3.28847			

Parameter Estimates

Variable	DF	Parameter Estimate	Standard Error	T for H0: Parameter=0	Prob > |T|
INTERCEP	1	-47.170477	28.82222878	-1.637	0.1092
LNP	1	35.662967	20.71258952	1.722	0.0925
LNY	1	-4.634014	2.69704943	-1.718	0.0931
LNCHAT2	1	2.841844	1.59062152	1.787	0.0812

TEST: LNCHAT=0

Dependent Variable: LNC
Test: Numerator: 0.0811   DF: 1   F value: 3.1920
       Denominator: 0.025415 DF: 42   Prob>F: 0.0812

The last regression gives the Thursby and Schmidt (1977) version of the RESET including $(\log P)^2$ and $(\log Y)^2$ and testing their joint significance with an F-test. This yields a t-value of 2.51 which is distributed as $F_{2,41}$ under the null hypothesis. This has a p-value of 0.093 and therefore insignificant at the 5 percent level.

THRUSBY AND SCHMIDT(1977) RESET TEST

Dependent Variable: LNC

Analysis of Variance

Source	DF	Sum of Squares	Mean Square	F Value	Prob>F
Model	4	0.62642	0.15660	6.276	0.0005
Error	41	1.02311	0.02495		
C Total	45	1.64953			

Root MSE	0.15797	R-square	0.3798	
Dep Mean	4.84784	Adj R-sq	0.3192	
C.V.	3.25852			

Parameter Estimates

Variable	DF	Parameter Estimate	Standard Error	T for H0: Parameter=0	Prob > |T|
INTERCEP	1	-33.873221	22.24434005	-1.523	0.1355
LNP	1	-2.656477	0.99215915	-2.677	0.0106
LNY	1	16.147638	9.24970948	1.746	0.0883
LNP2	1	3.372056	2.42490696	1.391	0.1719
LNY2	1	-1.665396	0.96108270	-1.733	0.0906

TEST: LNP2=LNY2=0

Dependent Variable: LNC

Test:	Numerator:	0.0627	DF:	2	F value:	2.5133
	Denominator:	0.024954	DF:	41	Prob>F:	0.0934

Hence, by the two versions of the RESET there is no evidence of misspecificiation unless we relax our type I error to the 10 percent level.

SAS PROGRAM

```
Data CIGARETT;
Input OBS STATE $ LNC LNP LNY;
Cards;

Data CIGA; set CIGARETT;
LNP2=LNP**2;
LNY2=LNY**2;

******* RAMSEY(1969) RESET TEST *******;

Proc reg data=CIGA;
    Model LNC=LNP LNY/COB;
  Output out=OUT1 P=LNCHAT H=LVG;
TITLE 'RAMSEY(1969) RESET TEST';

Data RESET; set OUT1;
LNCHAT2=LNCHAT**2;

Proc reg data=RESET;
    Model LNC=LNP LNY LNC_HAT2;
    TEST LNCHAT2=0;

******* THRUSBY AND SCHMIDT(1977) RESET TEST *******;

Proc reg data=CIGA;
    Model LNC=LNP LNY LNP2 LNY2;
    TEST LNP2=LNY2=0;
TITLE 'THRUSBY AND SCHMIDT(1977) RESET TEST';
RUN;
```

CHAPTER 9
Generalized Least Squares

9.1

a. Equation (7.5) of Chapter 7 gives $\hat{\beta}_{ols} = \beta + (X'X)^{-1}X'u$ so that $E(\hat{\beta}_{ols}) = \beta$ as long as X and u are uncorrelated and u has zero mean. Also,

$$\text{var}(\hat{\beta}_{ols}) = E(\hat{\beta}_{ols} - \beta)(\hat{\beta}_{ols} - \beta)' = E[(X'X)^{-1}X'uu'X(X'X)^{-1}]$$

$$= (X'X)^{-1}X'\,E(uu')X(X'X)^{-1} = \sigma^2(X'X)^{-1}X'\Omega X(X'X)^{-1}.$$

b. $\text{var}(\hat{\beta}_{ols}) - \text{var}(\hat{\beta}_{GLS}) = \sigma^2[(X'X)^{-1}X'\Omega X(X'X)^{-1} - (X'\Omega^{-1}X)^{-1}]$

$$= \sigma^2[(X'X)^{-1}X'\Omega X(X'X)^{-1} - (X'\Omega^{-1}X)^{-1}X'\Omega^{-1}\Omega\Omega^{-1}X(X'\Omega^{-1}X)^{-1}]$$

$$= \sigma^2[(X'X)^{-1}X' - (X'\Omega^{-1}X)^{-1}X'\Omega^{-1}]\,\Omega\,[X(X'X)^{-1} - \Omega^{-1}X(X'\Omega^{-1}X)^{-1}]$$

$$= \sigma^2\,A\Omega A'$$

where $A = [(X'X)^{-1}X' - (X'\Omega^{-1}X)^{-1}X'\Omega^{-1}]$. The second equality post multiplies $(X'\Omega^{-1}X)^{-1}$ by $(X'\Omega^{-1}X)(X'\Omega^{-1}X)^{-1}$ which is an identity of dimension K. The third equality follows since the cross-product terms give $-2(X'\Omega^{-1}X)^{-1}$. The difference in variances is positive semi-definite since Ω is positive definite.

9.2

a. From Chapter 7, we know that $s^2 = e'e/n-K = u'\bar{P}_X u/(n-K)$ or $(n-K)s^2 = u'\bar{P}_X u$. Hence,

$$(n-K)E(s^2) = E(u'\bar{P}_X u) = E[tr(u'\bar{P}_X u)] = tr[E(uu')\bar{P}_X] = tr(\Sigma\bar{P}_X) = \sigma^2 tr(\Omega\bar{P}_X)$$

and $E(s^2) = \sigma^2 tr(\Omega\bar{P}_X)/(n-K)$ which in general is not equal to σ^2.

b. From part (a),

$$(n-K)E(s^2) = tr(\Sigma\bar{P}_X) = tr(\Sigma) - tr(\Sigma P_X)$$

but, both Σ and P_X are non-negative definite. Hence, $tr(\Sigma P_X) \geq 0$ and $(n-K)E(s^2) \leq tr(\Sigma)$

which upon rearranging yields $E(s^2) \leq tr(\Sigma)/(n-K)$. Also, Σ and $\bar{P}_X$ are non-negative definite. Hence, $tr(\Sigma \bar{P}_X) \geq 0$. Therefore, $E(s^2) \geq 0$. This proves the bound derived by Dufour (1986):

$$0 \leq E(s^2) \leq tr(\Sigma)/(n-K)$$

where $tr(\Sigma) = \sum_{i=1}^{n} \sigma_i^2$. Under homoskedasticity $\sigma_i^2 = \sigma^2$ for $i = 1,2,..,n$. Hence, $tr(\Sigma) = n\sigma^2$ and the upper bound becomes $n\sigma^2/(n-K)$. A useful bound for $E(s^2)$ has been derived by Sathe and Vinod (1974) and Neudecker (1977, 1978). This is given by

$0 \leq$ mean of n-K smallest characteristic roots of $\Sigma \leq E(s^2)$

$\leq$ mean of n-K largest characteristic roots of $\Sigma \leq tr(\Sigma)/(n-K)$.

c. Using $s^2 = u'\bar{P}_X u/(n-K) = u'u/(n-K) - u'P_X u/(n-K)$

we have plim $s^2 =$ plim $u'u/(n-K) -$ plim $u'P_X u/(n-K)$.

By assumption plim $u'u/n = \sigma^2$. Hence, the first term tend in plim to σ^2 as $n \to \infty$. The second term has expectation $\sigma^2 tr(P_X \Omega)/(n-K)$. But, $P_X \Omega$ has rank K and therefore exactly K non-zero characteristic roots each of which cannot exceed λ_{max}. This means that

$E[u'P_X u/(n-K)] \leq \sigma^2 K \lambda_{max}/(n-K)$.

Using the condition that $\lambda_{max}/n \to 0$ as $n \to \infty$ proves that

$\lim E[u'P_X u/(n-K)] \to 0$

as $n \to \infty$. Hence, plim $[u'P_X u/(n-K)] \to 0$ as $n \to \infty$ and plim $s^2 = \sigma^2$. Therefore, a sufficient condition for s^2 to be consistent for σ^2 irrespective of X is that $\lambda_{max}/n \to 0$ and plim $(u'u/n) = \sigma^2$ as $n \to \infty$, see Krämer and Berghoff (1991).

d. From (9.6), $s^{*2} = e^{*'}e^*/(n-K)$ where $e^* = y^* - X^*\hat{\beta}_{GLS} = y^* - X^*(X^{*'}X^*)^{-1}X^{*'}y^*$
$= \bar{P}_{X^*} y^*$ using (9.4), where $\bar{P}_{X^*} = I_n - P_{X^*}$ and $P_{X^*} = X^*(X^{*'}X^*)^{-1}X^{*'}$. Substituting y^* from (9.3), we get $e^* = \bar{P}_{X^*} u^*$ where $\bar{P}_{X^*} X^* = 0$. Hence, $(n-K)s^{*2} = e^{*'}e^* = u^{*'}\bar{P}_{X^*} u^*$ with

$(n-K)E(s^{*2}) = E(u^{*'}\bar{P}_{X^*} u^*) = E[tr(u^* u^{*'} \bar{P}_{X^*})]$

Chapter 9: Generalized Least Squares

$$= \text{tr}[E(u^*u^{*\prime})\bar{P}_{X^*}] = \text{tr}(\sigma^2 \bar{P}_{X^*}) = \sigma^2(n-K)$$

from the fact that $\text{var}(u^*) = \sigma^2 I_n$. Hence, $E(s^{*2}) = \sigma^2$ and s^{*2} is unbiased for σ^2.

9.3 *The AR(1) Model.*

a. From (9.9) and (9.10) we get

$$\Omega\Omega^{-1} = \left[\frac{1}{1-\rho^2}\right] \begin{bmatrix} 1 & \rho & \rho^2 & \cdots & \rho^{T-1} \\ \rho & 1 & \rho & \cdots & \rho^{T-2} \\ \vdots & \vdots & \vdots & & \vdots \\ \rho^{T-1} & \rho^{T-2} & \rho^{T-3} & \cdots & 1 \end{bmatrix} \begin{bmatrix} 1 & -\rho & 0 & \cdots & 0 & 0 & 0 \\ -\rho & 1+\rho^2 & -\rho & \cdots & 0 & 0 & 0 \\ \vdots & \vdots & \vdots & & \vdots & \vdots & \vdots \\ 0 & 0 & 0 & \cdots & -\rho & 1+\rho^2 & -\rho \\ 0 & 0 & 0 & \cdots & 0 & -\rho & 1 \end{bmatrix}$$

$$= \left[\frac{1}{1-\rho^2}\right] \begin{bmatrix} (1-\rho^2) & 0 & 0 & \cdots & 0 \\ 0 & (1-\rho^2) & 0 & \cdots & 0 \\ \vdots & \vdots & \vdots & & \vdots \\ 0 & 0 & 0 & \cdots & (1-\rho^2) \end{bmatrix} = I_T$$

The multiplication is tedious but simple. The (1,1) element automatically gives $(1-\rho^2)$. The (1,2) element gives $-\rho + \rho(1+\rho^2) - \rho\rho^2 = -\rho + \rho + \rho^3 - \rho^3 = 0$. The (2,2) element gives $-\rho^2 + (1+\rho^2) - \rho\rho = 1 - \rho^2$ and so on.

b. For P^{-1} defined in (9.11), we get

$$P^{-1\prime} P^{-1} = \begin{bmatrix} \sqrt{1-\rho^2} & -\rho & 0 & \cdots & 0 & 0 \\ 0 & 1 & -\rho & \cdots & 0 & 0 \\ 0 & 0 & 1 & \cdots & 0 & 0 \\ \vdots & \vdots & \vdots & & \vdots & \vdots \\ 0 & 0 & 0 & \cdots & 1 & -\rho \\ 0 & 0 & 0 & \cdots & 0 & 1 \end{bmatrix} \begin{bmatrix} \sqrt{1-\rho^2} & 0 & 0 & \cdots & 0 & 0 & 0 \\ -\rho & 1 & 0 & \cdots & 0 & 0 & 0 \\ 0 & -\rho & 1 & \cdots & 0 & 0 & 0 \\ \vdots & \vdots & \vdots & & \vdots & \vdots & \vdots \\ 0 & 0 & 0 & \cdots & -\rho & 1 & 0 \\ 0 & 0 & 0 & \cdots & 0 & -\rho & 1 \end{bmatrix}$$

$$= \begin{bmatrix} 1 & -\rho & 0 & \cdots & 0 & 0 & 0 \\ -\rho & 1+\rho^2 & -\rho & \cdots & 0 & 0 & 0 \\ \vdots & \vdots & \vdots & & \vdots & \vdots & \vdots \\ 0 & 0 & 0 & \cdots & -\rho & 1+\rho^2 & -\rho \\ 0 & 0 & 0 & \cdots & 0 & -\rho & 1 \end{bmatrix} = (1-\rho^2)\Omega^{-1}$$

Again, the multiplication is simple but tedious. The (1,1) element gives

$$\sqrt{1-\rho^2}\sqrt{1-\rho^2} - \rho(-\rho) = (1-\rho^2) + \rho^2 = 1,$$

the (1,2) element gives $\sqrt{1-\rho^2}.0 - \rho.1 = -\rho$, the (2,2) element gives

$1 - \rho(-\rho) = 1+\rho^2$ and so on.

c. From part (b) we verified that $P^{-1'}P^{-1} = (1-\rho^2)\Omega^{-1}$. Hence, $\Omega/(1-\rho^2) = PP'$ or $\Omega = (1-\rho^2)PP'$. Therefore,

$$\text{var}(P^{-1}u) = P^{-1}\text{var}(u)P^{-1'} = \sigma_u^2 P^{-1}\Omega P^{-1'} = \sigma_u^2(1-\rho^2)P^{-1}PP'P^{-1'} = \sigma_\epsilon^2 I_T$$

since $\sigma_u^2 = \sigma_\epsilon^2/(1-\rho^2)$.

9.4 *Restricted GLS.* From Chapter 7, restricted least squares is given by $\hat{\beta}_{rls} = \hat{\beta}_{ols} + (X'X)^{-1}R'[R(X'X)^{-1}R']^{-1}(r-R\hat{\beta}_{ols})$. Applying the same analysis to the transformed model in (9.3) we get that $\hat{\beta}_{ols}^* = (X^{*'}X^*)^{-1}X^{*'}y^* = \hat{\beta}_{GLS}$.
From (9.4) and the above restricted estimator, we get

$$\hat{\beta}_{RGLS} = \hat{\beta}_{GLS} + (X^{*'}X^*)^{-1}R'[R(X^{*'}X^*)^{-1}R']^{-1}(r-R\hat{\beta}_{GLS})$$

where X^* now replaces X. But $X^{*'}X^* = X'\Omega^{-1}X$, hence,

$$\hat{\beta}_{RGLS} = \hat{\beta}_{GLS} + (X'\Omega^{-1}X)^{-1}R'[R(X'\Omega^{-1}X)^{-1}R']^{-1}(r-R\hat{\beta}_{GLS}).$$

9.5 *Best Linear Unbiased Prediction.* This is based on Goldberger (1962).

a. Consider linear predictors of the scalar y_{T+s} given by $\hat{y}_{T+s} = c'y$. From (9.1) we get $\hat{y}_{T+s} = c'X\beta + c'u$ and using the fact that $y_{T+s} = x'_{T+s}\beta + u_{T+s}$, we get

$$\hat{y}_{T+s} - y_{T+s} = (c'X - x'_{T+s})\beta + c'u - u_{T+s}.$$

The unbiased condition is given by $E(\hat{y}_{T+s} - y_{T+s}) = 0$. Since $E(u) = 0$ and $E(u_{T+s}) = 0$, this requires that $c'X = x'_{T+s}$ for this to hold for every β. Therefore, an unbiased predictor will have prediction error

$$\hat{y}_{T+s} - y_{T+s} = c'u - u_{T+s}.$$

Chapter 9: Generalized Least Squares

b. The prediction variance is given by

$$\text{var}(\hat{y}_{T+s}) = E(\hat{y}_{T+s} - y_{T+s})(\hat{y}_{T+s} - y_{T+s})' = E(c'u - u_{T+s})(c'u - u_{T+s})'$$

$$= c'E(uu')c + \text{var}(u_{T+s}) - 2c'E(u_{T+s}u) = c'\Sigma c + \sigma^2_{T+s} - 2c'\omega$$

using the definitions $\sigma^2_{T+s} = \text{var}(u_{T+s})$ and $\omega = E(u_{T+s}u)$.

c. Minimizing $\text{var}(\hat{y}_{T+s})$ subject to $c'X = x'_{T+s}$ sets up the following Lagrangian function

$$\psi(c,\lambda) = c'\Sigma c - 2c'\omega - 2\lambda'(X'c - x_{T+s})$$

where σ^2_{T+s} is fixed and where λ denotes the Kx1 vector of Lagrangian multipliers. The first order conditions of ψ with respect to c and λ yield

$$\partial\psi/\partial c = 2\Sigma c - 2\omega - 2X\lambda = 0 \quad \text{and} \quad \partial\psi/\partial\lambda = 2X'c - 2x_{T+s} = 0.$$

In matrix form, these two equations become

$$\begin{bmatrix} \Sigma & X \\ X' & 0 \end{bmatrix} \begin{pmatrix} \hat{c} \\ -\hat{\lambda} \end{pmatrix} = \begin{pmatrix} \omega \\ x_{T+s} \end{pmatrix}.$$

Using partitioned inverse matrix formulas one gets

$$\begin{pmatrix} \hat{c} \\ -\hat{\lambda} \end{pmatrix} = \begin{bmatrix} \Sigma^{-1}[I_T - X(X'\Sigma^{-1}X)^{-1}X'\Sigma^{-1}] & \Sigma^{-1}X(X'\Sigma^{-1}X)^{-1} \\ (X'\Sigma^{-1}X)^{-1}X'\Sigma^{-1} & -(X'\Sigma^{-1}X)^{-1} \end{bmatrix} \begin{pmatrix} \omega \\ x_{T+s} \end{pmatrix}$$

so that

$$\hat{c} = \Sigma^{-1}X(X'\Sigma^{-1}X)^{-1}x_{T+s} + \Sigma^{-1}[I_T - X(X'\Sigma^{-1}X)^{-1}X'\Sigma^{-1}]\omega.$$

Therefore, the BLUP is given by

$$\hat{y}_{T+s} = \hat{c}'y = x'_{T+s}(X'\Sigma^{-1}X)^{-1}X'\Sigma^{-1}y + \omega'\Sigma^{-1}y - \omega'\Sigma^{-1}X(X'\Sigma^{-1}X)^{-1}X'\Sigma^{-1}y$$

$$= x'_{T+s}\hat{\beta}_{GLS} + \omega'\Sigma^{-1}y - \omega'\Sigma^{-1}X\hat{\beta}_{GLS}$$

$$= x'_{T+s}\hat{\beta}_{GLS} + \omega'\Sigma^{-1}(y - X\hat{\beta}_{GLS})$$

$$= x'_{T+s}\hat{\beta}_{GLS} + \omega'\Sigma^{-1}e_{GLS}$$

where $e_{GLS} = y - X\hat{\beta}_{GLS}$. For $\Sigma = \sigma^2\Omega$, this can also be written as

$\hat{y}_{T+s} = x'_{T+s} \hat{\beta}_{GLS} + \omega'\Omega^{-1}e_{GLS}/\sigma^2.$

d. For the stationary AR(1) case

$u_t = \rho u_{t-1} + \epsilon_t$ with $\epsilon_t \sim IID(0,\sigma_\epsilon^2)$

$|\rho| < 1$ and $var(u_t) = \sigma_u^2 = \sigma_\epsilon^2/(1-\rho^2)$. In this case, $cov(u_t, u_{t-s}) = \rho^s \sigma_u^2$. Therefore, for s periods ahead forecast, we get

$$\omega = E(u_{T+s}u) = \begin{pmatrix} E(u_{T+s}u_1) \\ E(u_{T+s}u_2) \\ \vdots \\ E(u_{T+s}u_T) \end{pmatrix} = \sigma_u^2 \begin{pmatrix} \rho^{T+s-1} \\ \rho^{T+s-2} \\ \vdots \\ \rho^s \end{pmatrix}.$$

From Ω given in (9.9) we can deduce that $\omega = \rho^s \sigma_u^2$ (last column of Ω). But, $\Omega^{-1}\Omega = I_T$. Hence, Ω^{-1} (last column of Ω) = (last column of I_T) = $(0, 0, .., 1)'$. Substituting for the last column of Ω the expression $(\omega/\rho^s \sigma_u^2)$ yields

$\Omega^{-1} \omega/\rho^s \sigma_u^2 = (0,0,..,1)'$

which can be transposed and rewritten as

$\omega'\Omega^{-1}/\sigma_u^2 = \rho^s(0,0..,1).$

Substituting this expression in the BLUP for y_{T+s} in part (c) we get

$\hat{y}_{T+s} = x'_{T+s} \hat{\beta}_{GLS} + \omega'\Omega^{-1}e_{GLS}/\sigma = x'_{T+s}\hat{\beta}_{GLS} + \rho^s(0,0,..,1)e_{GLS}$

$= x'_{T+s} \hat{\beta}_{GLS} + \rho^s e_{T,GLS}$

where $e_{T,GLS}$ is the T-th GLS residual. For $T = 1$, this gives

$\hat{y}_{T+s} = x'_{T+s} \hat{\beta}_{GLS} + \rho e_{T,GLS}$

as shown in the text.

9.6 *The W, LR and LM Inequality.* From equation (9.27), the Wald statistic W can be interpreted as a LR statistic conditional on $\hat{\Sigma}$, the unrestricted MLE of Σ, i.e.,

$W = -2\log[\max_{R\beta=r} L(\beta/\hat{\Sigma})/\max_{\beta} L(\beta/\hat{\Sigma})]$. But, from (9.34), we know that the likelihood ratio statistic $LR = -2\log[\max_{R\beta=r,\Sigma} L(\beta,\Sigma)/\max_{\beta,\Sigma} L(\beta,\Sigma)]$. Using (9.33),

$\max_{R\beta=r} L(\beta/\hat{\Sigma}) \leq \max_{R\beta=r,\Sigma} L(\beta,\Sigma).$

Chapter 9: Generalized Least Squares   195

The right hand side term is an unconditional maximum over all Σ whereas the left hand side is a conditional maximum based on $\hat{\Sigma}$ under the null hypothesis H_o; $R\beta = r$. Also, from (9.32) $\max_{\beta,\Sigma} L(\beta,\Sigma) = \max_{\beta} L(\beta/\hat{\Sigma})$. Therefore, $W \geq LR$.

Similarly, from equation (9.31), the Lagrange Multiplier statistic can be interpreted as a LR statistic conditional on $\tilde{\Sigma}$, the restricted maximum likelihood of Σ, i.e., $LM = -2\log[\max_{R\beta=r} L(\beta/\tilde{\Sigma})/\max_{\beta} L(\beta/\tilde{\Sigma})]$.   Using   (9.33),

$\max_{R\beta=r} L(\beta/\tilde{\Sigma}) = \max_{R\beta=r,\Sigma} L(\beta,\Sigma)$ and from (9.32), we get $\max_{\beta} L(\beta/\tilde{\Sigma}) \leq \max_{\beta,\Sigma} L(\beta,\Sigma)$

because the latter is an unconditional maximum over all Σ. Hence, $LR \geq LM$. Therefore, $W \geq LR \geq LM$.

9.7 The W, LR and LM for this simple regression with H_o; $\beta = 0$ were derived in problem 7.16 in Chapter 7. Here, we follow the alternative derivation proposed by Breusch (1979) and considered in problem 9.6. From (9.34), the LR is given by $LR = -2\log[L(\tilde{\alpha}, \tilde{\beta}=0, \tilde{\sigma}^2)/L(\hat{\alpha}_{mle}, \hat{\beta}_{mle}, \hat{\sigma}^2_{mle})]$
where $\tilde{\alpha} = \bar{y}$, $\tilde{\beta} = 0$, $\tilde{\sigma}^2 = \sum_{i=1}^{n}(y_i-\bar{y})^2/n$ and
$\hat{\alpha}_{mle} = \hat{\alpha}_{ols} = \bar{y} - \hat{\beta}_{ols}\bar{X}$,
$\hat{\beta}_{mle} = \hat{\beta}_{ols} = \sum_{i=1}^{n} x_i y_i / \sum_{i=1}^{n} x_i^2$, $\hat{\sigma}^2_{mle} = \sum_{i=1}^{n} e_i^2/n$
and $e_i = y_i - \hat{\alpha}_{ols} - \hat{\beta}_{ols} X_i$,   see the solution to problem 7.16. But,
$\log L(\alpha,\beta,\sigma^2) = -\frac{n}{2}\log 2\pi - \frac{n}{2}\log\sigma^2 - \sum_{i=1}^{n}(y_i-\alpha-\beta X_i)^2/2\sigma^2$.

Therefore,
$\log L(\tilde{\alpha}, \tilde{\beta}=0, \tilde{\sigma}^2) = -\frac{n}{2}\log 2\pi - \frac{n}{2}\log\tilde{\sigma}^2 - \sum_{i=1}^{n}(y_i-\bar{y})^2/2\tilde{\sigma}^2$

$= -\frac{n}{2}\log 2\pi - \frac{n}{2}\log\tilde{\sigma}^2 - \frac{n}{2}$

and

$\log L(\hat{\alpha}_{mle}, \hat{\beta}_{mle}, \hat{\sigma}^2_{mle}) = -\frac{n}{2}\log 2\pi - \frac{n}{2}\log\hat{\sigma}^2_{mle} - \sum_{i=1}^{n} e_i^2/2\hat{\sigma}^2_{mle}$

$= -\frac{n}{2}\log 2\pi - \frac{n}{2}\log\hat{\sigma}^2_{mle} - \frac{n}{2}$.

Therefore,

$$LR = -2[-\frac{n}{2}\log\tilde{\sigma}^2 + \frac{n}{2}\log\hat{\sigma}^2_{mle}] = n\log(\tilde{\sigma}^2/\hat{\sigma}^2_{mle}) = n\log(TSS/RSS) = n\log(1/1-R^2)$$

where TSS = total sum of squares, and RSS = residual sum of squares for the simple regression. Of course, $R^2 = 1 - (RSS/TSS)$.

Similarly, from (9.31) we have

$$LM = -2\log[\max_{\alpha\beta=0} L(\alpha,\beta/\tilde{\sigma}^2)/\max_{\alpha,\beta} L(\alpha,\beta/\tilde{\sigma}^2)].$$

But, maximization of $L(\alpha,\beta/\tilde{\sigma}^2)$ gives $\hat{\alpha}_{ols}$ and $\hat{\beta}_{ols}$. Therefore,

$$\max_{\alpha,\beta} L(\alpha,\beta/\tilde{\sigma}^2) = L(\hat{\alpha},\hat{\beta},\tilde{\sigma}^2)$$

with

$$\log L(\hat{\alpha},\hat{\beta},\tilde{\sigma}^2) = -\frac{n}{2}\log 2\pi - \frac{n}{2}\log\tilde{\sigma}^2 - \sum_{i=1}^{n} e_i^2/2\tilde{\sigma}^2.$$

Also, restricted maximization of $L(\alpha,\beta/\tilde{\sigma}^2)$ under $H_o; \beta = 0$ gives $\tilde{\alpha} = \bar{y}$ and $\tilde{\beta} = 0$. Therefore, $\max_{\alpha\beta=0} L(\alpha,\beta/\tilde{\sigma}^2) = L(\tilde{\alpha},\tilde{\beta},\tilde{\sigma}^2)$. From this, we conclude that

$$LM = -2[-\frac{n}{2} + \frac{\sum_{i=1}^{n} e_i^2}{2\tilde{\sigma}^2}] = n - (\sum_{i=1}^{n} e_i^2/\tilde{\sigma}^2) = n - (n\sum_{i=1}^{n} e_i^2/\sum_{i=1}^{n} y_i^2) = n[1-(RSS/TSS)] = nR^2.$$

Finally, from (9.27), we have $W = 2\log[\max_{\alpha\beta=0} L(\alpha,\beta/\hat{\sigma}^2_{mle})/\max_{\alpha,\beta} L(\alpha,\beta/\hat{\sigma}^2_{mle})]$.

The maximization of $L(\alpha,\beta/\hat{\sigma}^2)$ gives $\hat{\alpha}_{ols}$ and $\hat{\beta}_{ols}$. Therefore,

$$\max_{\alpha,\beta} L(\alpha,\beta/\hat{\sigma}^2) = L(\hat{\alpha},\hat{\beta},\hat{\sigma}^2).$$

Also, restricted maximization of $L(\alpha,\beta/\hat{\sigma}^2)$ under $\beta = 0$ gives $\tilde{\alpha} = \bar{y}$ and $\tilde{\beta} = 0$.

Therefore, $\max_{\alpha\beta=0} L(\alpha,\beta/\tilde{\sigma}^2) = L(\tilde{\alpha},\tilde{\beta}=0,\hat{\sigma}^2)$ with

$$\log L(\tilde{\alpha},\tilde{\beta}=0,\hat{\sigma}^2) = -\frac{n}{2}\log 2\pi - \frac{n}{2}\log\hat{\sigma}^2 - \sum_{i=1}^{n}(y_i-\bar{y})^2/2\hat{\sigma}^2_{mle}.$$

Therefore,

$$W = -2[-\frac{\sum_{i=1}^{n}(y_i-\bar{y})^2}{2\hat{\sigma}^2_{mle}} + \frac{n}{2}] = \frac{TSS}{\hat{\sigma}^2_{mle}} - n = \frac{nTSS}{RSS} - n = n(\frac{TSS-RSS}{RSS}) = n(\frac{R^2}{1-R^2}).$$

This is exactly what we got in problem 7.16, but now from Breusch's (1979) alternative derivation. From problem 9.6, we infer using this LR interpretation of all three statistics that $W \geq LR \geq LM$.

9.8 *Sampling Distributions and Efficiency Comparison of OLS and GLS*. This is based on Baltagi (1992).

a. From the model it is clear that $\sum_{t=1}^{2} x_t^2 = 5$, $y_1 = 2 + u_1$, $y_2 = 4 + u_2$, and

$$\hat{\beta}_{ols} = \frac{\sum_{t=1}^{2} x_t y_t}{\sum_{t=1}^{2} x_t^2} = \beta + \frac{\sum_{t=1}^{2} x_t u_t}{\sum_{t=1}^{2} x_t^2} = 2 + 0.2u_1 + 0.4u_2$$

Let $u' = (u_1, u_2)$, then it is easy to verify that $E(u) = 0$ and

$$\Omega = \text{var}(u) = \begin{bmatrix} 1 & -1 \\ -1 & 4 \end{bmatrix}.$$

The disturbances have zero mean, are heteroskedastic and serially correlated with a correlation coefficient $\rho = -0.5$.

b. Using the joint probability function $P(u_1, u_2)$ and $\hat{\beta}_{ols}$ from part (a), one gets

$\hat{\beta}_{ols}$	Probability
1	1/8
1.4	3/8
2.6	3/8
3	1/8

Therefore, $E(\hat{\beta}_{ols}) = \beta = 2$ and $\text{var}(\hat{\beta}_{ols}) = 0.52$. These results can be also verified from $\hat{\beta}_{ols} = 2 + 0.2u_1 + 0.4u_2$. In fact, $E(\hat{\beta}_{ols}) = 2$ since $E(u_1) = E(u_2) = 0$ and

$\text{var}(\hat{\beta}_{ols}) = 0.04\,\text{var}(u_1) + 0.16\,\text{var}(u_2) + 0.16\,\text{cov}(u_1, u_2)$

$= 0.04 + 0.64 - 0.16 = 0.52$.

Also,

$\Omega^{-1} = \frac{1}{3}\begin{bmatrix} 4 & 1 \\ 1 & 1 \end{bmatrix}$ and $(x'\Omega^{-1}x)^{-1} = 1/4 = \text{var}(\hat{\beta}_{GLS})$

In fact, $\tilde{\beta}_{GLS} = (x'\Omega^{-1}x)^{-1}x'\Omega^{-1}y = 1/4(2y_1+y_2)$ which can be rewritten as $\tilde{\beta}_{GLS} = 2 + 1/4 [2u_1+u_2]$

Using $P(u_1,u_2)$ and this equation for $\tilde{\beta}_{GLS}$, one gets

$\tilde{\beta}_{GLS}$	Probability
1	1/8
2	3/4
3	1/8

Therefore, $E(\tilde{\beta}_{GLS}) = \beta = 2$ and $var(\tilde{\beta}_{GLS}) = 0.25$. This can also be verified from $\tilde{\beta}_{GLS} = 2 + 1/4 [2u_1+u_2]$. In fact, $E(\tilde{\beta}_{GLS}) = 2$ since $E(u_1) = E(u_2) = 0$ and

$$var(\tilde{\beta}_{GLS}) = \frac{1}{16}[4var(u_1) + var(u_2) + 4cov(u_1,u_2)] = \frac{1}{16}[4 + 4 - 4] = \frac{1}{4}.$$

This variance is approximately 48% of the variance of the OLS estimator.

c. The OLS predictions are given by $\hat{y}_t = \hat{\beta}_{ols}x_t$ which means that $\hat{y}_1 = \hat{\beta}_{ols}$ and $\hat{y}_2 = 2\hat{\beta}_{ols}$. The OLS residuals are given by $\hat{e}_t = y_t - \hat{y}_t$ and their probability function is given by

$(\hat{e}_1, \hat{e}_2)$	Probability
(0, 0)	1/4
(1.6, -0.8)	3/8
(-1.6, 0.8)	3/8

Now the estimated $var(\hat{\beta}_{ols}) = S(\hat{\beta}_{ols}) = \dfrac{s^2}{\sum\limits_{t=1}^{2} x_t^2} = \dfrac{\Sigma\hat{e}_t^2}{5}$, and this has probability function

Chapter 9: Generalized Least Squares

$S(\hat{\beta}_{ols})$	Probability
0	1/4
0.64	3/4

with $E[S(\hat{\beta}_{ols})] = 0.48 \neq var(\hat{\beta}_{ols}) = 0.52$.

Similarly, the GLS predictions are given by $\tilde{y}_t = \tilde{\beta}_{GLS} x_t$ which means that $\tilde{y}_1 = \tilde{\beta}_{GLS}$ and $\tilde{y}_2 = 2\tilde{\beta}_{GLS}$. The GLS residuals are given by $\tilde{e}_t = y_t - \tilde{y}_t$ and their probability function is given by

$(\tilde{e}_1, \tilde{e}_2)$	Probability
(0, 0)	1/4
(1, -2)	3/8
(-1, 2)	3/8

The MSE of the GLS regression is given by

$$\tilde{s}^2 = \tilde{e}'\Omega^{-1}\tilde{e} = 1/3\,[4\tilde{e}_1^2 + 2\tilde{e}_1\tilde{e}_2 + \tilde{e}_2^2]$$

and this has a probability function

$\tilde{s}^2$	Probability
0	1/4
4/3	3/4

with $E(\tilde{s}^2) = 1$. An alternative solution of this problem is given by Im and Snow (1993).

9.9 Equi-correlation.

a. For the regression with equi-correlated disturbances, OLS is equivalent to GLS as long as there is a constant in the regression model. Note that

$$\Omega = \begin{bmatrix} 1 & \rho & \rho & \cdots & \rho \\ \rho & 1 & \rho & \cdots & \rho \\ \vdots & \vdots & \vdots & & \vdots \\ \rho & \rho & \rho & \cdots & 1 \end{bmatrix}$$

so that u_t is homoskedastic and has constant serial correlation. In fact, correlation $(u_t, u_{t-s}) = \rho$ for $t \neq s$. Therefore, this is called equi-correlated. Zyskind's (1967) condition given in (9.8) yields

$P_X \Omega = \Omega P_X$.

In this case,

$P_X \Omega = (1-\rho) P_X + \rho P_X \iota_T \iota_T'$

and

$\Omega P_X = (1-\rho) P_X + \rho \iota_T \iota_T' P_X$.

But, we know that X contains a constant, i.e., a column of ones denoted by ι_T. Therefore, using $P_X X = X$ we get $P_X \iota_T = \iota_T$ since ι_T is a column of X. Substituting this in $P_X \Omega$ we get

$P_X \Omega = (1-\rho) P_X + \rho \iota_T \iota_T'$.

Similarly, substituting $\iota_T' P_X = \iota_T'$ in ΩP_X we get $\Omega P_X = (1-\rho) P_X + \rho \iota_T \iota_T'$. Hence, $\Omega P_X = P_X \Omega$ and OLS is equivalent to GLS for this model.

b. We know that $(T-K)s^2 = u' \bar{P}_X u$, see Chapter 7. Also that

$E(u' \bar{P}_X u) = E[\text{tr}(uu' \bar{P}_X)] = \text{tr}[E(uu' \bar{P}_X)] = \text{tr}(\sigma^2 \Omega \bar{P}_X)$

$\qquad = \sigma^2 \text{tr}[(1-\rho) \bar{P}_X + \rho \iota_T \iota_T' \bar{P}_X] = \sigma^2 (1-\rho) \text{tr}(\bar{P}_X)$

since $\iota_T' \bar{P}_X = \iota_T' - \iota_T' \bar{P}_X = \iota_T' - \iota_T' = 0$ see part (a). But, $\text{tr}(\bar{P}_X) = T-K$, hence, $E(u' \bar{P}_X u) = \sigma^2 (1-\rho)(T-K)$ and $E(s^2) = \sigma^2 (1-\rho)$.

Now for Ω to be positive semi-definite, it should be true that for every arbitrary non-zero vector a we have $a' \Omega a \geq 0$. In particular, for $a = \iota_T$, we get

$\iota_T' \Omega \iota_T = (1-\rho) \iota_T' \iota_T + \rho \iota_T' \iota_T \iota_T' \iota_T = T(1-\rho) + T^2 \rho$.

Chapter 9: Generalized Least Squares

This should be non-negative for every ρ. Hence, $(T^2-T)\rho + T \geq 0$ which gives $\rho \geq -1/(T-1)$. But, we know that $|\rho| \leq 1$. Hence, $-1/(T-1) \leq \rho \leq 1$ as required. This means that $0 \leq E(s^2) \leq [T/(T-1)]\sigma^2$ where the lower and upper bounds for $E(s^2)$ are attained at $\rho = 1$ and $\rho = -1/(T-1)$, respectively. These bounds were derived by Dufour (1986).

9.10

a. The model can be written in vector form as: $y = \alpha\iota_n + u$

where $y' = (y_1,..,y_n)$, ι_n is a vector of ones of dimension n, and $u' = (u_1,..,u_n)$.

Therefore, $\hat{\alpha}_{ols} = (\iota_n'\iota_n)^{-1} \iota_n'y = \sum_{i=1}^{n} y_i/n = \bar{y}$ and

$$\Sigma = \text{var}(u) = E(uu') = \sigma^2 \begin{bmatrix} 1 & \rho & .. & \rho \\ \rho & 1 & .. & \rho \\ \vdots & \vdots & .. & \vdots \\ \rho & \rho & .. & 1 \end{bmatrix} = \sigma^2[(1-\rho)I_n + \rho J_n]$$

where I_n is an identity matrix of dimension n and J_n is a matrix of ones of dimension n. Define $E_n = I_n - \bar{J}_n$ where $\bar{J}_n = J_n/n$, one can rewrite Σ as $\Sigma = \sigma^2[(1-\rho)E_n + (1+\rho(n-1))\bar{J}_n] = \sigma^2\Omega$ with

$$\Omega^{-1} = [\frac{1}{1-\rho}E_n + \frac{1}{1+\rho(n-1)}\bar{J}_n]$$

Therefore,

$$\hat{\alpha}_{GLS} = (\iota_n'\Sigma^{-1}\iota_n)^{-1} \iota_n'\Sigma^{-1}y = (\frac{\iota_n'\bar{J}_n\iota_n}{1+\rho(n-1)})^{-1} \frac{\iota_n'\bar{J}_n y}{1+\rho(n-1)} = \frac{\iota_n'\bar{J}_n y}{n} = \frac{\iota_n'y}{n} = \bar{y}$$

b. $s^2 = e'e/(n-1)$ where e is the vector of OLS residuals with typical element $e_i = y_i - \bar{y}$ for $i = 1,..,n$. In vector form, $e = E_n y$ and

$$s^2 = y'E_n y/(n-1) = u'E_n u/(n-1)$$

since $E_n \iota_n = 0$. But,

$$E(u'E_n u) = \text{tr}(\Sigma E_n) = \sigma^2 \text{tr}[(1-\rho)E_n] = \sigma^2(1-\rho)(n-1)$$

since $E_n\bar{J}_n = 0$ and $\text{tr}(E_n) = (n-1)$. Hence, $E(s^2) = \sigma^2(1-\rho)$ and $E(s^2) - \sigma^2 = -\rho\sigma^2$.

This bias is negative if $0 < \rho < 1$ and positive if $-1/(n-1) < \rho < 0$.

c. $s_*^2 = e'_{GLS}\Omega^{-1}e_{GLS}/(n-1) = e'\Omega^{-1}e/(n-1)$ where e_{GLS} denotes the vector of GLS residuals which in this case is identical to the OLS residuals. Substituting for $e = E_n y$ we get

$$s_*^2 = \frac{y'E_n\Omega^{-1}E_n y}{n-1} = \frac{u'E_n\Omega^{-1}E_n u}{(n-1)}$$

since $E_n \iota_n = 0$. Hence,

$$E(s_*^2) = \sigma^2 \text{tr}(\Omega E_n \Omega^{-1} E_n)/(n-1) = \frac{\sigma^2}{(n-1)} \text{tr}[(1-\rho)E_n][\frac{1}{1-\rho}E_n]$$
$$= \frac{\sigma^2}{(n-1)} \text{tr}(E_n) = \sigma^2$$

d. true $\text{var}(\hat{\alpha}_{ols}) = (\iota'_n \iota_n)^{-1} \iota'_n \Sigma \iota_n (\iota'_n \iota_n)^{-1} = \iota'_n \Sigma \iota_n / n^2$

$$= \sigma^2[(1+\rho(n-1))\iota'_n \bar{J}_n \iota_n]/n^2 = \sigma^2[1+\rho(n-1)]/n$$

which is equal to $\text{var}(\hat{\alpha}_{GLS}) = (\iota'_n \Sigma^{-1} \iota_n)^{-1}$ as it should be.

estimated $\text{var}(\hat{\alpha}_{ols}) = s^2(\iota'_n \iota_n)^{-1} = s^2/n$ so that

$E[\text{estimated var}(\hat{\alpha}_{ols}) - \text{true var}(\hat{\alpha}_{ols})] = E(s^2)/n - \sigma^2[1+\rho(n-1)]/n$

$$= \sigma^2[1-\rho-1-\rho(n-1)]/n = -\rho\sigma^2.$$

CHAPTER 10
Seemingly Unrelated Regressions

10.1 a. From (10.2), OLS on this system gives

$$\hat{\beta}_{ols} = \begin{pmatrix} \hat{\beta}_{1,ols} \\ \hat{\beta}_{2,ols} \end{pmatrix} = \begin{bmatrix} X_1'X_1 & 0 \\ 0 & X_2'X_2 \end{bmatrix}^{-1} \begin{bmatrix} X_1'y_1 \\ X_2'y_2 \end{bmatrix}$$

$$= \begin{bmatrix} (X_1'X_1)^{-1} & 0 \\ 0 & (X_2'X_2)^{-1} \end{bmatrix} \begin{pmatrix} X_1'y_1 \\ X_2'y_2 \end{pmatrix} = \begin{bmatrix} (X_1'X_1)^{-1}X_1'y_1 \\ (X_2'X_2)^{-1}X_2'y_2 \end{bmatrix}$$

This is OLS on each equation taken separately. For (10.2), the estimated var($\hat{\beta}_{ols}$) is given by

$$s^2 \begin{bmatrix} (X_1'X_1)^{-1} & 0 \\ 0 & (X_2'X_2)^{-1} \end{bmatrix}$$

where $s^2 = RSS/(2T-(K_1+K_2))$ and RSS denotes the residual sum of squares of this system. In fact, the $RSS = e_1'e_1 + e_2'e_2 = RSS_1 + RSS_2$ where

$$e_i = y_i - X_i\hat{\beta}_{i,ols} \qquad \text{for } i = 1,2.$$

If OLS was applied on each equation separately, then

$$\text{var}(\hat{\beta}_{1,ols}) = s_1^2(X_1'X_1)^{-1} \qquad \text{with} \qquad s_1^2 = RSS_1/(T-K_1)$$

and

$$\text{var}(\hat{\beta}_{2,ols}) = s_2^2(X_2'X_2)^{-1} \qquad \text{with} \qquad s_2^2 = RSS_2/(T-K_2).$$

Therefore, the estimates of the variance-covariance matrix of OLS from the system of two equations differs from OLS on each equation separately by a scalar, namely s^2 rather than s_i^2 for $i = 1,2$.

b. For the system of equations given in (10.2), $X' = \text{diag}[X_i']$, $\Omega^{-1} = \Sigma^{-1} \otimes I_T$ and

$$X'\Omega^{-1} = \begin{bmatrix} X_1' & 0 \\ 0 & X_2' \end{bmatrix} \begin{bmatrix} \sigma^{11}I_T & \sigma^{12}I_T \\ \sigma^{21}I_T & \sigma^{22}I_T \end{bmatrix} = \begin{bmatrix} \sigma^{11}X_1' & \sigma^{12}X_1' \\ \sigma^{21}X_2' & \sigma^{22}X_2' \end{bmatrix}.$$

Also, $X'X = \text{diag}[X_i'X_i]$ with $(X'X)^{-1} = \text{diag}[(X_i'X_i)^{-1}]$. Therefore,

$$P_X = \text{diag}[P_{X_i}] \quad \text{and} \quad \bar{P}_X = \text{diag}[\bar{P}_{X_i}].$$

Hence, $X'\Omega^{-1}\bar{P}_X = \begin{bmatrix} \sigma^{11}X_1'\bar{P}_{X_1} & \sigma^{12}X_1'\bar{P}_{X_2} \\ \sigma^{21}X_2'\bar{P}_{X_1} & \sigma^{22}X_2'\bar{P}_{X_2} \end{bmatrix}.$

But, $X_i'\bar{P}_{X_i} = 0$ for $i = 1,2$. Hence, $X'\Omega^{-1}\bar{P}_X = \begin{bmatrix} 0 & \sigma^{12}X_1'\bar{P}_{X_2} \\ \sigma^{21}X_2'\bar{P}_{X_1} & 0 \end{bmatrix}.$

and this is zero if $\sigma^{ij}X_i'\bar{P}_{X_j} = 0$ for $i \neq j$ with $i,j = 1,2$.

c. (i) If $\sigma_{ij} = 0$ for $i \neq j$, then Σ is diagonal. Hence, Σ^{-1} is diagonal and $\sigma^{ij} = 0$ for $i \neq j$. This automatically gives $\sigma^{ij}X_i'\bar{P}_{X_j} = 0$ for $i \neq j$ from part (b).
(ii) If all the X_i's are the same, then $X_1 = X_2 = X^*$ and $\bar{P}_{X_1} = \bar{P}_{X_2} = \bar{P}_{X^*}$. Hence,
$X_i'\bar{P}_{X_j} = X^{*'}\bar{P}_{X^*} = 0$ for $i,j = 1,2$.
This means that $\sigma^{ij}X_i'\bar{P}_{X_j} = 0$ from part (b) is satisfied for $i,j = 1,2$.

d. If $X_i = X_jC'$ where C is a non-singular matrix, then $X_i'\bar{P}_{X_j} = CX_j'\bar{P}_{X_j} = 0$. Alternatively, $X_i'X_i = CX_j'X_jC'$ and $(X_i'X_i)^{-1} = C'^{-1}(X_j'X_j)^{-1}C^{-1}$. Therefore,
$P_{X_i} = P_{X_j}$ and $\bar{P}_{X_i} = \bar{P}_{X_j}$.
Hence, $X_i'\bar{P}_{X_j} = X_i'\bar{P}_{X_i} = 0$. Note that when $X_i = X_j$ then $C = I_K$ where K is the number of regressors.

10.3 a. If X_1 and X_2 are orthogonal, then $X_1'X_2 = 0$. From (10.6) we get

$$\hat{\beta}_{GLS} = \begin{bmatrix} \sigma^{11}X_1'X_1 & 0 \\ 0 & \sigma^{22}X_2'X_2 \end{bmatrix}^{-1} \begin{bmatrix} \sigma^{11}X_1'y_1 & \sigma^{12}X_1'y_2 \\ \sigma^{21}X_2'y_1 & \sigma^{22}X_2'y_2 \end{bmatrix}$$

$$= \begin{bmatrix} (X_1'X_1)^{-1}/\sigma^{11} & 0 \\ 0 & (X_2'X_2)^{-1}/\sigma^{22} \end{bmatrix} \begin{bmatrix} \sigma^{11}X_1'y_1 & \sigma^{12}X_1'y_2 \\ \sigma^{21}X_2'y_1 & \sigma^{22}X_2'y_2 \end{bmatrix}$$

Chapter 10: Seemingly Unrelated Regressions

$$= \begin{bmatrix} (X_1'X_1)^{-1}X_1'y_1 + \sigma^{12}(X_1'X_1)^{-1}X_1'y_2/\sigma^{11} \\ \sigma^{21}(X_2'X_2)^{-1}X_2'y_1/\sigma^{22} + (X_2'X_2)^{-1}X_2'y_2 \end{bmatrix}$$

$$= \begin{bmatrix} \hat{\beta}_{1,ols} + (\sigma^{12}/\sigma^{11})(X_1'X_1)^{-1}X_1'y_2 \\ \hat{\beta}_{2,ols} + (\sigma^{21}/\sigma^{22})(X_2'X_2)^{-1}X_2'y_1 \end{bmatrix}$$

as required.

b. $\text{var}(\hat{\beta}_{GLS}) = (X'\Omega^{-1}X)^{-1} = \begin{bmatrix} (X_1'X_1)^{-1}/\sigma^{11} & 0 \\ 0 & (X_2'X_2)^{-1}/\sigma^{22} \end{bmatrix}$

If you have doubts, note that

$$\hat{\beta}_{1,GLS} = \beta_1 + (X_1'X_1)^{-1}X_1'u_1 + (\sigma^{12}/\sigma^{11})(X_1'X_1)^{-1}X_1'u_2$$

using $X_1'X_2 = 0$. Hence, $E(\hat{\beta}_{1,GLS}) = \beta_1$ and

$$\text{var}(\hat{\beta}_{1,GLS}) = E(\hat{\beta}_{1,GLS}-\beta_1)(\hat{\beta}_{1,GLS}-\beta_1)'$$

$$= \sigma_{11}(X_1'X_1)^{-1} + \sigma_{22}(\sigma^{12}/\sigma^{11})^2(X_1'X_1)^{-1} + 2\sigma_{12}\sigma^{12}/\sigma^{11}(X_1'X_1)^{-1}.$$

But, $\Sigma^{-1} = \begin{bmatrix} \sigma_{22} & -\sigma_{12} \\ -\sigma_{12} & \sigma_{11} \end{bmatrix}/(\sigma_{11}\sigma_{22}-\sigma_{12}^2).$

Hence, $\sigma^{12}/\sigma^{11} = -\sigma_{12}/\sigma_{22}$ substituting that in $\text{var}(\hat{\beta}_{1,GLS})$ we get

$$\text{var}(\hat{\beta}_{1,GLS}) = (X_1'X_1)^{-1}[\sigma_{11}+\sigma_{22}(\sigma_{12}^2/\sigma_{22}^2)-2\sigma_{12}^2/\sigma_{22}]$$

$$= (X_1'X_1)^{-1}[\sigma_{11}\sigma_{22}+\sigma_{12}^2-2\sigma_{12}^2]/\sigma_{22} = (X_1'X_1)^{-1}/\sigma^{11}$$

since $\sigma^{11} = \sigma_{22}/(\sigma_{11}\sigma_{22} - \sigma_{12}^2)$. Similarly, one can show that $\text{var}(\hat{\beta}_{2,GLS}) = (X_2'X_2)^{-1}/\sigma^{22}$.

c. We know that $\text{var}(\hat{\beta}_{1,ols}) = \sigma_{11}(X_1'X_1)^{-1}$ and $\text{var}(\hat{\beta}_{2,ols}) = \sigma_{22}(X_2'X_2)^{-1}$. If X_1 and X_2 are single regressors, then $(X_i'X_i)$ are scalars for $i = 1,2$. Hence,

$$\text{var}(\hat{\beta}_{1,GLS})/\text{var}(\hat{\beta}_{1,ols}) = \frac{(1/\sigma^{11})}{\sigma_{11}} = 1/\sigma_{11}\sigma^{11} = \frac{(\sigma_{11}\sigma_{22}-\sigma_{12}^2)}{\sigma_{11}\sigma_{22}} = 1 - \rho^2$$

where $\rho^2 = \sigma_{12}^2/\sigma_{22}\sigma_{11}$. Similarly,

$$\text{var}(\hat{\beta}_{2,GLS})/\text{var}(\hat{\beta}_{2,ols}) = \frac{(1/\sigma^{22})}{\sigma_{22}} = 1/\sigma_{22}\sigma^{22} = \frac{(\sigma_{11}\sigma_{22} - \sigma_{12}^2)}{\sigma_{22}\sigma_{11}} = 1 - \rho^2.$$

10.4 From (10.13), $\tilde{s}_{ij} = e_i'e_j/[T-K_i-K_j+\text{tr}(B)]$ for $i,j = 1,2$.

where $e_i = y_i - X_i\hat{\beta}_{i,ols}$ for $i = 1,2$. This can be rewritten as $e_i = \bar{P}_{X_i}y_i = \bar{P}_{X_i}u_i$ for $i = 1,2$. Hence,

$$E(e_i'e_j) = E(u_i'\bar{P}_{X_i}\bar{P}_{X_j}u_j) = E[\text{tr}(u_i'\bar{P}_{X_i}\bar{P}_{X_j}u_j)] = E[\text{tr}(u_ju_i'\bar{P}_{X_i}\bar{P}_{X_j})]$$

$$= \text{tr}[E(u_ju_i')\bar{P}_{X_i}\bar{P}_{X_j}] = \sigma_{ji}\text{tr}(\bar{P}_{X_i}\bar{P}_{X_j}).$$

But $\sigma_{ji} = \sigma_{ij}$ and $\text{tr}(\bar{P}_{X_i}\bar{P}_{X_j}) = \text{tr}(I_T - \bar{P}_{X_j} - \bar{P}_{X_i} + P_{X_i}P_{X_j}) = T - K_j - K_i + \text{tr}(B)$

where $B = P_{X_i}P_{X_j}$. Hence, $E(\tilde{s}_{ij}) = E(e_i'e_j)/\text{tr}(\bar{P}_{X_i}\bar{P}_{X_j}) = \sigma_{ij}$. This proves that $\tilde{s}_{ij}$ is unbiased for σ_{ij} for $i,j = 1,2$.

10.5 *Relative Efficiency of OLS in the Case of Simple Regressions.* This is based on Kmenta (1986, pp. 641-643).

a. Using the results in Chapter 3, we know that for a simple regression $Y_i = \alpha + \beta X_i + u_i$ that $\hat{\beta}_{ols} = \sum_{i=1}^{n}x_iy_i/\sum_{i=1}^{n}x_i^2$ and $\text{var}(\hat{\beta}_{ols}) = \sigma^2/\sum_{i=1}^{n}x_i^2$ where $x_i = X_i - \bar{X}$ and $\text{var}(u_i) = \sigma^2$. Hence, for the first equation in (10.15), we get $\text{var}(\hat{\beta}_{2,ols}) = \sigma_{11}/m_{x_1x_1}$ where $m_{x_1x_1} = \sum_{t=1}^{T}(X_{1t}-\bar{X}_1)^2$ and $\sigma_{11} = \text{var}(u_1)$. Similarly, for the second equation in (10.15), we get

$$\text{var}(\hat{\beta}_{22,ols}) = \sigma_{22}/m_{x_2x_2} \text{ where } m_{x_2x_2} = \sum_{t=1}^{T}(X_{2t}-\bar{X}_2)^2 \text{ and } \sigma_{22} = \text{var}(u_2).$$

b. In order to get rid of the constants in (10.15), one can premultiply (10.15) by $I_2 \otimes (I_T - \bar{J}_T)$ where $\bar{J}_T = J_T/T$ and $J_T = \iota_T\iota_T'$ with ι_T being a vector of ones. Recall, from the FWL Theorem in Chapter 7 that $I_T - \bar{J}_T$ is the orthogonal projection on the constant ι_T, see problem 7.2. This yields

$$\tilde{Y}_1 = \tilde{X}_1\beta_{12} + \tilde{u}_1$$
$$\tilde{Y}_2 = \tilde{X}_2\beta_{22} + \tilde{u}_2$$

where $\tilde{Y}_i = (I_T-\bar{J}_T)Y_i$, $\tilde{X}_i = (I_T-\bar{J}_T)X_i$ and $\tilde{u}_i = (I_T-\bar{J}_T)u_i$ for $i = 1,2$. Note that

$Y'_i = (Y_{i1},...,Y_{iT})$, $X'_i = (X_{i1},...,X_{iT})$ and $u'_i = (u_{i1},...,u_{iT})$ for $i = 1,2$. GLS on this system of two equations yields the same GLS estimators of β_{12} and β_{22} as in (10.15). Note that $\Omega = E\begin{pmatrix}\tilde{u}_1\\\tilde{u}_2\end{pmatrix}(\tilde{u}'_1,\tilde{u}'_2) = \Sigma \otimes (I_T - \bar{J}_T)$ where $\Sigma = [\sigma_{ij}]$ for $i,j = 1,2$. Also, for this transformed system

$$\tilde{X}'\Omega^{-1}\tilde{X} = \begin{bmatrix}\tilde{X}'_1 & 0\\ 0 & \tilde{X}'_2\end{bmatrix}(\Sigma^{-1}\otimes(I_T-\bar{J}_T))\begin{bmatrix}\tilde{X}_1 & 0\\ 0 & \tilde{X}_2\end{bmatrix} = \begin{bmatrix}\sigma^{11}\tilde{X}'_1\tilde{X}_1 & \sigma^{12}\tilde{X}'_1\tilde{X}_2\\ \sigma^{21}\tilde{X}'_2\tilde{X}_1 & \sigma^{22}\tilde{X}'_2\tilde{X}_2\end{bmatrix}$$

since $\Sigma^{-1} \otimes (I_T - \bar{J}_T)$ is the generalized inverse of $\Sigma \otimes (I_T - \bar{J}_T)$ and $(I_T - \bar{J}_T)\tilde{X}_i = \tilde{X}_i$ for $i = 1,2$. But, X_i and X_j are $T \times 1$ vectors, hence,

$$\tilde{X}'_i\tilde{X}_j = m_{x_i x_j} = \sum_{t=1}^{T}(X_{it} - \bar{X}_i)(X_{jt} - \bar{X}_j) \text{ for } i,j = 1,2. \text{ So}$$

$$\tilde{X}'\Omega^{-1}\tilde{X} = \begin{bmatrix}\sigma^{11}m_{x_1 x_1} & \sigma^{12}m_{x_1 x_2}\\ \sigma^{21}m_{x_2 x_1} & \sigma^{22}m_{x_2 x_2}\end{bmatrix}. \text{ But } \Sigma^{-1} = \begin{bmatrix}\sigma_{22} & -\sigma_{12}\\ -\sigma_{12} & \sigma_{11}\end{bmatrix}/(\sigma_{11}\sigma_{22}-\sigma_{12}^2).$$

Hence, $\text{var}\begin{pmatrix}\hat{\beta}_{12,GLS}\\ \hat{\beta}_{22,GLS}\end{pmatrix} = (\tilde{X}'\Omega^{-1}\tilde{X})^{-1} = (\sigma_{11}\sigma_{22}-\sigma_{12}^2)\begin{bmatrix}\sigma_{22}m_{x_1 x_1} & -\sigma_{12}m_{x_1 x_2}\\ -\sigma_{12}m_{x_2 x_1} & \sigma_{11}m_{x_2 x_2}\end{bmatrix}^{-1}.$

Simple inversion of this 2x2 matrix yields

$$\text{var}(\hat{\beta}_{12,GLS}) = (\sigma_{11}\sigma_{22}-\sigma_{12}^2)\sigma_{11}m_{x_2 x_2}/[\sigma_{11}\sigma_{22}m_{x_2 x_2}m_{x_1 x_1} - \sigma_{12}^2 m_{x_1 x_2}^2]$$

where the denominator is the determinant of the matrix to be inverted. Also,

$$\text{var}(\hat{\beta}_{22,GLS}) = (\sigma_{11}\sigma_{22}-\sigma_{12}^2)\sigma_{22}m_{x_1 x_1}/[\sigma_{11}\sigma_{22}m_{x_1 x_1}m_{x_2 x_2} - \sigma_{12}^2 m_{x_1 x_2}^2].$$

c. Defining $\rho = \text{correlation}(u_1,u_2) = \sigma_{12}/(\sigma_{11}\sigma_{22})^{1/2}$ and

$r = $ sample correlation coefficient between X_1 and $X_2 = m_{x_1 x_2}/(m_{x_2 x_2}m_{x_1 x_1})^{1/2}$,

then $\text{var}(\hat{\beta}_{12,GLS})/\text{var}(\hat{\beta}_{12,ols})$

$= (\sigma_{11}\sigma_{22} - \sigma_{12}^2)m_{x_1 x_1}m_{x_2 x_2}/[\sigma_{11}\sigma_{22}m_{x_2 x_2}m_{x_1 x_1} - \sigma_{12}^2 m_{x_1 x_2}^2]$

$= \sigma_{11}\sigma_{22}(1-\rho^2)/\sigma_{11}\sigma_{22}(1-\rho^2 r^2) = (1-\rho^2)/(1-\rho^2 r^2)$

similarly, $\text{var}(\hat{\beta}_{22,GLS})/\text{var}(\hat{\beta}_{22,ols})$

$= (\sigma_{11}\sigma_{22}-\sigma_{12}^2) m_{x_1x_1} m_{x_2x_2} / [\sigma_{11}\sigma_{22} m_{x_2x_2} m_{x_1x_1} - \sigma_{12}^2 m_{x_1x_2}^2]$

$= \sigma_{11}\sigma_{22}(1-\rho^2)/\sigma_{11}\sigma_{22}(1-\rho^2 r^2) = (1-\rho^2)/(1-\rho^2 r^2)$.

d. Call the relative efficiency ratio $E = (1-\rho^2)/(1-\rho^2 r^2)$. Let $\theta = \rho^2$, then $E = (1-\theta)/(1-\theta r^2)$. So that

$$\partial E/\partial \theta = \frac{-(1-\theta r^2)+r^2(1-\theta)}{(1-\theta r^2)^2} = \frac{-1+\theta r^2 + r^2 - \theta r}{(1-\theta r^2)^2} = \frac{-(1-r^2)}{(1-\theta r^2)^2} \leq 0$$

since $r^2 \leq 1$. Hence, the relative efficiency ratio is a non-increasing function of $\theta = \rho^2$. Similarly, let $\lambda = r^2$, then $E = (1-\theta)/(1-\theta\lambda)$ and $\partial E/\partial \lambda = \theta(1-\theta)/(1-\theta\lambda)^2 \geq 0$ since $0 \leq \theta \leq 1$. Hence, the relative efficiency ratio is a non-decreasing function of $\lambda = r^2$. This relative efficiency can be computed for various values of ρ^2 and r^2 between 0 and 1 at 0.1 intervals. See Kmenta (1986, Table 12-1, p. 642).

10.6 *Relative Efficiency of OLS in the Case of Multiple Regressions.* This is based on Binkley and Nelson (1988). Using partitioned inverse formulas, see the solution to problem 7.7(c), we get that the first block of the inverted matrix in (10.17) is

$\text{var}(\hat{\beta}_{1,GLS}) = A_{11} = [\sigma^{11} X_1' X_1 - \sigma^{12} X_1' X_2 (\sigma^{22} X_2' X_2)^{-1} \sigma^{21} X_2' X_1]^{-1}$.

But, $\Sigma^{-1} = \begin{bmatrix} \sigma_{22} & -\sigma_{12} \\ -\sigma_{12} & \sigma_{11} \end{bmatrix} / (\sigma_{11}\sigma_{22}-\sigma_{12}^2)$.

Divide both the matrix on the right hand side and the denominator by $\sigma_{11}\sigma_{22}$, we get $\Sigma^{-1} = [1/(1-\rho^2)] \begin{bmatrix} 1/\sigma_{11} & -\rho^2/\sigma_{12} \\ -\rho^2/\sigma_{21} & 1/\sigma_{22} \end{bmatrix}$

where $\rho^2 = \sigma_{12}^2/\sigma_{11}\sigma_{22}$. Hence,

Chapter 10: Seemingly Unrelated Regressions

$$\text{var}(\hat{\beta}_{1,GLS}) = [\frac{1}{\sigma_{11}(1-\rho^2)} X_1'X_1 - \sigma_{22}\left(\frac{\rho^2}{\sigma_{12}}\right)^2 (X_1' \bar{P}_{X_2} X_1)/(1-\rho^2)]^{-1}.$$

But $\rho^2 \sigma_{22}/\sigma_{12}^2 = 1/\sigma_{11}$, hence

$$\text{var}(\hat{\beta}_{1,GLS}) = [\frac{1}{\sigma_{11}(1-\rho^2)} X_1'X_1 - \frac{\rho^2}{\sigma_{11}(1-\rho^2)} X_1'X_1 \bar{P}_{X_2} X_1]^{-1}$$

$$= \sigma_{11}(1-\rho^2)[X_1'X_1 - \rho^2 X_1' \bar{P}_{X_2} X_1]^{-1}.$$

Add and subtract $\rho^2 X_1' X_1$ from the expression to be inverted, one gets

$$\text{var}(\hat{\beta}_{1,GLS}) = \sigma_{11}(1-\rho^2)[(1-\rho^2)X_1'X_1 + \rho^2 X_1' P_{X_2} X_1]^{-1}.$$

Factor out $(1-\rho^2)$ in the matrix to be inverted, one gets

$$\text{var}(\hat{\beta}_{1,GLS}) = \sigma_{11}\{X_1'X_1 + [\rho^2/(1-\rho^2)]E'E\}^{-1}$$

where $E = P_{X_2} X_1$ is the matrix whose columns are the OLS residuals of each variable in X_1 regressed on X_2.

10.7 When X_1 and X_2 are orthogonal, then $X_1'X_2 = 0$. R_q^2 is the R^2 of the regression of variable X_q on the other $(K_1 - 1)$ regressors in X_1. R_q^{*2} is the R^2 of the regression of $\begin{bmatrix} X_q \\ \theta e_q \end{bmatrix}$ on the other (K_1-1) regressors in $W = \begin{bmatrix} X_1 \\ \theta E \end{bmatrix}$. But $E = \bar{P}_{X_2} X_1 = X_1 - P_{X_2} X_1 = X_1$ since $P_{X_2} X_1 = X_2(X_2'X_2)^{-1}X_2'X_1 = 0$. So $e_q = X_q$ and $W = \begin{bmatrix} X_1 \\ \theta X_1 \end{bmatrix}$. Regressing $\begin{bmatrix} X_q \\ \theta X_q \end{bmatrix}$ on the other (K_1-1) regressors in $\begin{bmatrix} X_1 \\ \theta X_1 \end{bmatrix}$ yields the same R^2 as that of X_q on the other (K_1-1) regressors in X_1. Hence, $R_q^2 = R_q^{*2}$ and from (10.22) we get

$$\text{var}(\hat{\beta}_{q,SUR}) = \sigma_{11}/\{\sum_{t=1}^{T} x_{tq}^2(1-R_q^2) + \theta^2 \sum_{t=1}^{T} e_{tq}^2(1-R_q^2)\}$$

$$= \sigma_{11}/\{\sum_{t=1}^{T} x_{tq}^2(1-R_q^2)\}(1+\theta^2) = \sigma_{11}(1-\rho^2)/\sum_{t=1}^{T} x_{tq}^2(1-R_q^2)$$

since $1 + \theta^2 = 1/(1-\rho^2)$.

10.8 *SUR With Unequal Number of Observations.* This is based on Schmidt (1977).

a. Ω given in (10.25) is block-diagonal. Therefore, Ω^{-1} is block-diagonal:

$$\Omega^{-1} = \begin{bmatrix} \sigma^{11}I_T & \sigma^{12}I_T & 0 \\ \sigma^{12}I_T & \sigma^{22}I_T & 0 \\ 0 & 0 & \frac{1}{\sigma_{22}}I_N \end{bmatrix} \text{ with } \Sigma^{-1} = [\sigma^{ij}] \text{ for i,j} = 1,2.$$

$$X = \begin{bmatrix} X_1 & 0 \\ 0 & X_2^* \\ 0 & X_2^o \end{bmatrix} \quad \text{and} \quad y = \begin{pmatrix} y_1 \\ y_2^* \\ y_2^o \end{pmatrix}$$

where X_1 is $T \times K_1$, X_2^* is $T \times K_2$ and X_2^o is $N \times K_2$. Similarly, y_1 is $T \times 1$, y_2^* is $T \times 1$ and y_2^o is $N \times 1$.

$$X'\Omega^{-1} = \begin{bmatrix} X_1' & 0 & 0 \\ 0 & X_2^{*\prime} & X_2^{o\prime} \end{bmatrix} \begin{bmatrix} \sigma^{11}I_T & \sigma^{12}I_T & 0 \\ \sigma^{12}I_T & \sigma^{22}I_T & 0 \\ 0 & 0 & \frac{1}{\sigma_{22}}I_N \end{bmatrix} = \begin{bmatrix} \sigma^{11}X_1' & \sigma^{12}X_1' & 0 \\ \sigma^{12}X_2^{*\prime} & \sigma^{22}X_2^{*\prime} & \frac{1}{\sigma_{22}}X_2^{o\prime} \end{bmatrix}.$$

Therefore,

$$\hat{\beta}_{GLS} = (X'\Omega^{-1}X)^{-1}X'\Omega^{-1}y$$

$$= \begin{bmatrix} \sigma^{11}X_1'X_1 & \sigma^{12}X_1'X_2^* \\ \sigma^{12}X_2^{*\prime}X_1 & \sigma^{22}X_2^{*\prime}X_2^* + (X_2^{o\prime}X_2^o/\sigma_{22}) \end{bmatrix}^{-1}$$

$$\begin{bmatrix} \sigma^{11}X_1'y_1 + \sigma^{12}X_1'y_2^* \\ \sigma^{12}X_2^{*\prime}y_1 + \sigma^{22}X_2^{*\prime}y_2^* + (X_2^{o\prime}y_2^o/\sigma_{22}) \end{bmatrix}$$

b. If $\sigma_{12} = 0$, then from (10.25)

$$\Omega = \begin{bmatrix} \sigma_{11}I_T & 0 & 0 \\ 0 & \sigma_{22}I_T & 0 \\ 0 & 0 & \sigma_{22}I_N \end{bmatrix} \quad \text{and} \quad \Omega^{-1} = \begin{bmatrix} \dfrac{1}{\sigma_{11}}I_T & 0 & 0 \\ 0 & \dfrac{1}{\sigma_{22}}I_T & 0 \\ 0 & 0 & \dfrac{1}{\sigma_{22}}I_N \end{bmatrix}$$

so that $\sigma^{12} = 0$ and $\sigma^{ii} = 1/\sigma_{ii}$ for $i = 1,2$. From (10.26)

$$\hat{\beta}_{GLS} = \begin{bmatrix} \dfrac{X_1'X_1}{\sigma_{11}} & 0 \\ 0 & \dfrac{X_2^{*'}X_2^{*}}{\sigma_{22}} + \dfrac{X_2^{o'}X_2^{o}}{\sigma_{22}} \end{bmatrix}^{-1} \begin{bmatrix} (X_1'y_1/\sigma_{11}) \\ (X_2^{*'}y_2^{*} + X_2^{o'}y_2^{o})/\sigma_{22} \end{bmatrix}$$

and

$$\hat{\beta}_{GLS} = \begin{bmatrix} (X_1'X_1)^{-1}X_1'y_1 \\ [X_2^{*'}X_2^{*} + X_2^{o'}X_2^{o}]^{-1}[X_2^{*'}y_2^{*} + X_2^{o'}y_2^{o}] \end{bmatrix} = \begin{pmatrix} \hat{\beta}_{1,ols} \\ \hat{\beta}_{2,ols} \end{pmatrix}$$

Therefore, SUR with unequal number of observations reduces to OLS on each equation separately if $\sigma_{12} = 0$.

10.9 Grunfeld (1958) investment equation.

a. OLS on each firm.

```
Firm 1
Autoreg Procedure
Dependent Variable = INVEST
Ordinary Least Squares Estimates
SSE            275298.9    DFE                   16
MSE            17206.18    Root MSE        131.1723
SBC            244.7953    AIC              241.962
Reg Rsq          0.8411    Total Rsq         0.8411
Durbin-Watson    1.3985

        Godfrey's Serial Correlation Test
        Alternative         LM      Prob>LM
           AR(+  1)      2.6242      0.1052
           AR(+  2)      2.9592      0.2277
           AR(+  3)      3.8468      0.2785
```

Variable	DF	B Value	Std Error	t Ratio	Approx Prob
Intercept	1	-72.906480	154.0	-0.473	0.6423
VALUE1	1	0.101422	0.0371	2.733	0.0147
CAPITAL1	1	0.465852	0.0623	7.481	0.0001

Covariance of B-Values

	Intercept	VALUE1	CAPITAL1
Intercept	23715.391439	-5.465224899	0.9078247414
VALUE1	-5.465224899	0.0013768903	-0.000726424
CAPITAL1	0.9078247414	-0.000726424	0.0038773611

Firm 2

Autoreg Procedure

Dependent Variable = INVEST

Ordinary Least Squares Estimates

SSE	224688.6	DFE	16
MSE	14043.04	Root MSE	118.5033
SBC	240.9356	AIC	238.1023
Reg Rsq	0.1238	Total Rsq	0.1238
Durbin-Watson	1.1116		

Godfrey's Serial Correlation Test

Alternative	LM	Prob>LM
AR(+ 1)	4.6285	0.0314
AR(+ 2)	10.7095	0.0047
AR(+ 3)	10.8666	0.0125

Variable	DF	B Value	Std Error	t Ratio	Approx Prob
Intercept	1	306.273712	185.2	1.654	0.1177
VALUE1	1	0.015020	0.0913	0.165	0.8713
CAPITAL1	1	0.309876	0.2104	1.473	0.1602

Covariance of B-Values

	Intercept	VALUE1	CAPITAL1
Intercept	34309.129267	-15.87143322	-8.702731269
VALUE1	-15.87143322	0.0083279981	-0.001769902
CAPITAL1	-8.702731269	-0.001769902	0.0442679729

Firm 3

Autoreg Procedure

Dependent Variable = INVEST

Ordinary Least Squares Estimates

SSE	14390.83	DFE	16
MSE	899.4272	Root MSE	29.99045
SBC	188.7212	AIC	185.8879
Reg Rsq	0.6385	Total Rsq	0.6385
Durbin-Watson	1.2413		

Chapter 10: Seemingly Unrelated Regressions

```
         Godfrey's Serial Correlation Test
       Alternative        LM     Prob>LM
          AR(+ 1)      2.7742     0.0958
          AR(+ 2)      8.6189     0.0134
          AR(+ 3)     11.2541     0.0104
```

Variable	DF	B-Value	Std Error	t Ratio	Approx Prob
Intercept	1	-14.649578	39.6927	-0.369	0.7169
VALUE1	1	0.031508	0.0189	1.665	0.1154
CAPITAL1	1	0.162300	0.0311	5.213	0.0001

Covariance of B-Values

	Intercept	VALUE1	CAPITAL1
Intercept	1575.5078379	-0.706680779	-0.498664487
VALUE1	-0.706680779	0.0003581721	0.00007153
CAPITAL1	-0.498664487	0.00007153	0.0009691442

Plot of RESID*YEAR.

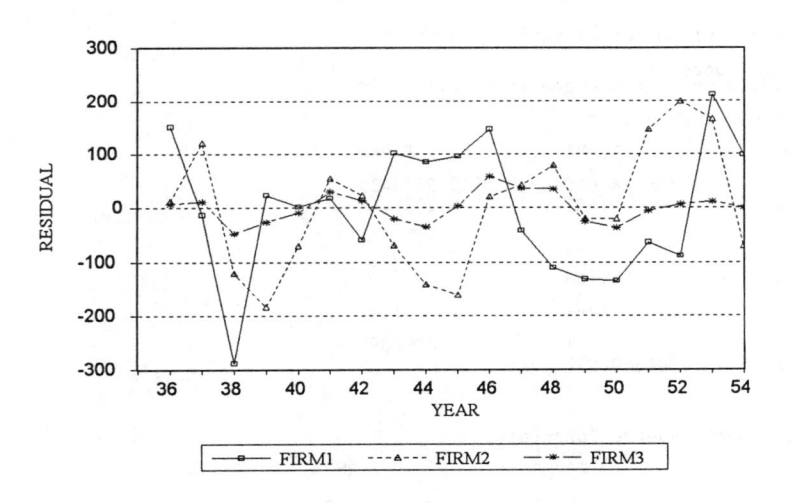

c. SUR model using the first 2 firms

```
Model: FIRM1
Dependent variable: FIRM1_I
```
Parameter Estimates

Variable	DF	Parameter Estimate	Standard Error	T for H0: Parameter=0	Prob > |T|
INTERCEP	1	-74.776254	153.974556	-0.486	0.6338
FIRM1_F1	1	0.101594	0.037101	2.738	0.0146
FIRM1_C1	1	0.467861	0.062263	7.514	0.0001

```
Model: FIRM2
Dependent variable: FIRM2_I
```
Parameter Estimates

Variable	DF	Parameter Estimate	Standard Error	T for H0: Parameter=0	Prob > |T|
INTERCEP	1	309.978987	185.198638	1.674	0.1136
FIRM2_F1	1	0.012508	0.091244	0.137	0.8927
FIRM2_C1	1	0.314339	0.210382	1.494	0.1546

SUR model using the first 2 firms

SYSLIN Procedure
Seemingly Unrelated Regression Estimation

Cross Model Covariance

Sigma	FIRM1	FIRM2
FIRM1	17206.183816	-355.9331435
FIRM2	-355.9331435	14043.039401

Cross Model Correlation

Corr	FIRM1	FIRM2
FIRM1	1	-0.022897897
FIRM2	-0.022897897	1

Cross Model Inverse Correlation

Inv Corr	FIRM1	FIRM2
FIRM1	1.0005245887	0.0229099088
FIRM2	0.0229099088	1.0005245887

Cross Model Inverse Covariance

Inv Sigma	FIRM1	FIRM2
FIRM1	0.0000581491	1.4738406E-6
FIRM2	1.4738406E-6	0.000071247

System Weighted MSE: 0.99992 with 32 degrees of freedom.
System Weighted R-Square: 0.7338

Chapter 10: Seemingly Unrelated Regressions

d. SUR model using the first 3 firms

```
Model: FIRM1
Dependent variable: FIRM1_I
```
Parameter Estimates

Variable	DF	Parameter Estimate	Standard Error	T for H0: Parameter=0	Prob > |T|
INTERCEP	1	-27.616240	147.621696	-0.187	0.8540
FIRM1_F1	1	0.088732	0.035366	2.509	0.0233
FIRM1_C1	1	0.481536	0.061546	7.824	0.0001

```
Model: FIRM2
Dependent variable: FIRM2_I
```
Parameter Estimates

Variable	DF	Parameter Estimate	Standard Error	T for H0: Parameter=0	Prob > |T|
INTERCEP	1	255.523339	167.011628	1.530	0.1456
FIRM2_F1	1	0.034710	0.081767	0.425	0.6769
FIRM2_C1	1	0.353757	0.195529	1.809	0.0892

```
Model: FIRM3
Dependent variable: FIRM3_I
```
Parameter Estimates

Variable	DF	Parameter Estimate	Standard Error	T for H0: Parameter=0	Prob > |T|
INTERCEP	1	-27.024792	35.444666	-0.762	0.4569
FIRM3_F1	1	0.042147	0.016579	2.542	0.0217
FIRM3_C1	1	0.141415	0.029134	4.854	0.0002

SUR model using the first 3 firms

SYSLIN Procedure
Seemingly Unrelated Regression Estimation

Cross Model Covariance

Sigma	FIRM1	FIRM2	FIRM3
FIRM1	17206.183816	-355.9331435	1432.3838645
FIRM2	-355.9331435	14043.039401	1857.4410783
FIRM3	1432.3838645	1857.4410783	899.4271732

Cross Model Correlation

Corr	FIRM1	FIRM2	FIRM3
FIRM1	1	-0.022897897	0.3641112847
FIRM2	-0.022897897	1	0.5226386059
FIRM3	0.3641112847	0.5226386059	1

Cross Model Inverse Correlation

Inv Corr	FIRM1	FIRM2	FIRM3
FIRM1	1.2424073467	0.3644181288	-0.642833518
FIRM2	0.3644181288	1.4826915084	-0.907600576
FIRM3	-0.642833518	-0.907600576	1.7084100377

Cross Model Inverse Covariance

Inv Sigma	FIRM1	FIRM2	FIRM3
FIRM1	0.000072207	0.0000234438	-0.000163408
FIRM2	0.0000234438	0.000105582	-0.000255377
FIRM3	-0.000163408	-0.000255377	0.0018994423

System Weighted MSE: 0.95532 with 48 degrees of freedom.
System Weighted R-Square: 0.6685

SAS PROGRAM

```
data A; infile 'B:\DATA\grunfeld.dat';
input firm year invest value capital;

data aa; set A; keep firm year invest value capital;
if firm>3 then delete;

data A1; set aa; keep firm year invest value capital;
if firm>1 then delete;
data AA1; set A1;
value1=lag(value);
capital1=lag(capital);

Proc autoreg data=AA1;
    model invest=value1 capital1/ godfrey=3 covb;
    output out=E1 r=resid;
title 'Firm 1';
proc plot data=E1;
    plot resid*year='*';
Title 'Firm 1';

run;

************************************************************;

data A2; set aa; keep firm year invest value capital;
if firm=1 or firm=3 then delete;
data AA2; set A2;
value1=lag(value);
capital1=lag(capital);

Proc autoreg data=AA2;
    model invest=value1 capital1/ godfrey=3 covb;
    output out=E2 r=resid;
title 'Firm 2';
proc plot data=E2;
    plot resid*year='*';
Title 'Firm 2';

run;

data A3; set aa;  keep firm year invest value capital;
if firm<=2 then delete;
```

Chapter 10: Seemingly Unrelated Regressions   217

```
data AA3; set A3;
value1=lag(value);
capital1=lag(capital);

Proc autoreg data=AA3;
    model invest=value1 capital1/ godfrey=3 covb;
    output out=E3 r=resid;
title 'Firm 3';
proc plot data=E3;
    plot resid*year='*';
title 'Firm 3';
run;

Proc iml;
use aa;
read all into temp;
sur=temp[1:20,3:5]||temp[21:40,3:5]||temp[41:60,3:5];
c={"F1_i" "F1_f" "F1_c" "F2_i" "F2_f" "F2_c" "F3_i" "F3_f"
"F3_c"};
create sur_data from sur [colname=c];
append from sur;

data surdata; set sur_data;
firm1_i=f1_i; firm1_f1=lag(f1_f); firm1_c1=lag(f1_c);
firm2_i=f2_i; firm2_f1=lag(f2_f); firm2_c1=lag(f2_c);
firm3_i=f3_i; firm3_f1=lag(f3_f); firm3_c1=lag(f3_c);

proc syslin sur data=surdata;
    Firm1: model firm1_i=firm1_f1 firm1_c1;
    Firm2: model firm2_i=firm2_f1 firm2_c1;
title 'SUR model using the first 2 firms';
run;

proc syslin sur data=surdata;
    Firm1: model firm1_i=firm1_f1 firm1_c1;
    Firm2: model firm2_i=firm2_f1 firm2_c1;
    Firm3: model firm3_i=firm3_f1 firm3_c1;
title 'SUR model using the first 3 firms';
run;
```

10.11 Grunfeld (1958) Data -Unequal Observations. The SAS output is given below along with the program.

```
       Ignoring the extra OBS
        BETA      STD_BETA
      48.910297  116.04249
       0.0834951  0.0257922
       0.2419409  0.0654934

      -58.84498  125.53869
       0.1804117  0.0629202
       0.3852054  0.1213539
```

```
              WILKS METHOD
         BETA      STD_BETA
      47.177574  116.24131
       0.0835601  0.0258332
       0.245064   0.0656327

     -58.93095   135.57919
       0.1803626  0.0679503
       0.3858253  0.1309581
```

```
      SRIVASTAVA & ZAATAR METHOD
         BETA      STD_BETA
      47.946223  116.04249
       0.0835461  0.0257922
       0.2435492  0.0654934

     -59.66478   135.42233
       0.1808292  0.067874
       0.3851937  0.1309081
```

```
      HOCKING & SMITH METHOD
         BETA      STD_BETA

      48.769108  116.12029
       0.083533   0.0258131
       0.2419114  0.0655068

     -60.39359   135.24547
       0.1813054  0.0677879
       0.384481   0.1308514
```

SAS PROGRAM

```
data AA; infile 'a:\ch10\grunfeld.dat';
     input firm year invest value capital;

data AAA; set AA;
     keep firm year invest value capital;
     if firm>=3 then delete;

Proc IML;
     use AAA; read all into temp;
     Y1=temp[1:15,3];    Y2=temp[21:40,3];   Y=Y1//Y2;
     X1=temp[1:15,4:5];  X2=temp[21:40,4:5];
     N1=15; N2=20; NN=5; K=3;

     *---------- SCHMIDT'S FEASIBLE GLS ESTIMATORS ----------*;

     X1=J(N1,1,1)||X1;
     X2=J(N2,1,1)||X2;
     X=(X1//J(N2,K,0))||(J(N1,K,0)//X2);
     BT1=INV(X1`*X1)*X1`*Y1;
     BT2=INV(X2`*X2)*X2`*Y2;
     e1=Y1-X1*BT1; e2=Y2-X2*BT2;
     e2_15=e2[1:N1,]; ee=e2[16:N2,];
     S11=e1`*e1/N1; S12=e1`*e2_15/N1;
     S22_15=e2_15`*e2_15/N1;
     S22_4=ee`*ee/NN; S22=e2`*e2/N2;
```

Chapter 10: Seemingly Unrelated Regressions 219

```
       S_12=S12*SQRT(S22/S22_15);
       S_11=S11-(NN/N2)*((S12/S22_15)**(2))*(S22_15-S22_4);
       ST_2=S12*S22/S22_15; ZERO=J(NN,2*N1,0);
       OMG1=((S11||S12)//(S12||S22_15))@I(N1);
       OMG2=((S11||S12)//(S12||S22))@I(N1);
       OMG3=((S11||S_12)//(S_12||S22))@I(N1);
       OMG4=((S_11||S1_2)//(S1_2||S22))@I(N1);
       OMEGA1=(OMG1//ZERO)||(ZERO`//(S22_15@I(NN)));
       OMEGA2=(OMG2//ZERO)||(ZERO`//(S22@I(NN)));
       OMEGA3=(OMG3//ZERO)||(ZERO`//(S22@I(NN)));
       OMEGA4=(OMG4//ZERO)||(ZERO`//(S22@I(NN)));
       OMG1_INV=INV(OMEGA1); OMG2_INV=INV(OMEGA2);
       OMG3_INV=INV(OMEGA3); OMG4_INV=INV(OMEGA4);

       ********** Ignoring the extra OBS **********;
       BETA1=INV(X`*OMG1_INV*X)*X`*OMG1_INV*Y;
       VAR_BT1=INV(X`*OMG1_INV*X);
       STD_BT1=SQRT(VECDIAG(VAR_BT1));
       OUT1=BETA1||STD_BT1; C={"BETA" "STD_BETA"};
       PRINT 'Ignoring the extra OBS',,OUT1(|COLNAME=C|);

       ********** WILKS METHOD **********;
       BETA2=INV(X`*OMG2_INV*X)*X`*OMG2_INV*Y;
       VAR_BT2=INV(X`*OMG2_INV*X);
       STD_BT2=SQRT(VECDIAG(VAR_BT2));
       OUT2=BETA2||STD_BT2;
       PRINT 'WILKS METHOD',,OUT2(|COLNAME=C|);

       ********** SRIVASTAVA & ZAATAR METHOD **********;
       BETA3=INV(X`*OMG3_INV*X)*X`*OMG3_INV*Y;
       VAR_BT3=INV(X`*OMG3_INV*X);
       STD_BT3=SQRT(VECDIAG(VAR_BT3));
       OUT3=BETA3||STD_BT3;
       PRINT 'SRIVASTAVA & ZAATAR METHOD',,OUT3(|COLNAME=C|);

       ********** HOCKING & SMITH METHOD **********;
       BETA4=INV(X`*OMG4_INV*X)*X`*OMG4_INV*Y;
       VAR_BT4=INV(X`*OMG4_INV*X);
       STD_BT4=SQRT(VECDIAG(VAR_BT4));
       OUT4=BETA4||STD_BT4;
       PRINT 'HOCKING & SMITH METHOD',,OUT4(|COLNAME=C|);
```

10.12 Baltagi and Griffin (1983) Gasoline Data.

a. SUR Model with the first 2 countries

SYSLIN Procedure
Seemingly Unrelated Regression Estimation

Model: AUSTRIA
Dependent variable: AUS_Y

Parameter Estimates

Variable	DF	Parameter Estimate	Standard Error	T for H0: Parameter=0	Prob > |T|
INTERCEP	1	3.713252	0.371877	9.985	0.0001
AUS_X1	1	0.721405	0.208790	3.455	0.0035

AUS_X2	1	-0.753844	0.146377	-5.150	0.0001
AUS_X3	1	-0.496348	0.111424	-4.455	0.0005

Model: BELGIUM
Dependent variable: BEL_Y

Parameter Estimates

Variable	DF	Parameter Estimate	Standard Error	T for H0: Parameter=0	Prob > |T|
INTERCEP	1	2.843323	0.445235	6.386	0.0001
BEL_X1	1	0.835168	0.169508	4.927	0.0002
BEL_X2	1	-0.130828	0.153945	-0.850	0.4088
BEL_X3	1	-0.686411	0.092805	-7.396	0.0001

b. SUR Model with the first three countries

SYSLIN Procedure
Seemingly Unrelated Regression Estimation

Model: AUSTRIA
Dependent variable: AUS_Y

Parameter Estimates

Variable	DF	Parameter Estimate	Standard Error	T for H0: Parameter=0	Prob > |T|
INTERCEP	1	3.842153	0.369141	10.408	0.0001
AUS_X1	1	0.819302	0.205608	3.985	0.0012
AUS_X2	1	-0.786415	0.145256	-5.414	0.0001
AUS_X3	1	-0.547701	0.109719	-4.992	0.0002

Model: BELGIUM
Dependent variable: BEL_Y

Parameter Estimates

Variable	DF	Parameter Estimate	Standard Error	T for H0: Parameter=0	Prob > |T|
INTERCEP	1	2.910755	0.440417	6.609	0.0001
BEL_X1	1	0.887054	0.167952	5.282	0.0001
BEL_X2	1	-0.128480	0.151578	-0.848	0.4100
BEL_X3	1	-0.713870	0.091902	-7.768	0.0001

Model: CANADA
Dependent variable: CAN_Y

Parameter Estimates

Variable	DF	Parameter Estimate	Standard Error	T for H0: Parameter=0	Prob > |T|
INTERCEP	1	3.001741	0.272684	11.008	0.0001
CAN_X1	1	0.410169	0.076193	5.383	0.0001
CAN_X2	1	-0.390490	0.086275	-4.526	0.0004
CAN_X3	1	-0.462567	0.070169	-6.592	0.0001

c. Cross Model Correlation

Corr	AUSTRIA	BELGIUM	CANADA
AUSTRIA	1	0.2232164952	-0.211908098
BELGIUM	0.2232164952	1	-0.226527683
CANADA	-0.211908098	-0.226527683	1

Breusch and Pagan (1980) diagonality LM test statistic for the first three countries yields $\lambda_{LM} = T(r_{12}^2 + r_{31}^2 + r_{32}^2) = 2.77$ which is asympotically distributed as χ_3^2 under the null hypothesis. This does not reject the diagonality of the variance-covariance matrix across the three countries.

SAS PROGRAM

```
Data gasoline;
Input COUNTRY $ YEAR Y X1 X2 X3;
CARDS;

Proc IML;
use GASOLINE;
read all into temp;
sur=temp[1:19,2:5]||temp[20:38,2:5]||temp[39:57,2:5];
c={"AUS_Y" "AUS_X1" "AUS_X2" "AUS_X3" "BEL_Y" "BEL_X1" "BEL_X2"
"BEL_X3" "CAN_Y" "CAN_X1" "CAN_X2" "CAN_X3"};
create SUR_DATA from SUR [colname=c];
append from SUR;
proc syslin sur data=SUR_DATA;
      AUSTRIA: model AUS_Y=AUS_X1 AUS_X2 AUS_X3;
      BELGIUM: model BEL_Y=BEL_X1 BEL_X2 BEL_X3;
title 'SUR MODEL WITH THE FIRST 2 COUNTRIES';
proc syslin sur data=SUR_DATA;
      AUSTRIA: model AUS_Y=AUS_X1 AUS_X2 AUS_X3;
      BELGIUM: model BEL_Y=BEL_X1 BEL_X2 BEL_X3;
      CANADA:  model CAN_Y=CAN_X1 CAN_X2 CAN_X3;
title 'SUR MODEL WITH THE FIRST 3 COUNTRIES';
run;
```

10.13 *Trace Minimization of Singular Systems with Cross-Equation Restrictions.* This is based on Baltagi (1993).

a. Note that $\sum_{i=1}^{3} y_{it} = 1$, implies that $\sum_{i=1}^{3} \bar{y}_i = 1$, where $\bar{y}_i = \sum_{t=1}^{T} y_{it}/T$, for i=1,2,3. This means that $\sum_{t=1}^{T} x_t(y_{3t}-\bar{y}_3) = -\sum_{t=1}^{T} x_t(y_{2t}-\bar{y}_2) - \sum_{t=1}^{T} x_t(y_{1t}-\bar{y}_1)$

or $b_3 = -b_2 - b_1$. This shows that the b_i's satisfy the adding up restriction $\beta_1 + \beta_2 + \beta_3 = 0$.

b. Note also that $\beta_1 = \beta_2$ and $\beta_1 + \beta_2 + \beta_3 = 0$ imply $\beta_3 = -2\beta_1$. If we ignore the first equation, and impose $\beta_1 = \beta_2$, we get

$$\begin{bmatrix} y_2 \\ y_3 \end{bmatrix} = \begin{bmatrix} \iota & 0 & X \\ 0 & \iota & -2X \end{bmatrix} \begin{bmatrix} \alpha_2 \\ \alpha_3 \\ \beta_1 \end{bmatrix} + \begin{bmatrix} \epsilon_1 \\ \epsilon_2 \end{bmatrix}$$

which means that the OLS normal equations yield

$$T\hat{\alpha}_2 + \hat{\beta}_1 \sum_{t=1}^{T} x_t = \sum_{t=1}^{T} y_{2t}$$

$$T\hat{\alpha}_3 - 2\hat{\beta}_1 \sum_{t=1}^{T} x_t = \sum_{t=1}^{T} y_{3t}$$

$$\hat{\alpha}_2 \sum_{t=1}^{T} x_t - 2\hat{\alpha}_3 \sum_{t=1}^{T} x_t + 5\hat{\beta}_1 \sum_{t=1}^{T} x_t^2 = \sum_{t=1}^{T} x_t y_{2t} - 2 \sum_{t=1}^{T} x_t y_{3t}.$$

Substituting the expressions for $\hat{\alpha}_2$ and $\hat{\alpha}_3$ from the first two equations into the third equation, we get $\hat{\beta}_1 = 0.2b_2 - 0.4b_3$. Using $b_1 + b_2 + b_3 = 0$ from part (a), one gets $\hat{\beta}_1 = 0.4b_1 + 0.6b_2$.

c. Similarly, deleting the second equation and imposing $\beta_1 = \beta_2$ one gets

$$\begin{bmatrix} y_1 \\ y_3 \end{bmatrix} = \begin{bmatrix} \iota & 0 & X \\ 0 & \iota & -2X \end{bmatrix} \begin{bmatrix} \alpha_1 \\ \alpha_3 \\ \beta_1 \end{bmatrix} + \begin{bmatrix} \epsilon_1 \\ \epsilon_3 \end{bmatrix}.$$

Forming the OLS normal equations and solving for $\hat{\beta}_1$, one gets $\hat{\beta}_1 = 0.2b_1 - 0.4b_3$. Using $b_1 + b_2 + b_3 = 0$ gives $\hat{\beta}_1 = 0.6b_1 + 0.4b_2$.

d. Finally, deleting the third equation and imposing $\beta_1 = \beta_2$ one gets

$$\begin{bmatrix} y_1 \\ y_2 \end{bmatrix} = \begin{bmatrix} \iota & 0 & X \\ 0 & \iota & X \end{bmatrix} \begin{bmatrix} \alpha_1 \\ \alpha_2 \\ \beta_1 \end{bmatrix} + \begin{bmatrix} \epsilon_1 \\ \epsilon_2 \end{bmatrix}.$$

Forming the OLS normal equations and solving for $\hat{\beta}_1$ one gets

$$\hat{\beta}_1 = 0.5b_1 + 0.5b_2.$$

Therefore, the estimate of β_1 is not invariant to the deleted equation. Also, this non-invariancy affects Zellner's SUR estimation if the restricted least squares residuals are used rather than the unrestricted least squares residuals in estimating the variance covariance matrix of the disturbances. An alternative solution is given by Im (1994).

10.14 Natural Gas Data.

a. SUR Model with the first two states.

Model: NEW_YORK
Dependent variable: NY_Y

SYSLIN Procedure
Seemingly Unrelated Regression Estimation

Parameter Estimates

Variable	DF	Parameter Estimate	Standard Error	T for H0: Parameter=0	Prob > |T|
INTERCEP	1	5.141425	1.883616	2.730	0.0143
NY_PG	1	-0.185166	0.098068	-1.888	0.0762
NY_PE	1	-0.033223	0.153473	-0.216	0.8312
NY_PO	1	0.127063	0.066114	1.922	0.0715
NY_HD	1	0.228977	0.169892	1.348	0.1954
NY_PI	1	0.581221	0.139036	4.180	0.0006

Model: FLORIDA
Dependent variable: FL_Y

Parameter Estimates

Variable	DF	Parameter Estimate	Standard Error	T for H0: Parameter=0	Prob > |T|
INTERCEP	1	-5.882191	3.910058	-1.504	0.1508
FL_PG	1	-0.847736	0.230939	-3.671	0.0019
FL_PE	1	0.194490	0.307218	0.633	0.5351
FL_PO	1	0.182123	0.158025	1.152	0.2651
FL_HD	1	0.316060	0.096603	3.272	0.0045
FL_PI	1	1.481960	0.435257	3.405	0.0034

Cross Model Correlation

Corr	NEW_YORK	FLORIDA	MICHIGAN
NEW_YORK	1	0.4262967404	0.3245924668
FLORIDA	0.4262967404	1	-0.402537885
MICHIGAN	0.3245924668	-0.402537885	1
TEXAS	-0.232095809	0.1579786634	-0.271254696
UTAH	0.1075654549	-0.287458971	0.5053131494
CALIFOR	0.4807552233	0.0321358445	0.2799913878

Corr	TEXAS	UTAH	CALIFOR
NEW_YORK	-0.232095809	0.1075654549	0.4807552233
FLORIDA	0.1579786634	-0.287458971	0.0321358445
MICHIGAN	-0.271254696	0.5053131494	0.2799913878
TEXAS	1	-0.472397852	-0.486656417
UTAH	-0.472397852	1	0.3653875996
CALIFOR	-0.486656417	0.3653875996	1

b. SUR Model with all six states

Model: NEW_YORK
Dependent variable: NY_Y

SYSLIN Procedure
Seemingly Unrelated Regression Estimation

Parameter Estimates

Variable	DF	Parameter Estimate	Standard Error	T for H0: Parameter=0	Prob > |T|
INTERCEP	1	5.672111	1.562787	3.629	0.0021
NY_PG	1	-0.184625	0.077028	-2.397	0.0283
NY_PE	1	0.012393	0.124361	0.100	0.9218
NY_PO	1	0.098964	0.055543	1.782	0.0927
NY_HD	1	0.223733	0.143076	1.564	0.1363
NY_PI	1	0.530281	0.117275	4.522	0.0003

Model: FLORIDA
Dependent variable: FL_Y

SUR MODEL WITH ALL 6 STATES

SYSLIN Procedure
Seemingly Unrelated Regression Estimation

Parameter Estimates

Variable	DF	Parameter Estimate	Standard Error	T for H0: Parameter=0	Prob > |T|
INTERCEP	1	-9.016811	3.557765	-2.534	0.0214
FL_PG	1	-0.967320	0.199515	-4.848	0.0002
FL_PE	1	0.168007	0.259203	0.648	0.5255
FL_PO	1	0.210304	0.138594	1.517	0.1475
FL_HD	1	0.277781	0.076030	3.654	0.0020
FL_PI	1	1.859821	0.395134	4.707	0.0002

Model: MICHIGAN
Dependent variable: MI_Y

SUR MODEL WITH ALL 6 STATES

SYSLIN Procedure
Seemingly Unrelated Regression Estimation

Parameter Estimates

Variable	DF	Parameter Estimate	Standard Error	T for H0: Parameter=0	Prob > |T|
INTERCEP	1	5.428544	2.562203	2.119	0.0491
MI_PG	1	0.120098	0.095555	1.257	0.2258
MI_PE	1	-0.578227	0.190784	-3.031	0.0075
MI_PO	1	0.207916	0.054188	3.837	0.0013
MI_HD	1	0.238018	0.186806	1.274	0.2198
MI_PI	1	0.569369	0.232354	2.450	0.0254

Chapter 10: Seemingly Unrelated Regressions   225

Model: TEXAS
Dependent variable: TX_Y

                    SUR MODEL WITH ALL 6 STATES

                          SYSLIN Procedure
                Seemingly Unrelated Regression Estimation

                          Parameter Estimates

                    Parameter   Standard   T for H0:
    Variable   DF   Estimate   Error   Parameter=0   Prob > |T|

    INTERCEP   1   15.434300   4.178850   3.693   0.0018
    TX_PG   1   1.052946   0.177247   5.941   0.0001
    TX_PE   1   -1.569572   0.195742   -8.019   0.0001
    TX_PO   1   0.074450   0.054124   1.376   0.1868
    TX_HD   1   0.238954   0.107118   2.231   0.0395
    TX_PI   1   -0.424107   0.439970   -0.964   0.3486

Model: UTAH
Dependent variable: UT_Y

                    SUR MODEL WITH ALL 6 STATES

                          SYSLIN Procedure
                Seemingly Unrelated Regression Estimation

                          Parameter Estimates

                    Parameter   Standard   T for H0:
    Variable   DF   Estimate   Error   Parameter=0   Prob > |T|

    INTERCEP   1   8.344877   10.272148   0.812   0.4278
    UT_PG   1   0.174993   0.325348   0.538   0.5976
    UT_PE   1   -0.300111   0.414779   -0.724   0.4792
    UT_PO   1   0.169002   0.096725   1.747   0.0986
    UT_HD   1   0.629814   0.494090   1.275   0.2196
    UT_PI   1   -0.360010   0.893025   -0.403   0.6919

Model: CALIFOR
Dependent variable: CA_Y
                    SUR MODEL WITH ALL 6 STATES

                          SYSLIN Procedure
                Seemingly Unrelated Regression Estimation

                          Parameter Estimates

                    Parameter   Standard   T for H0:
    Variable   DF   Estimate   Error   Parameter=0   Prob > |T|

    INTERCEP   1   -5.361533   4.353498   -1.232   0.2349
    CA_PG   1   -0.170188   0.095825   -1.776   0.0936
    CA_PE   1   -0.251253   0.165693   -1.516   0.1478
    CA_PO   1   0.142102   0.058672   2.422   0.0269
    CA_HD   1   0.833115   0.146974   5.668   0.0001
    CA_PI   1   1.280422   0.400995   3.193   0.0053

SAS PROGRAM

```
Data NATURAL;
infile 'a:\data\natural.asc' firstobs=2;
Input STATE $ SCODE YEAR Cons Pg Pe Po LPgas HDD Pi;

Proc iml;
use NATURAL;
read all into temp;
sur=LOG(temp[1:23,3:9]||temp[24:46,3:9]||temp[47:69,3:9]||
    temp[70:92,3:9]||temp[93:115,3:9]||temp[116:138,3:9]);

c={"NY_Y" "NY_PG" "NY_PE" "NY_PO" "NY_LPG" "NY_HD" "NY_PI"
   "FL_Y" "FL_PG" "FL_PE" "FL_PO" "FL_LPG" "FL_HD" "FL_PI"
   "MI_Y" "MI_PG" "MI_PE" "MI_PO" "MI_LPG" "MI_HD" "MI_PI"
   "TX_Y" "TX_PG" "TX_PE" "TX_PO" "TX_LPG" "TX_HD" "TX_PI"
   "UT_Y" "UT_PG" "UT_PE" "UT_PO" "UT_LPG" "UT_HD" "UT_PI"
   "CA_Y" "CA_PG" "CA_PE" "CA_PO" "CA_LPG" "CA_HD" "CA_PI"};

create SUR_DATA from SUR [colname=c];
append from SUR;

proc syslin sur data=SUR_DATA;
    NEW_YORK: model NY_Y=NY_PG NY_PE NY_PO NY_HD NY_PI ;
    FLORIDA:  model FL_Y=FL_PG FL_PE FL_PO FL_HD FL_PI;
title 'SUR MODEL WITH THE FIRST 2 STATES';

proc syslin sur data=SUR_DATA;
    NEW_YORK:   model NY_Y=NY_PG NY_PE NY_PO NY_HD NY_PI;
    FLORIDA:    model FL_Y=FL_PG FL_PE FL_PO FL_HD FL_PI;
    MICHIGAN:   model MI_Y=MI_PG MI_PE MI_PO MI_HD MI_PI;
    TEXAS:      model TX_Y=TX_PG TX_PE TX_PO TX_HD TX_PI;
    UTAH:       model UT_Y=UT_PG UT_PE UT_PO UT_HD UT_PI;
    CALIFOR:    model CA_Y=CA_PG CA_PE CA_PO CA_HD CA_PI;
title 'SUR MODEL WITH ALL 6 STATES';
RUN;
```

CHAPTER 11
Simultaneous Equations Model

11.1 The OLS estimator from equation (11.14) yields $\hat{\delta}_{ols} = \sum_{t=1}^{T} p_t q_t / \sum_{t=1}^{T} p_t^2$

where $p_t = P_t - \bar{P}$ and $q_t = Q_t - \bar{Q}$. Substituting $q_t = \delta p_t + (u_{2t} - \bar{u}_2)$ from (11.14)

we get $\hat{\delta}_{ols} = \delta + \sum_{t=1}^{T} p_t (u_{2t} - \bar{u}_2) / \sum_{t=1}^{T} p_t^2$. Using (11.18), we get

plim $\sum_{t=1}^{T} p_t(u_{2t} - \bar{u}_2)/T = (\sigma_{12} - \sigma_{22})/(\delta - \beta)$ where $\sigma_{ij} = \text{cov}(u_{it}, u_{jt})$ for i,j = 1,2 and

t = 1,2,..,T. Using (11.20) we get

plim $\hat{\delta}_{ols} = \delta + [(\sigma_{12} - \sigma_{22})/(\delta - \beta)]/[(\sigma_{11} + \sigma_{22} - 2\sigma_{12})/(\delta - \beta)^2]$

$= \delta + (\sigma_{12} - \sigma_{22})(\delta - \beta)/(\sigma_{11} + \sigma_{22} - 2\sigma_{12})$.

11.2

a. For equation (11.30) $y_1 = \alpha_{12} y_2 + \alpha_{13} y_3 + \beta_{11} X_1 + \beta_{12} X_2 + u_1$

when we regress y_2 on X_1, X_2 and X_3 to get $y_2 = \hat{y}_2 + \hat{v}_2$, the residuals satisfy

$\sum_{t=1}^{T} \hat{y}_{2t} \hat{v}_{2t} = 0$ and $\sum_{t=1}^{T} \hat{v}_{2t} X_{1t} = \sum_{t=1}^{T} \hat{v}_{2t} X_{2t} = \sum_{t=1}^{T} \hat{v}_{2t} X_{3t} = 0$.

Similarly, when we regress y_3 on X_1, X_2 and X_4 to get $y_3 = \hat{y}_3 + \hat{v}_3$, the

residuals satisfy $\sum_{t=1}^{T} \hat{y}_{3t} \hat{v}_{3t} = 0$ and $\sum_{t=1}^{T} \hat{v}_{3t} X_{1t} = \sum_{t=1}^{T} \hat{v}_{3t} X_{2t} = \sum_{t=1}^{T} \hat{v}_{3t} X_{4t} = 0$.

In the second stage regression $y_1 = \alpha_{12} \hat{y}_2 + \alpha_{13} \hat{y}_3 + \beta_{11} X_1 + \beta_{12} X_2 + \epsilon_1$

where $\epsilon_1 = \alpha_{12}(y_2 - \hat{y}_2) + \alpha_{13}(y_3 - \hat{y}_3) + u_1 = \alpha_{12} \hat{v}_2 + \alpha_{13} \hat{v}_3 + u_1$. For this to yield

consistent estimates, we need

$\sum_{t=1}^{T} \hat{y}_{2t} \epsilon_{1t} = \sum_{t=1}^{T} \hat{y}_{2t} u_{1t}$; $\sum_{t=1}^{T} \hat{y}_{3t} \epsilon_{1t} = \sum_{t=1}^{T} \hat{y}_{3t} u_{1t}$; $\sum_{t=1}^{T} X_{1t} \epsilon_{1t} = \sum_{t=1}^{T} X_{1t} u_{1t}$ and

$\sum_{t=1}^{T} X_{2t} \epsilon_{1t} = \sum_{t=1}^{T} X_{2t} u_{1t}$. But $\sum_{t=1}^{T} \hat{y}_{2t} \epsilon_{1t} = \alpha_{12} \sum_{t=1}^{T} \hat{v}_{2t} \hat{y}_{2t} + \alpha_{13} \sum_{t=1}^{T} \hat{v}_{3t} \hat{y}_{2t} + \sum_{t=1}^{T} \hat{y}_{2t} u_{1t}$

with $\sum_{t=1}^{T} \hat{v}_{2t} \hat{y}_{2t} = 0$ because $\hat{v}_2$ are the residuals and $\hat{y}_2$ are the predicted values from the same regression. However, $\sum_{t=1}^{T} \hat{v}_{3t}\hat{y}_{2t}$ is not necessarily zero because $\hat{v}_3$ is only orthogonal to X_1, X_2 and X_4, while $\hat{y}_2$ is a perfect linear combination of X_1, X_2 and X_3. Hence, $\sum_{t=1}^{T} \hat{v}_{3t} X_{3t}$ is not necessarily zero, which makes $\sum_{t=1}^{T} \hat{v}_{3t}\hat{y}_{2t}$ not necessarily zero.

Similarly, $\sum_{t=1}^{T} \hat{y}_{3t}\epsilon_{1t} = \alpha_{12} \sum_{t=1}^{T} \hat{v}_{2t}\hat{y}_{3t} + \alpha_{13} \sum_{t=1}^{T} \hat{v}_{3t}\hat{y}_{3t} + \sum_{t=1}^{T} \hat{y}_{3t} u_{1t}$ with $\sum_{t=1}^{T} \hat{v}_{3t}\hat{y}_{3t} = 0$ and $\sum_{t=1}^{T} \hat{v}_{2t}\hat{y}_{3t}$ not necessarily zero. This is because $\hat{v}_{2t}$ is only orthogonal to X_1, X_2 and X_3, while $\hat{y}_3$ is a perfect linear combination of X_1, X_2 and X_4. Hence, $\sum_{t=1}^{T} \hat{v}_{2t} X_{4t}$ is not necessarily zero, which makes $\sum_{t=1}^{T} \hat{v}_{2t}\hat{y}_{3t}$ not necessarily zero.

Since both X_1 and X_2 are included in the first stage regressions, we have $\sum_{t=1}^{T} X_{1t}\epsilon_{1t} = \sum_{t=1}^{T} X_{1t} u_{1t}$ using the fact that $\sum_{t=1}^{T} X_{1t}\hat{v}_{2t} = \sum_{t=1}^{T} X_{1t}\hat{v}_{3t} = 0$. Also, $\sum_{t=1}^{T} X_{2t}\epsilon_{1t} = \sum_{t=1}^{T} X_{2t} u_{1t}$ since $\sum_{t=1}^{T} X_{2t}\hat{v}_{2t} = \sum_{t=1}^{T} X_{2t}\hat{v}_{3t} = 0$.

b. The first stage regressions regress y_2 and y_3 on X_2, X_3 and X_4 to get
$$y_2 = \hat{y}_2 + \hat{v}_2 \quad \text{and} \quad y_3 = \hat{y}_3 + \hat{v}_3$$
with the residuals satisfying $\sum_{t=1}^{T} \hat{v}_{2t}\hat{y}_{2t} = \sum_{t=1}^{T} \hat{v}_{2t} X_{2t} = \sum_{t=1}^{T} \hat{v}_{2t} X_{3t} = \sum_{t=1}^{T} \hat{v}_{2t} X_{4t} = 0$ and $\sum_{t=1}^{T} \hat{v}_{3t}\hat{y}_{3t} = \sum_{t=1}^{T} \hat{v}_{3t} X_{2t} = \sum_{t=1}^{T} \hat{v}_{3t} X_{3t} = \sum_{t=1}^{T} \hat{v}_{3t} X_{4t} = 0$. In the second stage regression $y_1 = \alpha_{12} \hat{y}_2 + \alpha_{13} \hat{y}_3 + \beta_{11} X_1 + \beta_{12} X_2 + \epsilon_1$ where $\epsilon_1 = \alpha_{12}(y_2 - \hat{y}_2) + \alpha_{13}(y_3 - \hat{y}_3) + u_1 = \alpha_{12}\hat{v}_2 + \alpha_{13}\hat{v}_3 + u_1$. For this to yield consistent estimates, we need

$$\sum_{t=1}^{T} \hat{y}_{2t}\epsilon_{1t} = \sum_{t=1}^{T} \hat{y}_{2t} u_{1t} \; ; \; \sum_{t=1}^{T} \hat{y}_{3t}\epsilon_{1t} = \sum_{t=1}^{T} \hat{y}_{3t} u_{1t} \; ; \; \sum_{t=1}^{T} X_{1t}\epsilon_{1t} = \sum_{t=1}^{T} X_{1t} u_{1t}$$

and $\sum_{t=1}^{T} X_{2t}\epsilon_{1t} = \sum_{t=1}^{T} X_{2t} u_{1t}$.

The first two equalities are satisfied because

$$\sum_{t=1}^{T} \hat{y}_{2t}\hat{v}_{2t} = \sum_{t=1}^{T} \hat{y}_{2t}\hat{v}_{3t} = 0 \quad \text{and} \quad \sum_{t=1}^{T} \hat{y}_{3t}\hat{v}_{2t} = \sum_{t=1}^{T} \hat{y}_{3t}\hat{v}_{3t} = 0$$

since the *same* set of X's are included in both first stage regressions.

Chapter 11: Simultaneous Equations Model   229

$\sum_{t=1}^{T} X_{2t}\hat{\epsilon}_{1t} = \sum_{t=1}^{T} X_{2t}u_{1t}$ because $\sum_{t=1}^{T} X_{2t}\hat{v}_{2t} = \sum_{t=1}^{T} X_{2t}\hat{v}_{3t} = 0$ since X_2 is included in both first-stage regressions. However, $\sum_{t=1}^{T} X_{1t}\hat{\epsilon}_{1t} \neq \sum_{t=1}^{T} X_{1t}u_{1t}$ since $\sum_{t=1}^{T} X_{1t}\hat{v}_{2t} \neq 0$ and $\sum_{t=1}^{T} X_{1t}\hat{v}_{3t} \neq 0$ because X_1 was not included in the first stage regressions.

11.3 If equation (11.34) is just-identified, then X_2 is of the same dimension as Y_1, i.e., both are Txg_1. Hence, Z_1 is of the same dimension as X, both of dimension $Tx(g_1+k_1)$. Therefore, $X'Z_1$ is a square non-singular matrix of dimension (g_1+k_1). Hence, $(Z_1'X)^{-1}$ exists. But from (11.36), we know that

$$\hat{\delta}_{1,2SLS} = (Z_1'P_XZ_1)^{-1}Z_1'P_Xy_1 = [Z_1'X(X'X)^{-1}X'Z_1]^{-1}Z_1'X(X'X)^{-1}X'y_1$$
$$= (X'Z_1)^{-1}(X'X)(Z_1'X)^{-1}(Z_1'X)(X'X)^{-1}X'y_1 = (X'Z_1)^{-1}X'y_1$$

as required. This is exactly the IV estimator with $W = X$ given in (11.41). It is important to emphasize that this is only feasible if $X'Z_1$ is square and non-singular.

11.4 Premultiplying (11.34) by X' we get $X'y_1 = X'Z_1\delta_1 + X'u_1$ with $X'u_1$ having mean zero and $var(X'u_1) = \sigma_{11}X'X$ since $var(u_1) = \sigma_{11}I_T$. Performing GLS on this transformed equation yields

$$\hat{\delta}_{1,GLS} = [Z_1'X(\sigma_{11}X'X)^{-1}X'Z_1]^{-1}Z_1'X(\sigma_{11}X'X)^{-1}X'y_1 = (Z_1'P_XZ_1)^{-1}Z_1'P_Xy_1$$

as required.

11.5 *The Equivalence of 3SLS and 2SLS.*

a. From (11.46), (i) if Σ is diagonal then Σ^{-1} is diagonal and $\hat{\Sigma}^{-1} \otimes P_X$ is block-diagonal with the i-th block consisting of $P_X/\hat{\sigma}_{ii}$. Also, Z is block-diagonal, therefore, $\{Z'[\hat{\Sigma}^{-1} \otimes P_X]Z\}^{-1}$ is block-diagonal with the i-th block consisting of $\hat{\sigma}_{ii}(Z_i'P_XZ_i)^{-1}$. In other words,

$$Z'[\hat{\Sigma}^{-1} \otimes P_X]Z = \begin{pmatrix} Z_1' & 0 & .. & 0 \\ 0 & Z_2' & .. & 0 \\ \vdots & \vdots & & \vdots \\ 0 & 0 & .. & Z_G' \end{pmatrix} \begin{pmatrix} P_X/\hat{\sigma}_{11} & 0 & .. & 0 \\ 0 & P_X/\hat{\sigma}_{22} & .. & 0 \\ \vdots & \vdots & & \vdots \\ 0 & 0 & .. & P_X/\hat{\sigma}_{GG} \end{pmatrix} \begin{pmatrix} Z_1 & 0 & .. & 0 \\ 0 & Z_2 & .. & 0 \\ \vdots & \vdots & & \vdots \\ 0 & 0 & .. & Z_G \end{pmatrix}$$

$$= \begin{pmatrix} Z_1'P_XZ_1/\hat{\sigma}_{11} & 0 & \cdots & 0 \\ 0 & Z_2'P_XZ_2/\hat{\sigma}_{22} & \cdots & 0 \\ \vdots & \vdots & & \vdots \\ 0 & 0 & \cdots & Z_G'P_XZ_G/\hat{\sigma}_{GG} \end{pmatrix}$$

and $\quad Z'[\hat{\Sigma}^{-1} \otimes P_X]y = \begin{pmatrix} Z_1'P_Xy_1/\hat{\sigma}_{11} \\ Z_2'P_Xy_2/\hat{\sigma}_{22} \\ \vdots \\ Z_G'P_Xy_G/\hat{\sigma}_{GG} \end{pmatrix}$

Hence, from (11.46) $\hat{\delta}_{3SLS} = \{Z'[\hat{\Sigma}^{-1} \otimes P_X]Z\}^{-1}\{Z'[\hat{\Sigma}^{-1} \otimes P_X]y\}$

$$= \begin{pmatrix} \hat{\sigma}_{11}[Z_1'P_XZ_1]^{-1} & 0 & \cdots & 0 \\ 0 & \hat{\sigma}_{22}[Z_2'P_XZ_2]^{-1} & \cdots & 0 \\ \vdots & \vdots & \cdots & \vdots \\ 0 & 0 & \cdots & \hat{\sigma}_{GG}[Z_G'P_XZ_G]^{-1} \end{pmatrix} \begin{pmatrix} Z_1'P_Xy_1/\hat{\sigma}_{11} \\ Z_2'P_Xy_2/\hat{\sigma}_{22} \\ \vdots \\ Z_G'P_Xy_G/\hat{\sigma}_{GG} \end{pmatrix}$$

$$= \begin{pmatrix} (Z_1'P_XZ_1)^{-1}Z_1'P_Xy_1 \\ (Z_2'P_XZ_2)^{-1}Z_2'P_Xy_2 \\ \vdots \\ (Z_G'P_XZ_G)^{-1}Z_G'P_Xy_G \end{pmatrix} = \begin{pmatrix} \hat{\delta}_{1,2SLS} \\ \hat{\delta}_{2,2SLS} \\ \vdots \\ \hat{\delta}_{G,2SLS} \end{pmatrix}$$

(ii) If every equation in the system is just-identified, then $Z_i'X$ is square and non-singular for $i = 1,2,..,G$. In problem 3, we saw that for say the first equation, both Z_1 and X have the same dimension $Tx(g_1+k_1)$. Hence, $Z_1'X$ is square and of dimension (g_1+k_1). This holds for every equation of the just-identified system. In fact, problem 3 proved that $\hat{\delta}_{i,2SLS} = (X'Z_i)^{-1}X'y_i$ for $i = 1,2,..,G$. Also, from (11.44), we get

$\hat{\delta}_{3SLS} = \{\text{diag}[Z_i'X][\hat{\Sigma}^{-1} \otimes (X'X)^{-1}] \text{diag}[X'Z_i]\}^{-1}$

$\quad \{\text{diag}[Z_i'X] [\hat{\Sigma}^{-1} \otimes (X'X)^{-1}](I_G \otimes X')y\}$.

But each $Z_i'X$ is square and non-singular, hence

$\hat{\delta}_{3SLS} = [\text{diag}(X'Z_i)^{-1}(\hat{\Sigma} \otimes (X'X)) \text{diag}(Z_i'X)^{-1}]$

$\quad \{\text{diag}(Z_i'X) [\hat{\Sigma}^{-1} \otimes (X'X)^{-1}](I_G \otimes X')y\}$.

Chapter 11: Simultaneous Equations Model

which upon cancellation yields

$$\hat{\delta}_{3SLS} = \text{diag}(X'Z_i)^{-1} (I_G \otimes X')y$$

$$= \begin{pmatrix} (X'Z_1)^{-1} & 0 & \cdots & 0 \\ 0 & (X'Z_2)^{-1} & \cdots & 0 \\ \vdots & \vdots & \ddots & \vdots \\ 0 & 0 & \cdots & (X'Z_G)^{-1} \end{pmatrix} \begin{pmatrix} X'y_1 \\ X'y_2 \\ \vdots \\ X'y_G \end{pmatrix}$$

$$= \begin{pmatrix} (X'Z_1)^{-1}X'y_1 \\ (X'Z_2)^{-1}X'y_2 \\ \vdots \\ (X'Z_G)^{-1}X'y_G \end{pmatrix} = \begin{pmatrix} \hat{\delta}_{1,2SLS} \\ \hat{\delta}_{2,2SLS} \\ \vdots \\ \hat{\delta}_{G,2SLS} \end{pmatrix} \quad \text{as required.}$$

b. Premultiplying (11.43) by $(I_G \otimes P_X)$ yields $y^* = Z^*\delta + u^*$

where $y^* = (I_G \otimes P_X)y$, $Z^*(I_G \otimes P_X)Z$ and $u^* = (I_G \otimes P_X)u$. OLS on this transformed model yields

$$\hat{\delta}_{ols} = (Z^{*\prime}Z^*)^{-1}Z^{*\prime}y^* = [Z'(I_G \otimes P_X)Z]^{-1} [Z'(I_G \otimes P_X)y]$$

$$= \begin{pmatrix} (Z_1'P_XZ_1)^{-1} & 0 & \cdots & 0 \\ 0 & (Z_2'P_XZ_2)^{-1} & \cdots & 0 \\ \vdots & \vdots & \ddots & \vdots \\ 0 & 0 & \cdots & (Z_G'P_XZ_G)^{-1} \end{pmatrix} \begin{pmatrix} Z_1'P_Xy_1 \\ Z_2'P_Xy_2 \\ \vdots \\ Z_G'P_Xy_G \end{pmatrix}$$

$$= \begin{pmatrix} (Z_1'P_XZ_1)^{-1}Z_1'P_Xy_1 \\ (Z_2'P_XZ_2)^{-1}Z_2'P_Xy_2 \\ \vdots \\ (Z_G'P_XZ_G)^{-1}Z_G'P_Xy_G \end{pmatrix} = \begin{pmatrix} \hat{\delta}_{1,2SLS} \\ \hat{\delta}_{2,2SLS} \\ \vdots \\ \hat{\delta}_{G,2SLS} \end{pmatrix}$$

which is the 2SLS estimator of each equation in (11.43). Note that u^* has mean zero and $\text{var}(u^*) = \Sigma \otimes P_X$ since $\text{var}(u) = \Sigma \otimes I_T$. The generalized inverse of $\text{var}(u^*)$ is $\Sigma^{-1} \otimes P_X$. Hence, GLS on this transformed model yields

$$\hat{\delta}_{GLS} = [Z^{*\prime}(\Sigma^{-1} \otimes P_X)Z^*]^{-1} Z^{*\prime}(\Sigma^{-1} \otimes P_X)y^*$$
$$= [Z'(I_G \otimes P_X)(\Sigma^{-1} \otimes P_X)(I_G \otimes P_X)Z]^{-1} Z'(I_G \otimes P_X)(\Sigma^{-1} \otimes P_X)(I_G \otimes P_X)y$$
$$= [Z'(\Sigma^{-1} \otimes P_X)Z]^{-1}Z'[\Sigma^{-1} \otimes P_X]y.$$

Using the Milliken and Albohali condition for the equivalence of OLS and GLS given in equation (9.7) of Chapter 9, we can obtain a necessary and sufficient condition for 2SLS to be equivalent to 3SLS. After all, the last two estimators are respectively OLS and GLS on the transformed * system. This condition gives $Z^{*\prime}(\Sigma^{-1} \otimes P_X)\bar{P}_Z. = 0$ since Z^* is the matrix of regressors and $\Sigma \otimes P_X$ is the corresponding variance-covariance matrix. Note that $Z^* = \text{diag}[P_X Z_i] = \text{diag}[\hat{Z}_i]$ or a block-diagonal matrix with the i-th block being the matrix of regressors of the second-stage regression of 2SLS. Also, $P_Z. = \text{diag}[P_{\hat{Z}_i}]$ and $\bar{P}_Z. = \text{diag}[\bar{P}_{\hat{Z}_i}]$. Hence, the above condition reduces to $\sigma^{ij} \hat{Z}_i' \bar{P}_{\hat{Z}_j} = 0$ for $i \neq j$ and $i,j = 1,2,..,G$.

If Σ is diagonal, this automatically satisfies this condition since $\sigma_{ij} = \sigma^{ij} = 0$ for $i \neq j$. Also, if each equation in the system is just-identified, then $Z_i'X$ and $X'Z_j$ are square and non-singular. Hence,

$$\hat{Z}_i' P_{\hat{Z}_j} = \hat{Z}_i' \hat{Z}_j (\hat{Z}_j' \hat{Z}_j)^{-1} \hat{Z}_j' = (Z_i'X(X'X)^{-1}X'Z_j[Z_j'X(X'X)^{-1}X'Z_j]^{-1}$$

$$Z_j'X(X'X)^{-1}X') = Z_i'X(X'X)^{-1}X' = Z_i'P_X = \hat{Z}_i'$$

after some cancellations. Hence, $\hat{Z}_i' \bar{P}_{\hat{Z}_j} = \hat{Z}_i' - \hat{Z}_i' P_{\hat{Z}_j} = \hat{Z}_i' - \hat{Z}_i' = 0$, under just-identification of each equation in the system.

11.6 a. Writing this system of two equations in the matrix form given by (11.47) we get

$$B = \begin{bmatrix} 1 & b \\ 1 & -d \end{bmatrix} ; \Gamma = \begin{bmatrix} -a & 0 & 0 \\ -c & -e & -f \end{bmatrix} ; y_t = \begin{pmatrix} Q_t \\ P_t \end{pmatrix}$$

$x_t' = [1, W_t, L_t]$ and $u_t' = (u_{1t}, u_{2t})$.

b. There are two zero restrictions on Γ. These restrictions state that W_t and L_t do not appear in the demand equation. Therefore, the first equation is over-identified by the order condition of identification. There are two excluded exogenous variables and only one right hand side included endogenous variable. However, the supply equation is not identified by the order condition of identification. There are no excluded exogenous variables, but there is one right hand side included endogenous variable.

c. The transformed matrix FB should satisfy the following normalization restrictions:

$f_{11} + f_{12} = 1$ and $f_{21} + f_{22} = 1$.

These are the same normalization restrictions given in (11.52). Also, FΓ must satisfy the following zero restrictions:

$f_{11}0 - f_{12} e = 0$ and $f_{11}0 - f_{12}f = 0$.

If either e or f are different from zero, so that at least W_t or L_t appear in the supply equation, then $f_{12} = 0$. This means that $f_{11} + 0 = 1$ or $f_{11} = 1$, from the first normalization equation. Hence, the first row of F is indeed the first row of an identity matrix and the demand equation is identified. However, there are no zero restrictions on the second equation and in the absence of any additional restrictions $f_{21} \neq 0$ and the second row of F is not necessarily the second row of an identity matrix. Hence, the supply equation is not identified.

11.7 a. In this case,

$$B = \begin{bmatrix} 1 & b \\ 1 & -f \end{bmatrix} ; \Gamma = \begin{bmatrix} -a & -c & -d & 0 & 0 \\ -e & 0 & 0 & -g & -h \end{bmatrix} ; y_t = \begin{pmatrix} Q_t \\ P_t \end{pmatrix}$$

$x'_t = [1, Y_t, A_t, W_t, L_t]$ and $u'_t = (u_{1t}, u_{2t})$.

b. There are four zero restrictions on Γ. These restrictions state that Y_t and A_t are absent from the supply equation whereas W_t and L_t are absent from the demand equation. Therefore, both equations are over-identified. For each equation, there is one right hand sided included endogenous variable and two excluded exogenous variables.

c. The transformed matrix FB should satisfy the following normalization restrictions:

$f_{11} + f_{12} = 1$ and $f_{21} + f_{22} = 1$.

Also, FΓ must satisfy the following zero restrictions:

$$-f_{21}c + f_{22}0 = 0 \qquad\qquad -f_{21}d + f_{22}0 = 0.$$

and

$$f_{11}0 - f_{12}g = 0 \qquad\qquad f_{11}0 - f_{12}h = 0.$$

If either c or d are different from zero, so that at least Y_t or A_t appear in the demand equation, then $f_{21} = 0$ and from the second normalization equation we deduce that $f_{22} = 1$. Hence, the second row of F is indeed the second row of an identity matrix and the supply equation is identified. Similarly, if either g or h are different from zero, so that at least W_t or L_t appear in the supply equation, then $f_{12} = 0$ and from the first normalization equation we deduce that $f_{11} = 1$. Hence, the first row of F is indeed the first row of an identity matrix and the demand equation is identified.

11.8 a. From example 3 and equation (11.51), we get

$$A = [B,\Gamma] = \begin{bmatrix} \beta_{11} & \beta_{12} & \gamma_{11} & \gamma_{12} & \gamma_{13} \\ \beta_{21} & \beta_{22} & \gamma_{21} & \gamma_{22} & \gamma_{23} \end{bmatrix} = \begin{bmatrix} 1 & b & -a & -c & 0 \\ 1 & -e & -d & 0 & -f \end{bmatrix}$$

Therefore, ϕ for the first equation consists of only one restriction, namely that W_t is not in that equation, or $\gamma_{13} = 0$. This makes $\phi' = (0,0,0,0,1)$, since $\alpha_1' \phi = 0$ gives $\gamma_{13} = 0$. Therefore,

$$A\phi = \begin{pmatrix} \gamma_{13} \\ \gamma_{23} \end{pmatrix} = \begin{pmatrix} 0 \\ -f \end{pmatrix}.$$

This is of rank one as long as $f \neq 0$. Similarly, for ϕ the second equation consists of only one restriction, namely that Y_t is not in that equation, or $\gamma_{22} = 0$. This makes $\phi = (0,0,0,1,0)$, since $\alpha_1' \phi = 0$ gives $\gamma_{22} = 0$. Therefore,

$$A\phi = \begin{pmatrix} \gamma_{12} \\ \gamma_{22} \end{pmatrix} = \begin{pmatrix} -c \\ 0 \end{pmatrix}. \text{ This is of rank one as long as } c \neq 0.$$

b. For the supply and demand model given by (11.49) and (11.50), the reduced form is given by

$$\Pi = \begin{bmatrix} \pi_{11} & \pi_{12} & \pi_{13} \\ \pi_{21} & \pi_{22} & \pi_{23} \end{bmatrix} = \begin{bmatrix} ea+bd & ec & bf \\ a-d & c & -f \end{bmatrix} / (e+b).$$

Chapter 11: Simultaneous Equations Model

This can be easily verified by solving the two equations in (11.49) and (11.50) for P_t and Q_t in terms of the constant, Y_t and W_t.

From part (a) and equation (11.63), we can show that the parameters of the first structural equation can be obtained from $\alpha'_1[W, \phi] = 0$ where $\alpha'_1 = (\beta_{11}, \beta_{12}, \gamma_{11}, \gamma_{12}, \gamma_{13})$ represents the parameters of the first structural equation. $W' = [\Pi, I_3]$ and $\phi' = (0,0,0,0,1)$ as seen in part (a).

$\beta_{11} \pi_{11} + \beta_{12} \pi_{21} + \gamma_{11} = 0$

$\beta_{11} \pi_{12} + \beta_{12} \pi_{22} + \gamma_{12} = 0$

$\beta_{11} \pi_{13} + \beta_{12} \pi_{23} + \gamma_{13} = 0$

$$\gamma_{13} = 0$$

If we normalize by setting $\beta_{11} = 1$, we get $\beta_{12} = -\pi_{13}/\pi_{23}$ also

$$\gamma_{12} = -\pi_{12} + \frac{\pi_{13}}{\pi_{23}} \pi_{22} = \frac{\pi_{22}\pi_{13} - \pi_{12}\pi_{23}}{\pi_{23}} \text{ and }$$

$$\gamma_{11} = -\pi_{11} + \frac{\pi_{13}}{\pi_{23}} \pi_{21} = \frac{\pi_{21}\pi_{13} - \pi_{11}\pi_{23}}{\pi_{23}}.$$

One can easily verify from the reduced form parameters that

$\beta_{12} = -\pi_{13}/\pi_{23} = -bf/-f = b$.

Similarly, $\gamma_{12} = (\pi_{22}\pi_{13} - \pi_{12}\pi_{23})/\pi_{23} = (cbf+ecf)/-f(e+b) = -c$. Finally,

$\gamma_{11} = (\pi_{21}\pi_{13} - \pi_{11}\pi_{23})/\pi_{23} = (abf+aef)/-f(e+b) = -a$.

Similarly, for the second structural equation, we get,

$$(\beta_{21}, \beta_{22}, \gamma_{21}, \gamma_{22}, \gamma_{23}) \begin{bmatrix} \pi_{11} & \pi_{12} & \pi_{13} & 0 \\ \pi_{21} & \pi_{22} & \pi_{23} & 0 \\ 1 & 0 & 0 & 0 \\ 0 & 1 & 0 & 1 \\ 0 & 0 & 1 & 0 \end{bmatrix} = (0,0,0,0,0)$$

which can be rewritten as

$\beta_{21}\pi_{11} + \beta_{22}\pi_{21} + \gamma_{21} = 0$

$\beta_{21}\pi_{12} + \beta_{22}\pi_{22} + \gamma_{22} = 0$

$\beta_{21}\pi_{13} + \beta_{22}\pi_{23} + \gamma_{23} = 0$

$$\gamma_{22} = 0.$$

If we normalize by setting $\beta_{21} = 1$, we get $\beta_{22} = -\pi_{12}/\pi_{22}$ also

$$\gamma_{23} = -\pi_{13} + \frac{\pi_{12}}{\pi_{22}}\pi_{23} = \frac{\pi_{23}\pi_{12} - \pi_{13}\pi_{22}}{\pi_{22}} \quad \text{and}$$

$$\gamma_{21} = -\pi_{11} + \frac{\pi_{12}}{\pi_{22}}\pi_{21} = \frac{\pi_{12}\pi_{21} - \pi_{11}\pi_{22}}{\pi_{22}}.$$

One can easily verify from the reduced form parameters that

$\beta_{22} = -\pi_{12}/\pi_{22} = -ec/c = -e$. Similarly,

$\gamma_{23} = (\pi_{23}\pi_{13} - \pi_{13}\pi_{22})/\pi_{22} = (-ecf - bdf)/c(e+b) = -f$. Finally,

$\gamma_{21} = (\pi_{12}\pi_{21} - \pi_{11}\pi_{22})/\pi_{22} = (-dec - bdc)/c(e+b) = -d$.

11.9 For the simultaneous model in problem 6, the reduced form parameters are given

by $\pi = \begin{bmatrix} \pi_{11} & \pi_{12} & \pi_{13} \\ \pi_{21} & \pi_{22} & \pi_{23} \end{bmatrix} = \begin{bmatrix} ab+bc & be & bf \\ a-c & -e & -f \end{bmatrix} / (b+d)$

This can be easily verified by solving for P_t and Q_t in terms of the constant, W_t and L_t. From the solution to 6, we get

$$A = [B, \Gamma] = \begin{bmatrix} \beta_{11} & \beta_{12} & \gamma_{11} & \gamma_{12} & \gamma_{13} \\ \beta_{21} & \beta_{22} & \gamma_{21} & \gamma_{22} & \gamma_{23} \end{bmatrix} = \begin{bmatrix} 1 & b & -a & 0 & 0 \\ 1 & -d & -c & -e & -f \end{bmatrix}$$

For the first structural equation,

$$\phi = \begin{bmatrix} 0 & 0 \\ 0 & 0 \\ 0 & 0 \\ 1 & 0 \\ 0 & 1 \end{bmatrix}$$

with $\alpha_1' \phi = 0$ yielding $\gamma_{12} = \gamma_{13} = 0$. Therefore, (11.63) reduces to

$$(\beta_{11}, \beta_{12}, \gamma_{11}, \gamma_{12}, \gamma_{13}) \begin{bmatrix} \pi_{11} & \pi_{12} & \pi_{13} & 0 & 0 \\ \pi_{21} & \pi_{22} & \pi_{23} & 0 & 0 \\ 1 & 0 & 0 & 0 & 0 \\ 0 & 1 & 0 & 1 & 0 \\ 0 & 0 & 1 & 0 & 1 \end{bmatrix} = (0,0,0,0,0)$$

This can be rewritten as

$\beta_{11}\pi_{11} + \beta_{12}\pi_{21} + \gamma_{11} = 0$

$\beta_{11}\pi_{12} + \beta_{12}\pi_{22} + \gamma_{12} = 0$

$\beta_{11}\pi_{13} + \beta_{12}\pi_{23} + \gamma_{13} = 0$

$\gamma_{12} = 0$

$\gamma_{13} = 0$

If we normalize by setting $\beta_{11} = 1$, we get $\beta_{12} = -\pi_{13}/\pi_{23}$; $\beta_{12} = -\pi_{12}/\pi_{22}$.

Also, $\gamma_{11} = -\pi_{11} + (\pi_{12}/\pi_{22})\pi_{21} = \dfrac{\pi_{12}\pi_{21} - \pi_{11}\pi_{22}}{\pi_{22}}$ and

$\gamma_{11} = -\pi_{11} + (\pi_{13}/\pi_{23})\pi_{21} = \dfrac{\pi_{21}\pi_{13} - \pi_{11}\pi_{23}}{\pi_{23}}$

one can easily verify from the reduced form parameters that

$\beta_{12} = -\pi_{13}/\pi_{23} = -bf/-f = b$. Also,

$\beta_{12} = -\pi_{12}/\pi_{22} = -be/-e = b$. Similarly,

$\gamma_{11} = \dfrac{\pi_{12}\pi_{21} - \pi_{11}\pi_{22}}{\pi_{22}} = \dfrac{abe - bce + ead + ebc}{-e(b+d)} = \dfrac{a(be+ed)}{-(be+ed)} = -a$

Also,

$\gamma_{11} = \dfrac{\pi_{21}\pi_{13} - \pi_{11}\pi_{23}}{\pi_{23}} = \dfrac{abf - bcf + adf + bcf}{-f(b+d)} = \dfrac{a(bf+df)}{-(bf+df)} = -a$

For the second structural equation there are no zero restrictions. Therefore, (11.61) reduces to

$(\beta_{21}, \beta_{22}, \gamma_{21}, \gamma_{22}, \gamma_{23}) \begin{bmatrix} \pi_{11} & \pi_{12} & \pi_{13} \\ \pi_{21} & \pi_{22} & \pi_{23} \\ 1 & 0 & 0 \\ 0 & 1 & 0 \\ 0 & 0 & 1 \end{bmatrix} = (0,0,0,0,0)$

This can be rewritten as

$\beta_{21}\pi_{11} + \beta_{22}\pi_{21} + \gamma_{21} = 0$

$\beta_{21}\pi_{12} + \beta_{22}\pi_{22} + \gamma_{22} = 0$

$\beta_{21}\pi_{13} + \beta_{22}\pi_{23} + \gamma_{23} = 0$

Three equations in five unknowns. When we normalize by setting $\beta_{21} = 1$, we still get three equations in four unknowns for which there is no unique solution.

11.10 *Just-Identified Model.*

  a. The *generalized instrumental variable* estimator for δ_1 based on W is given below (11.41)

$$\hat{\delta}_{1,IV} = (Z_1'P_W Z_1)^{-1} Z_1' P_W y_1 = [Z_1' W(W'W)^{-1} W' Z_1]^{-1} Z_1' W(W'W)^{-1} W' y_1$$

$$= (W'Z_1)^{-1}(W'W)(Z_1'W)^{-1}(Z_1'W)(W'W)^{-1}W'y_1 = (W'Z_1)^{-1}W'y_1$$

since $W'Z_1$ is square and non-singular under just-identification. This is exactly the expression for the *simple instrumental variable* estimator for δ_1 given in (11.38).

b. Let the minimized value of the criterion function be

$$Q = (y_1 - Z_1 \hat{\delta}_{1,IV})' P_W (y_1 - Z_1 \hat{\delta}_{1,IV}).$$

Substituting the expression for $\hat{\delta}_{1,IV}$ and expanding terms, we get

$$Q = y_1' P_W y_1 - y_1' W(Z_1'W)^{-1} Z_1' P_W y_1 - y_1' P_W Z_1 (W'Z_1)^{-1} W' y_1$$

$$+ y_1' W(Z_1'W)^{-1} Z_1' P_W Z_1 (W'Z_1)^{-1} W' y_1$$

$$Q = y_1' P_W y_1 - y_1' W(Z_1'W)^{-1} Z_1' W(W'W)^{-1} W' y_1$$

$$- y_1' W(W'W)^{-1} W' Z_1 (W_1'Z_1)^{-1} W' y_1$$

$$+ y_1' W(Z_1'W)^{-1} Z_1' W(W'W)^{-1} W' Z_1 (W'Z_1)^{-1} W' y_1$$

$$Q = y_1' P_W y_1 - y_1' P_W y_1 - y_1' P_W y_1 + y_1' P_W y_1 = 0$$

as required.

c. Let $\hat{Z}_1 = P_W Z_1$ be the set of second stage regressors obtained from regressing each variable in Z_1 on the set of instrumental variables W. For the just-identified case

$$P_{\hat{Z}_1} = P_{P_W Z_1} = \hat{Z}_1 (\hat{Z}_1' \hat{Z}_1)^{-1} \hat{Z}_1' = P_W Z_1 (Z_1' P_W Z_1)^{-1} Z_1' P_W$$

$$= W(W'W)^{-1} W' Z_1 [Z_1' W(W'W)^{-1} W' Z_1]^{-1} Z_1' W(W'W)^{-1} W'$$

$$= W(W'W)^{-1} W' Z_1 (W'Z_1)^{-1} (W'W)(Z_1'W)^{-1} Z_1' W(W'W)^{-1} W'$$

$$= W(W'W)^{-1} W' = P_W.$$

Hence, the residual sum of squares of the second stage regression for the just-identified model, yields $y_1' \bar{P}_{\hat{Z}_1} y_1 = y_1' \bar{P}_W y_1$ since $P_{\hat{Z}_1} = P_W$. This is exactly the residual sum of squares from regression y_1 on the matrix of instruments W.

Chapter 11: Simultaneous Equations Model

11.11 If the set of instruments W_1 is spanned by the space of the set of instruments W_2, then $P_{W_2} W_1 = W_1$. The corresponding two instrumental variables estimators are

$$\hat{\delta}_{1,W_i} = (Z_1' P_{W_i} Z_1)^{-1} Z_1' P_{W_i} y_1 \quad \text{for} \quad i = 1,2$$

with asymptotic covariance matrices

$$\sigma_{11} \, \text{plim} \, (Z_1' P_{W_i} Z_1/T)^{-1} \quad \text{for} \quad i = 1,2.$$

The difference between the asymptotic covariance matrix of $\hat{\delta}_{1,W_1}$ and that of $\hat{\delta}_{1,W_2}$ is positive semi-definite if

$$\sigma_{11} \left[\text{plim} \, \frac{Z_1' P_{W_1} Z_1}{T} \right]^{-1} - \sigma_{11} \left[\text{plim} \, \frac{Z_1' P_{W_2} Z_1}{T} \right]^{-1}$$

is positive semi-definite. This holds, if $Z_1' P_{W_2} Z_1 - Z_1' P_{W_1} Z_1$ is positive semi-definite. To show this, we prove that $P_{W_2} - P_{W_1}$ is idempotent.

$$(P_{W_2} - P_{W_1})(P_{W_2} - P_{W_1}) = P_{W_2} - P_{W_2} P_{W_1} - P_{W_1} P_{W_2} + P_{W_1}$$

$$= P_{W_2} - P_{W_1} - P_{W_1} + P_{W_1} = P_{W_2} - P_{W_1}.$$

This uses the fact that $P_{W_2} P_{W_1} = P_{W_2} W_1 (W_1' W_1)^{-1} W_1' = W_1 (W_1' W_1)^{-1} W_1' = P_{W_1}$. Since $P_{W_2} W_1 = W_1$.

11.12 *Testing for Over-Identification.* The main point here is that W spans the same space as $[\hat{Z}_1, W^*]$ and that both are of full rank ℓ. In fact, W^* is a subset of instruments W, of dimension $(\ell - k_1 - g_1)$, that are linearly independent of $\hat{Z}_1 = P_W Z_1$.

a. Using instruments W, the second stage regression on $y_1 = Z_1 \delta_1 + W^* \gamma + u_1$ regresses y_1 on $P_W[Z_1, W^*] = [\hat{Z}_1, W^*]$ since $\hat{Z}_1 = P_W Z_1$ and $W^* = P_W W^*$. But the matrix of instruments W is of the same dimension as the matrix of regressors $[\hat{Z}_1, W^*]$. Hence, this equation is just-identified, and from problem 10, we deduce that the residual sum of squares of this second stage regression is exactly the residual sum of squares obtained by regressing y_1 on the matrix of instruments W, i.e.,

$$\text{URSS}^* = y_1' \bar{P}_W y_1 = y_1' y_1 - y_1' P_W y_1.$$

b. Using instruments W, the second stage regression on $y_1 = Z_1\delta_1 + u_1$ regresses y_1 on $\hat{Z}_1 = P_W Z_1$. Hence, the $\text{RRSS}^* = y_1'\bar{P}_{\hat{Z}_1} y_1 = y_1'y_1 - y_1'P_{\hat{Z}_1}y_1$ where $\bar{P}_{\hat{Z}_1} = I - P_{\hat{Z}_1}$ and $P_{\hat{Z}_1} = \hat{Z}_1(\hat{Z}_1'\hat{Z}_1)^{-1}\hat{Z}_1' = P_W Z_1(Z_1'P_W Z_1)^{-1}Z_1'P_W$. Therefore, using the results in part (a) we get

$$\text{RRSS}^* - \text{URSS}^* = y_1' P_W y_1 - y_1' P_{\hat{Z}_1} y_1$$

as given in (11.77).

c. Hausman (1983) proposed regressing the 2SLS residuals $(y_1 - Z_1\hat{\delta}_{1,2SLS})$ on the set of instruments W and obtaining nR_u^2 where R_u^2 is the uncentered R^2 of this regression. Note that the regression sum of squares yields

$$(y_1 - Z_1\hat{\delta}_{1,2SLS})' P_W (y_1 - Z_1\hat{\delta}_{1,2SLS})$$
$$= y_1' P_W y_1 - y_1' P_W Z_1(Z_1'P_W Z_1)^{-1} Z_1' P_W y_1 - y_1' P_W Z_1(Z_1'P_W Z_1)^{-1} Z_1' P_W y_1$$
$$+ y_1' P_W Z_1(Z_1'P_W Z_1)^{-1} Z_1' P_W Z_1(Z_1'P_W Z_1)^{-1} Z_1' P_W y_1$$
$$= y_1' P_W y_1 - y_1' P_W Z_1(Z_1'P_W Z_1)^{-1} Z_1' P_W y_1$$
$$= y_1' P_W y_1 - y_1' P_{\hat{Z}_1} y_1 = \text{RRSS}^* - \text{URSS}^*$$

as given in part (b). The total sum of squares of this regression, uncentered, is the 2SLS residuals sum of squares given by

$$(y_1 - Z_1\hat{\delta}_{1,2SLS})'(y_1 - Z_1\hat{\delta}_{1,2SLS}) = T\hat{\sigma}_{11}$$

where $\hat{\sigma}_{11}$ is given in (11.78). Hence, the test statistic

$$\frac{\text{RRSS}^* - \text{URSS}^*}{\hat{\sigma}_{11}} = \frac{\text{Regression Sum of Squares}}{\text{Uncentered Total Sum of Squares}/T}$$

$= T \text{ (uncentered } R^2) = T R_u^2$ as required.

This is asymptotically distributed as χ^2 with $\ell - (g_1 + k_1)$ degrees of freedom. Large values of this test statistic reject the null hypothesis.

d. The GNR for the unrestricted model given in (11.75) computes the residuals from the restricted model under $H_o; \gamma = 0$. These are the 2SLS residuals $(y_1 - Z_1\hat{\delta}_{1,2SLS})$ based on the set of instruments W. Next, one differentiates the model with respect to its parameters and evaluates these derivatives at the

Chapter 11: Simultaneous Equations Model   241

restricted estimates. For instrumental variables W one has to premultiply the right hand side of the GNR by P_W, this yields

$$(y_1 - Z_1\hat{\delta}_{1,2SLS}) = P_W Z_1 b_1 + P_W W^* b_2 + \text{residuals}.$$

But $\hat{Z}_1 = P_W Z_1$ and $P_W W^* = W^*$ since W^* is a subset of W. Hence, the GNR becomes $(y_1 - Z_1\hat{\delta}_{1,2SLS}) = \hat{Z}_1 b_1 + W^* b_2 + \text{residuals}$. However, $[\hat{Z}_1, W^*]$ spans the same space as W, see part (a). Hence, this GNR regresses 2SLS residuals on the matrix of instruments W and computes TR_u^2 where R_u^2 is the uncentered R^2 of this regression. This is Hausman's (1983) test statistic derived in part (c).

11.15

a. The artificial regression in (11.83) is given by

$$y_1 = Z_1\delta_1 + (Y_1 - \hat{Y}_1)\eta + \text{residuals}$$

This can be rewritten as $y_1 = Z_1\delta_1 + \bar{P}_W Y_1 \eta + \text{residuals}$
where we made use of the fact that $\hat{Y}_1 = P_W Y_1$ with $Y_1 - \hat{Y}_1 = Y_1 - P_W Y_1 = \bar{P}_W Y_1$.
Note that to residual out the matrix of regressors $\bar{P}_W Y_1$, we need

$$\bar{P}_{\bar{P}_W Y_1} = I - \bar{P}_W Y_1 (Y_1'\bar{P}_W Y_1)^{-1} Y_1'\bar{P}_W$$

so that, by the FWL Theorem, the OLS estimate of δ_1 can also be obtained from the following regression

$$\bar{P}_{\bar{P}_W Y_1} y_1 = \bar{P}_{\bar{P}_W Y_1} Z_1 \delta_1 + \text{residuals}.$$

Note that

$$\bar{P}_{\bar{P}_W Y_1} Z_1 = Z_1 - \bar{P}_W Y_1 (Y_1'\bar{P}_W Y_1)^{-1} Y_1'\bar{P}_W Z_1$$

with $\bar{P}_W Z_1 = [\bar{P}_W Y_1, 0]$ and $\bar{P}_W X_1 = 0$ since X_1 is part of the instruments in W.
Hence, $\bar{P}_{\bar{P}_W Y_1} Z_1 = [Y_1, X_1] - [\bar{P}_W Y_1, 0] = [P_W Y_1, X_1]$
$= [\hat{Y}, X_1] = P_W[Y_1, X_1] = P_W Z_1 = \hat{Z}_1$

In this case,

$$\hat{\delta}_{1,ols} = (Z_1'\bar{P}_{\bar{P}_W Y_1} Z_1)^{-1} Z_1'\bar{P}_{\bar{P}_W Y_1} y_1 = (\hat{Z}_1'\hat{Z}_1)^{-1}\hat{Z}_1' y_1 = (Z_1' P_W Z_1)^{-1} Z_1' P_W y_1 = \hat{\delta}_{1,IV}$$

as required.

b. The FWL Theorem also states that the residuals from the regressions in part (a) are the same. Hence, their residuals sum of squares are the same. The last regression computes an estimate of the var($\hat{\delta}_{1,ols}$) as $\tilde{s}^2 (Z_1' P_W Z_1)^{-1}$ where $\tilde{s}^2$ is the residual sum of squares divided by the degrees of freedom of the regression. In (11.83), this is the MSE of that regression, since it has the same residuals sum of squares. Note that when $\eta \neq 0$ in (11.83), IV estimation is necessary and $\tilde{s}_{11}$ underestimates σ_{11} and will have to be replaced by

$$(y_1 - Z_1 \hat{\delta}_{1,IV})'(y_1 - Z_1 \hat{\delta}_{1,IV})/T.$$

11.16 Recursive Systems.

a. The order condition of identification yields one excluded exogenous variable (x_3) from the first equation and no right hand side endogenous variable i.e., $k_2 = 1$ and $g_1 = 0$. Therefore, the first equation is over-identified with degree of over-identification equal to one. The second equation has two excluded exogenous variables (x_1 and x_2) and one right hand side endogenous variable (y_1) so that $k_2 = 2$ and $g_1 = 1$. Hence, the second equation is over-identified of degree one. The rank condition of identification can be based on

$$A = [B, \Gamma] = \begin{bmatrix} 1 & 0 & \gamma_{11} & \gamma_{12} & 0 \\ \beta_{21} & 1 & 0 & 0 & \gamma_{23} \end{bmatrix}$$

with $y_t' = (y_{1t}, y_{2t})$ and $x_t' = (x_{1t}, x_{2t}, x_{3t})$. For the first equation, the set of zero restrictions yield

$$\phi_1 = \begin{bmatrix} 0 & 0 \\ 1 & 0 \\ 0 & 0 \\ 0 & 0 \\ 0 & 1 \end{bmatrix} \text{ so that } \alpha_1' \phi_1 = (\beta_{12}, \gamma_{13}) = 0$$

where α_1' is the first row of A. For the second equation, the set of zero restrictions yield

$$\phi_2 = \begin{bmatrix} 0 & 0 \\ 0 & 0 \\ 1 & 0 \\ 0 & 1 \\ 0 & 0 \end{bmatrix} \text{ so that } \alpha_2' \phi_2 = (\gamma_{21}, \gamma_{22}) = 0$$

where α_2' is the second row of A. Hence, for the first equation

$$A\phi_1 = \begin{bmatrix} 0 & 0 \\ 1 & \gamma_{23} \end{bmatrix}$$

which is of rank 1 in general. Hence, the first equation is identified by the rank condition of identification. Similarly, for the second equation

$$A\phi_2 = \begin{bmatrix} \gamma_{11} & \gamma_{12} \\ 0 & 0 \end{bmatrix}$$

which is of rank 1 as long as either γ_{11} or γ_{12} are different from zero. This ensures that either x_1 or x_2 appear in the first equation to identify the second equation.

b. The first equation is already in reduced form format

$$y_{1t} = -\gamma_{11} x_{1t} - \gamma_{12} x_{2t} + u_{1t}$$

y_{1t} is a function of x_{1t}, x_{2t} and only u_{1t}, not u_{2t}. For the second equation, one can substitute for y_{1t} to get

$$y_{2t} = -\gamma_{23} x_{3t} - \beta_{21}(-\gamma_{11} x_{1t} - \gamma_{12} x_{2t} + u_{1t}) + u_{2t}$$

$$= \beta_{21}\gamma_{11} x_{1t} + \beta_{21}\gamma_{12} x_{2t} - \gamma_{23} x_{3t} + u_{2t} - \beta_{21} u_{1t}$$

so that y_{2t} is a function of x_{1t}, x_{2t}, x_{3t} and the error is a linear combination of u_{1t} and u_{2t}.

c. OLS on the first structural equation is also OLS on the first reduced form equation with only exogenous variables on the right hand side. Since x_1 and x_2 are not correlated with u_{1t}, this yields consistent estimates for γ_{11} and γ_{12}. OLS on the second equation regresses y_{2t} on y_{1t} and x_{3t}. Since x_{3t} is not correlated with u_{2t}, the only possible correlation is from y_{1t} and u_{2t}. However, from the first reduced form equation y_{1t} is only a function of exogenous variables and u_{1t}. Since u_{1t} and u_{2t} are not correlated because Σ is diagonal for a recursive system, OLS estimates of β_{21} and γ_{23} are consistent.

d. Assuming that $u_t \sim N(0,\Sigma)$ where $u'_t = (u_{1t}, u_{2t})$, this exercise shows that OLS is MLE for this recursive system. Conditional on the set of x's, the likelihood function is given by

$$L(B, \Gamma, \Sigma) = (2\pi)^{-T/2} |B|^T |\Sigma|^{-T/2} \exp\left(-\frac{1}{2} \sum_{t=1}^{T} u'_t \Sigma^{-1} u_t\right)$$

so that

$$\log L = -(T/2)\log 2\pi + T\log|B| - (T/2)\log|\Sigma| - \frac{1}{2} \sum_{t=1}^{T} u'_t \Sigma^{-1} u_t.$$

Since B is triangular, $|B| = 1$ so that $\log|B| = 0$. Therefore, the only place in $\log L$ where B and Γ parameters appear is in $\sum_{t=1}^{T} u'_t \Sigma^{-1} u_t$. Maximizing $\log L$ with respect to B and Γ is equivalent to minimizing $\sum_{t=1}^{T} u'_t \Sigma^{-1} u_t$ with respect to B and Γ. But Σ is diagonal, so Σ^{-1} is diagonal and

$$\sum_{t=1}^{T} u'_t \Sigma^{-1} u_t = \sum_{t=1}^{T} u_{1t}^2/\sigma_{11} + \sum_{t=1}^{T} u_{2t}^2/\sigma_{22}.$$

The partial derivatives of $\log L$ with respect to the coefficients of the first structural equation are simply the partial derivatives of $\sum_{t=1}^{T} u_{1t}^2/\sigma_{11}$ with respect to those coefficients. Setting these partial derivatives equal to zero yields the OLS estimates of the first structural equation. Similarly, the partial derivatives of $\log L$ with respect to the coefficients of the second structural equation are simply the derivatives of $\sum_{t=1}^{T} u_{2t}^2/\sigma_{22}$ with respect to those coefficients. Setting these partial derivatives equal to zero yields the OLS estimates of the second structural equation. Hence, MLE is equivalent to OLS for a recursive system.

11.17 *Hausman's Specification Test: 2SLS Versus 3SLS.* This is based on Baltagi (1989).

a. The two equations model considered is

$$y_1 = Z_1 \delta_1 + u_1 \quad \text{and} \quad y_2 = Z_2 \delta_2 + u_2 \quad (1)$$

where y_1 and y_2 are endogenous; $Z_1 = [y_2, x_1, x_2]$, $Z_2 = [y_1, x_3]$; and x_1, x_2 and x_3 are exogenous (the y's and x's are Tx1 vectors). As noted in the problem,

Chapter 11: Simultaneous Equations Model   245

$\tilde{\delta}$ = 2SLS, $\tilde{\tilde{\delta}}$ = 3SLS, and the corresponding residuals are denoted by $\tilde{u}$ and $\tilde{\tilde{u}}$. $\delta'_1 = (\alpha, \beta_1, \beta_2)$ and $\delta'_2 = (\gamma, \beta_3)$ with $\alpha\gamma \neq 1$, so that the system is complete and we can solve for the reduced form.

Let $X = [x_1, x_2, x_3]$ and $P_X = X(X'X)^{-1}X'$, the 3SLS equations for model (1) are given by

$$Z'(\tilde{\Sigma}^{-1} \otimes P_X)Z\tilde{\tilde{\delta}} = Z'(\tilde{\Sigma}^{-1} \otimes P_X)y \tag{2}$$

where $Z = \text{diag}[Z_i]$, $y' = (y'_1, y'_2)$, and $\tilde{\Sigma}^{-1} = [\tilde{\sigma}^{ij}]$ is the inverse of the estimated variance-covariance matrix obtained from 2SLS residuals. Equation (2) can also be written as

$$Z'(\tilde{\Sigma}^{-1} \otimes P_X)\tilde{\tilde{u}} = 0, \tag{3}$$

or more explicitly as

$$\tilde{\sigma}^{11}Z'_1 P_X \tilde{\tilde{u}}_1 + \tilde{\sigma}^{12}Z'_1 P_X \tilde{\tilde{u}}_2 = 0 \tag{4}$$

$$\tilde{\sigma}^{12}Z'_2 P_X \tilde{\tilde{u}}_1 + \tilde{\sigma}^{22}Z'_2 P_X \tilde{\tilde{u}}_2 = 0 \tag{5}$$

Using the fact that $Z'_1 X$ is a square non-singular matrix, one can premultiply (4) by $(X'X)(Z'_1 X)^{-1}$ to get

$$\tilde{\sigma}^{11} X' \tilde{\tilde{u}}_1 + \tilde{\sigma}^{12} X' \tilde{\tilde{u}}_2 = 0. \tag{6}$$

b. Writing equations (5) and (6) in terms of $\tilde{\tilde{\delta}}_1$ and $\tilde{\tilde{\delta}}_2$, one gets

$$\tilde{\sigma}^{12}(Z'_2 P_X Z_1)\tilde{\tilde{\delta}}_1 + \tilde{\sigma}^{22}(Z'_2 P_X Z_2)\tilde{\tilde{\delta}}_2 = \tilde{\sigma}^{12}(Z'_2 P_X y_1) + \tilde{\sigma}^{22}(Z'_2 P_X y_2) \tag{7}$$

and

$$\tilde{\sigma}^{11} X' Z_1 \tilde{\tilde{\delta}}_1 + \tilde{\sigma}^{12} X' Z_2 \tilde{\tilde{\delta}}_2 = \tilde{\sigma}^{11} X' y_1 + \tilde{\sigma}^{12} X' y_2. \tag{8}$$

Premultiply (8) by $\tilde{\sigma}^{12} Z'_2 X (X'X)^{-1}$ and subtract it from $\tilde{\sigma}^{11}$ times (7). This eliminates $\tilde{\tilde{\delta}}_1$, and solves for $\tilde{\tilde{\delta}}_2$:

$$\tilde{\tilde{\delta}}_2 = (Z'_2 P_X Z_2)^{-1} Z'_2 P_X y_2 = \tilde{\delta}_2. \tag{9}$$

Therefore, the 3SLS estimator of the over-identified equation is equal to its 2SLS counterpart.

Substituting (9) into (8), and rearranging terms, one gets

$$\tilde{\sigma}^{11}X'Z_1\tilde{\tilde{\delta}}_1 = \tilde{\sigma}^{11}X'y_1 + \tilde{\sigma}^{12}X'\tilde{u}_2.$$

Premultiplying by $Z_1'X(X'X)^{-1}/\tilde{\sigma}^{11}$, and solving for $\tilde{\tilde{\delta}}_1$, one gets

$$\tilde{\tilde{\delta}}_1 = (Z_1'P_XZ_1)^{-1}Z_1'P_Xy_1 + (\tilde{\sigma}^{12}/\tilde{\sigma}^{11})(Z_1'P_XZ_1)^{-1}Z_1'P_X\tilde{u}_2. \tag{10}$$

Using the fact that $\tilde{\sigma}^{12} = -\tilde{\sigma}_{12}/|\tilde{\Sigma}|$, and $\tilde{\sigma}^{11} = \tilde{\sigma}_{22}/|\tilde{\Sigma}|$, (10) becomes

$$\tilde{\tilde{\delta}}_1 = \tilde{\delta}_1 - (\tilde{\sigma}_{12}/\tilde{\sigma}_{22})(Z_1'P_XZ_1)^{-1}Z_1'P_X\tilde{u}_2. \tag{11}$$

Therefore, the 3SLS estimator of the just-identified equation differs from its 2SLS (or indirect least squares) counterpart by a linear combination of the 2SLS (or 3SLS) residuals of the over-identified equation; see Theil (1971).

c. A Hausman-type test based on $\tilde{\tilde{\delta}}_1$ and $\tilde{\delta}_1$ is given by

$$m = (\tilde{\delta}_1 - \tilde{\tilde{\delta}}_1)'[V(\tilde{\delta}_1) - V(\tilde{\tilde{\delta}}_1)]^{-1}(\tilde{\delta}_1 - \tilde{\tilde{\delta}}_1), \tag{12}$$

where $V(\tilde{\delta}_1) = \tilde{\sigma}_{11}(Z_1'P_XZ_1)^{-1}$

and $V(\tilde{\tilde{\delta}}_1) = (1/\tilde{\sigma}^{11})(Z_1'P_XZ_1)^{-1} + (\tilde{\sigma}_{12}^2/\tilde{\sigma}_{22})(Z_1'P_XZ_1)^{-1}$
$(Z_1'P_XZ_2)(Z_2'P_XZ_2)^{-1}(Z_2'P_XZ_1)(Z_1'P_XZ_1)^{-1}$.

The latter term is obtained by using the partitioned inverse of $Z'(\tilde{\Sigma}^{-1}\otimes P_X)Z$.

Using (11), the Hausman test statistic becomes

$$m = (\tilde{\sigma}_{12}^2/\tilde{\sigma}_{22}^2)(\tilde{u}_2'P_XZ_1)[(\tilde{\sigma}_{11} - 1/\tilde{\sigma}^{11})(Z_1'P_XZ_1)$$
$$- (\tilde{\sigma}_{12}^2/\tilde{\sigma}_{22})(Z_1'P_XZ_2)(Z_2'P_XZ_2)^{-1}(Z_2'P_XZ_1)]^{-1}(Z_1'P_X\tilde{u}_2).$$

However, $(\tilde{\sigma}_{11} - 1/\tilde{\sigma}^{11}) = (\tilde{\sigma}_{12}^2/\tilde{\sigma}_{22}^2)$; this enables us to write

$$m = \tilde{u}_2'\hat{Z}_1[\hat{Z}_1'\hat{Z}_1 - \hat{Z}_1'\hat{Z}_2(\hat{Z}_2'\hat{Z}_2)^{-1}\hat{Z}_2'\hat{Z}_1]^{-1}\hat{Z}_1'\tilde{u}_2/\tilde{\sigma}_{22}, \tag{13}$$

where $\hat{Z}_i = P_XZ_i$ is the matrix of second stage regressors of 2SLS, for i = 1,2. This statistic is asymptotically distributed as χ_3^2, and can be given the following interpretation:

Claim: $m = TR^2$, where R^2 is the R-squared of the regression of $\tilde{u}_2$ on the set of second stage regressors of both equations, $\hat{Z}_1$ and $\hat{Z}_2$.

Chapter 11: Simultaneous Equations Model

Proof: This result follows immediately using the fact that
$\tilde{\sigma}_{22} = \tilde{u}_2'\tilde{u}_2/T =$ total sums of squares/T,
and the fact that the regression sum of squares is given by

$$RSS = \tilde{u}_2'[\hat{Z}_1 \ \hat{Z}_2] \begin{bmatrix} \hat{Z}_1'\hat{Z}_1 & \hat{Z}_1'\hat{Z}_2 \\ \hat{Z}_2'\hat{Z}_1 & \hat{Z}_2'\hat{Z}_2 \end{bmatrix}^{-1} \begin{bmatrix} \hat{Z}_1' \\ \hat{Z}_2' \end{bmatrix} \tilde{u}_2, \qquad (14)$$

$$= \tilde{u}_2'\hat{Z}_1[\hat{Z}_1'\hat{Z}_1 - \hat{Z}_1'\hat{Z}_2(\hat{Z}_2'\hat{Z}_2)^{-1}\hat{Z}_2'\hat{Z}_1]^{-1}\hat{Z}_1'\tilde{u}_2,$$

where the last equality follows from partitioned inverse and the fact that
$\hat{Z}_2'\tilde{u}_2 = 0$.

11.18

a. Using the order condition of identification, the first equation has two excluded exogenous variables x_2 and x_3 and only one right hand sided included endogenous variable y_2. Hence,

$$k_2 = 2 > g_1 = 1$$

and the first equation is over-identified with degree of over-identification equal to one. Similarly, the second equation has one excluded exogenous variable x_1 and only one right hand side included endogenous variable y_1. Hence, $k_2 = 1 = g_1$ and the second equation is just-identified. Using the rank condition of identification

$$A = [B \ \Gamma] = \begin{bmatrix} 1 & -\beta_{12} & \gamma_{11} & 0 & 0 \\ -\beta_{21} & 1 & 0 & \gamma_{22} & \gamma_{23} \end{bmatrix}$$

with $y_t' = (y_{1t}, y_{2t})$ and $x_t' = (x_{1t}, x_{2t}, x_{3t})$. The first structural equation has the following zero restrictions:

$$\phi_1 = \begin{bmatrix} 0 & 0 \\ 0 & 0 \\ 0 & 0 \\ 1 & 0 \\ 0 & 1 \end{bmatrix} \qquad \text{so that } \alpha_1'\phi_1 = (\gamma_{12}, \gamma_{13}) = 0$$

where α_1' is the first row of A. Similarly, the second structural equation has the following zero restriction:

$$\phi_2 = \begin{bmatrix} 0 \\ 0 \\ 1 \\ 0 \\ 0 \end{bmatrix} \text{ so that } \alpha_2' \phi_2 = \gamma_{21} = 0$$

where α_2' is the second row of A. Hence, $A\phi_1 = \begin{bmatrix} 0 & 0 \\ \gamma_{22} & \gamma_{23} \end{bmatrix}$ which has rank one provided either γ_{22} or γ_{23} is different from zero. Similarly, $A\phi_2 = \begin{pmatrix} \gamma_{11} \\ 0 \end{pmatrix}$ which has rank one provided $\gamma_{11} \neq 0$. Hence, both equations are identified by the rank condition of identification.

b. For the first equation, the OLS normal equations are given by

$$(Z_1'Z_1)\begin{pmatrix} \hat{\beta}_{12} \\ \hat{\gamma}_{11} \end{pmatrix}_{ols} = Z_1'y_1$$

where $Z_1 = [y_2, x_1]$. Hence,

$$Z_1'Z_1 = \begin{bmatrix} y_2'y_2 & y_2'x_1 \\ x_1'y_2 & x_1'x_1 \end{bmatrix} = \begin{bmatrix} 8 & 10 \\ 10 & 1 \end{bmatrix}$$

$$Z_1'y_1 = \begin{pmatrix} y_2'y_1 \\ x_1'y_1 \end{pmatrix} = \begin{pmatrix} 4 \\ 5 \end{pmatrix}$$

obtained from the matrices of cross-products provided by problem 18. Hence, the OLS normal equations are

$$\begin{bmatrix} 8 & 10 \\ 10 & 1 \end{bmatrix}\begin{pmatrix} \hat{\beta}_{12} \\ \hat{\gamma}_{11} \end{pmatrix}_{ols} = \begin{pmatrix} 4 \\ 5 \end{pmatrix}$$

solving for the OLS estimators, we get

$$\begin{pmatrix} \hat{\beta}_{12} \\ \hat{\gamma}_{11} \end{pmatrix}_{ols} = \frac{-1}{92}\begin{bmatrix} 1 & -10 \\ -10 & 8 \end{bmatrix}\begin{pmatrix} 4 \\ 5 \end{pmatrix} = \begin{pmatrix} 0.5 \\ 0 \end{pmatrix}$$

OLS on the second structural equation can be deduced in a similar fashion.

c. For the first equation, the 2SLS normal equations are given by

$$(Z_1'P_XZ_1)\begin{pmatrix} \hat{\beta}_{12} \\ \hat{\gamma}_{11} \end{pmatrix}_{2SLS} = Z_1'P_Xy_1$$

Chapter 11: Simultaneous Equations Model

where $Z_1 = [y_2, x_1]$ and $X = [x_1, x_2, x_3]$. Hence,

$$Z_1'X = \begin{bmatrix} y_2'x_1 & y_2'x_2 & y_2'x_3 \\ x_1'x_1 & x_1'x_2 & x_1'x_3 \end{bmatrix} = \begin{bmatrix} 8 & 20 & 30 \\ 1 & 0 & 0 \end{bmatrix}$$

$$(X'X)^{-1} = \begin{bmatrix} 1 & 0 & 0 \\ 0 & 1/20 & 0 \\ 0 & 0 & 1/10 \end{bmatrix} \text{ and } X'y_1 = \begin{bmatrix} x_1'y_1 \\ x_2'y_1 \\ x_3'y_1 \end{bmatrix} = \begin{pmatrix} 5 \\ 40 \\ 20 \end{pmatrix}$$

$$Z_1'X(X'X)^{-1} = \begin{bmatrix} 8 & 1 & 3 \\ 1 & 0 & 0 \end{bmatrix} \text{ and } Z_1'P_X Z_1 = \begin{bmatrix} 8 & 1 & 3 \\ 1 & 0 & 0 \end{bmatrix} \begin{bmatrix} 8 & 1 \\ 20 & 0 \\ 30 & 0 \end{bmatrix} = \begin{bmatrix} 174 & 8 \\ 8 & 1 \end{bmatrix}$$

with

$$Z_1'P_X y_1 = \begin{bmatrix} 8 & 1 & 3 \\ 1 & 0 & 0 \end{bmatrix} \begin{pmatrix} 5 \\ 40 \\ 20 \end{pmatrix} = \begin{pmatrix} 140 \\ 5 \end{pmatrix}.$$

Therefore, the 2SLS normal equations are

$$\begin{bmatrix} 174 & 8 \\ 8 & 1 \end{bmatrix} \begin{pmatrix} \hat{\beta}_{12} \\ \hat{\gamma}_{11} \end{pmatrix}_{2SLS} = \begin{pmatrix} 140 \\ 5 \end{pmatrix}$$

solving for the 2SLS estimates, we get

$$\begin{pmatrix} \hat{\beta}_{12} \\ \hat{\gamma}_{11} \end{pmatrix}_{2SLS} = \frac{1}{110} \begin{bmatrix} 1 & -8 \\ -8 & 174 \end{bmatrix} \begin{pmatrix} 140 \\ 5 \end{pmatrix} = \frac{1}{110} \begin{pmatrix} 100 \\ -250 \end{pmatrix} = \begin{pmatrix} 10/11 \\ -25/11 \end{pmatrix}$$

2SLS on the second structural equation can be deduced in a similar fashion.

d. The first equation is over-identified and the reduced form parameter estimates will give more than one solution to the structural form parameters. However, the second equation is just-identified. Hence, 2SLS reduces to indirect least squares which in this case solves uniquely for the structural parameters from the reduced form parameters. ILS on the second equation yields

$$\begin{pmatrix} \hat{\beta}_{21} \\ \hat{\gamma}_{22} \\ \hat{\gamma}_{23} \end{pmatrix}_{ILS} = (X'Z_1)^{-1} X'y_2$$

where $Z_1 = [y_1, x_2, x_3]$ and $X = [x_1, x_2, x_3]$. Therefore,

$$X'Z_1 = \begin{bmatrix} x_1'y_1 & x_1'x_2 & x_1'x_3 \\ x_2'y_1 & x_2'x_2 & x_2'x_3 \\ x_3'y_1 & x_3'x_2 & x_3'x_3 \end{bmatrix} = \begin{bmatrix} 5 & 0 & 0 \\ 40 & 20 & 0 \\ 20 & 0 & 10 \end{bmatrix} \quad \text{and} \quad X'y_2 = \begin{pmatrix} x_1'y_2 \\ x_2'y_2 \\ x_3'y_2 \end{pmatrix} = \begin{pmatrix} 10 \\ 20 \\ 30 \end{pmatrix}$$

so that

$$\begin{pmatrix} \hat{\beta}_{21} \\ \hat{\gamma}_{22} \\ \hat{\gamma}_{23} \end{pmatrix}_{ILS} = \begin{bmatrix} 5 & 0 & 0 \\ 40 & 20 & 0 \\ 20 & 0 & 10 \end{bmatrix}^{-1} \begin{pmatrix} 10 \\ 20 \\ 30 \end{pmatrix}$$

Note that $X'Z_1$ for the first equation is of dimension 3 x 2 and is not even square.

11.19 Supply and Demand Equations for Traded Money. This is based on Laffer (1970).

b. OLS Estimation

SYSLIN Procedure
Ordinary Least Squares Estimation

Model: SUPPLY
Dependent variable: LNTM_P

Analysis of Variance

Source	DF	Sum of Squares	Mean Square	F Value	Prob>F
Model	2	0.28682	0.14341	152.350	0.0001
Error	18	0.01694	0.00094		
C Total	20	0.30377			

Root MSE	0.03068	R-Square	0.9442	
Dep Mean	5.48051	Adj R-SQ	0.9380	
C.V.	0.55982			

Parameter Estimates

Variable	DF	Parameter Estimate	Standard Error	T for H0: Parameter=0	Prob > |T|
INTERCEP	1	3.661183	0.272547	13.433	0.0001
LNRM_P	1	0.553762	0.089883	6.161	0.0001
LNI	1	0.165917	0.012546	13.225	0.0001

OLS ESTIMATION OF MONEY DEMAND FUNCTION

SYSLIN Procedure
Ordinary Least Squares Estimation

Model: DEMAND
Dependent variable: LNTM_P

Chapter 11: Simultaneous Equations Model

Analysis of Variance

Source	DF	Sum of Squares	Mean Square	F Value	Prob>F
Model	4	0.29889	0.07472	245.207	0.0001
Error	16	0.00488	0.00030		
C Total	20	0.30377			

Root MSE	0.01746	R-Square	0.9839	
Dep Mean	5.48051	Adj R-SQ	0.9799	
C.V.	0.31852			

Parameter Estimates

Variable	DF	Parameter Estimate	Standard Error	T for H0: Parameter=0	Prob > |T|
INTERCEP	1	3.055597	0.647464	4.719	0.0002
LNY_P	1	0.618770	0.055422	11.165	0.0001
LNI	1	-0.015724	0.021951	-0.716	0.4841
LNS1	1	-0.305535	0.101837	-3.000	0.0085
LNS2	1	0.147360	0.202409	0.728	0.4771

c. 2SLS Estimation

SYSLIN Procedure
Two-Stage Least Squares Estimation

Model: SUPPLY
Dependent variable: LNTM_P

Analysis of Variance

Source	DF	Sum of Squares	Mean Square	F Value	Prob>F
Model	2	0.29742	0.14871	150.530	0.0001
Error	18	0.01778	0.00099		
C Total	20	0.30377			

Root MSE	0.03143	R-Square	0.9436	
Dep Mean	5.48051	Adj R-SQ	0.9373	
C.V.	0.57350			

Parameter Estimates

Variable	DF	Parameter Estimate	Standard Error	T for H0: Parameter=0	Prob > |T|
INTERCEP	1	3.741721	0.280279	13.350	0.0001
LNRM_P	1	0.524645	0.092504	5.672	0.0001
LNI	1	0.177757	0.013347	13.318	0.0001

Test for Overidentifying Restrictions
Numerator: 0.001342 DF: 2 F Value: 1.4224
Denominator: 0.000944 DF: 16 Prob>F: 0.2700

2SLS ESTIMATION OF MONEY DEMAND FUNCTION

SYSLIN Procedure
Two-Stage Least Squares Estimation

Model: DEMAND
Dependent variable: LNTM_P

Analysis of Variance

Source	DF	Sum of Squares	Mean Square	F Value	Prob>F
Model	4	0.30010	0.07503	154.837	0.0001
Error	16	0.00775	0.00048		
C Total	20	0.30377			

Root MSE	0.02201	R-Square	0.9748	
Dep Mean	5.48051	Adj R-SQ	0.9685	
C.V.	0.40165			

Parameter Estimates

Variable	DF	Parameter Estimate	Standard Error	T for H0: Parameter=0	Prob > |T|
INTERCEP	1	1.559954	1.222958	1.276	0.2203
LNY_P	1	0.761774	0.111639	6.824	0.0001
LNI	1	-0.083170	0.049518	-1.680	0.1125
LNS1	1	-0.187052	0.147285	-1.270	0.2222
LNS2	1	0.228561	0.259976	0.879	0.3923

The total number of instruments equals the number of parameters in the equation. The equation is just identified, and the test for over identification is not computed.

d. 3SLS Estimation

SYSLIN Procedure
Three-Stage Least Squares Estimation

Cross Model Covariance

Sigma	SUPPLY	DEMAND
SUPPLY	0.0009879044	-0.000199017
DEMAND	-0.000199017	0.000484546

Cross Model Correlation

Corr	SUPPLY	DEMAND
SUPPLY	1	-0.287650003
DEMAND	-0.287650003	1

                  Cross Model Inverse Correlation

        Inv Corr   SUPPLY   DEMAND

        SUPPLY   1.0902064321   0.3135978834
        DEMAND   0.3135978834   1.0902064321

                  Cross Model Inverse Covariance

        Inv Sigma   SUPPLY   DEMAND

        SUPPLY   1103.5545963   453.26080422
        DEMAND   453.26080422   2249.9545257

     System Weighted MSE: 0.53916 with 34 degrees of freedom.
     System Weighted R-Square: 0.9858

Model: SUPPLY
Dependent variable: LNTM_P
              3SLS ESTIMATION OF MODEY SUPPLY AND DEMAND FUNCTION

                           SYSLIN Procedure
                   Three-Stage Least Squares Estimation

                           Parameter Estimates

                       Parameter   Standard   T for H0:
     Variable   DF   Estimate   Error   Parameter=0   Prob > |T|

     INTERCEP   1   3.741721   0.280279   13.350   0.0001
     LNRM_P   1   0.524645   0.092504   5.672   0.0001
     LNI   1   0.177757   0.013347   13.318   0.0001

Model: DEMAND
Dependent variable: LNTM_P

                           Parameter Estimates

                       Parameter   Standard   T for H0:
     Variable   DF   Estimate   Error   Parameter=0   Prob > |T|

     INTERCEP   1   1.887370   1.186847   1.590   0.1313
     LNY_P   1   0.764657   0.110168   6.941   0.0001
     LNI   1   -0.077564   0.048797   -1.590   0.1315
     LNS1   1   -0.234061   0.141070   -1.659   0.1165
     LNS2   1   0.126953   0.250346   0.507   0.6190

e. HAUSMAN TEST

                              Hausman Test
                  2SLS vs OLS for SUPPLY Equation

                                HAUSMAN
                               6.7586205

                  2SLS vs OLS for DEMAND Equation

                                HAUSMAN
                               2.3088302

SAS PROGRAM

```
Data Laffer1; Input TM RM Y S2 i S1 P;
Cards;

Data Laffer; Set Laffer1;

LNTM_P=LOG(TM/P);
LNRM_P=LOG(RM/P);
LNi=LOG(i);
LNY_P=LOG(Y/P);
LNS1=LOG(S1);
LNS2=LOG(S2);
T=21;

Proc syslin data=Laffer outest=ols_s outcov;
 SUPPLY: MODEL LNTM_P=LNRM_P LNi;
TITLE ' OLS ESTIMATION OF MONEY SUPPLY FUNCTION';

Proc syslin data=Laffer outest=ols_d outcov;
 DEMAND: MODEL LNTM_P=LNY_P  LNi LNS1 LNS2;
TITLE ' OLS ESTIMATION OF MONEY DEMAND FUNCTION';

Proc syslin 2SLS data=Laffer outest=TSLS_s outcov;
 ENDO LNTM_P LNi;
 INST LNRM_P LNY_P LNS1 LNS2;
 SUPPLY: MODEL LNTM_P=LNRM_P LNi/overid;
TITLE ' 2SLS ESTIMATION OF MONEY SUPPLY FUNCTION';

Proc syslin 2SLS data=Laffer outest=TSLS_d outcov;
 ENDO LNTM_P LNi;
 INST LNRM_P LNY_P LNS1 LNS2;
 DEMAND: MODEL LNTM_P=LNY_P LNi LNS1 LNS2/overid;
TITLE ' 2SLS ESTIMATION OF MONEY DEMAND FUNCTION';

Proc SYSLIN 3SLS data=Laffer outest=S_3SLS outcov;
 ENDO LNTM_P LNi;
 INST LNRM_P LNY_P LNS1 LNS2;
 SUPPLY: MODEL LNTM_P=LNRM_P LNi;
 DEMAND: MODEL LNTM_P=LNY_P  LNi LNS1 LNS2;
TITLE ' 3SLS ESTIMATION';
RUN;

PROC IML;
TITLE 'HAUSMAN TEST';
```

```
use ols_s; read all var {intercep lnrm_p lni} into ols1;
olsbt_s=ols1[1,]; ols_v_s=ols1[2:4,];
use ols_d; read all var {intercep lny_p lni lns1 lns2} into ols2;
olsbt_d=ols2[1,]; ols_v_d=ols2[2:6,];

use tsls_s; read all var {intercep lnrm_p lni} into tsls1;
tslsbt_s=tsls1[13,]; tsls_v_s=tsls1[14:16,];
use tsls_d; read all var {intercep lny_p lni lns1 lns2} into
tsls2;
tslsbt_d=tsls2[13,]; tsls_v_d=tsls2[14:18,];

d=tslsbt_s`-olsbt_s`; varcov=tsls_v_s-ols_v_s;
Hausman=d`*inv(varcov)*d;
print 'Hausman Test',, '2SLS vs OLS for SUPPLY Equation',,
Hausman;

d=tslsbt_d`-olsbt_d`; varcov=tsls_v_d-ols_v_d;
Hausman=d`*inv(varcov)*d;
print '2SLS vs OLS for DEMAND Equation',, Hausman;
```

11.20

a. Writing this system of two equations in matrix form, as described in (11.47) we get

$$B = \begin{bmatrix} 1 & \alpha_1 \\ 1 & -\beta_1 \end{bmatrix}; \Gamma = \begin{bmatrix} -\alpha_0 & -\alpha_2 \\ -\beta_0 & 0 \end{bmatrix}; y_t = \begin{pmatrix} Q_t \\ P_t \end{pmatrix}$$

$x'_t = [1, X_t]$ and $u'_t = (u_{1t}, u_{2t})$.

There are no zero restrictions for the demand equation and only one zero restriction for the supply equation (X_t does not appear in the supply equation). Therefore, the demand equation is not identified by the order condition since the number of excluded exogenous variables $k_2 = 0$ is less than the number of right hand side included endogenous variables $g_1 = 1$. Similarly, the supply equation is just-identified by the order condition since $k_2 = g_1 = 1$.

By the rank condition, only the supply equation need be checked for identification. In this case, $\phi' = (0,1)$ since $\gamma_{22} = 0$. Therefore,

$$A\phi = [B, \Gamma]\phi = \begin{pmatrix} \gamma_{12} \\ \gamma_{22} \end{pmatrix} = \begin{pmatrix} -\alpha_2 \\ 0 \end{pmatrix}.$$

This is of rank one as long as $\alpha_2 \neq 0$, i.e., as long as X_t is present in the demand equation. One can also premultiply this system of two equations by

$$F = \begin{bmatrix} f_{11} & f_{12} \\ f_{21} & f_{22} \end{bmatrix},$$

the transformed matrix FB should satisfy the following normalization restrictions:

$$f_{11} + f_{12} = 1 \quad \text{and} \quad f_{21} + f_{22} = 1.$$

Also, FΓ must satisfy the following zero restriction:

$$-\alpha_2 f_{21} + 0 f_{22} = 0.$$

If $\alpha_2 \neq 0$, then $f_{21} = 0$ which also gives from the second normalization equation that $f_{22} = 1$. Hence, the second row of F is indeed the second row of an identity matrix and the supply equation is identified. However, for the demand equation, the only restriction is the normalization restriction $f_{11} + f_{12} = 1$ which is not enough for identification.

b. The OLS estimator of the supply equation yields $\hat{\beta}_{1,ols} = \sum_{t=1}^{T} p_t q_t / \sum_{t=1}^{T} p_t^2$

where $p_t = P_t - \bar{P}$ and $q_t = Q_t - \bar{Q}$. Substituting $q_t = \beta_1 p_t + (u_{2t} - \bar{u}_2)$ we get

$$\hat{\beta}_{1,ols} = \beta_1 + \sum_{t=1}^{T} p_t(u_{2t} - \bar{u}_2) / \sum_{t=1}^{T} p_t^2.$$

The reduced form equation for P_t yields $P_t = \dfrac{\alpha_o - \beta_o}{\alpha_1 + \beta_1} + \dfrac{\alpha_2}{\alpha_1 + \beta_1} X_t + \dfrac{u_{1t} - u_{2t}}{\alpha_1 + \beta_1}$

Hence, $P_t = \dfrac{\alpha_2}{\alpha_1 + \beta_1} x_t + \dfrac{(u_{1t} - \bar{u}_1) - (u_{2t} - \bar{u}_2)}{(\alpha_1 + \beta_1)}$

where $x_t = X_t - \bar{X}$. Defining $m_{xx} = \text{plim} \sum_{t=1}^{T} x_t^2 / T$, we get

$$m_{pp} = \text{plim}_{T \to \infty} \sum_{t=1}^{T} p_t^2 / T = \left(\dfrac{\alpha_2}{\alpha_1 + \beta_1}\right)^2 m_{xx} + \dfrac{m_{u_1 u_1} + m_{u_2 u_2} - 2m_{u_1 u_2}}{(\alpha_1 + \beta_1)^2}$$

using the fact that $m_{xu_1} = m_{xu_2} = 0$. Also,

$$m_{pu_2} = \text{plim} \sum_{t=1}^{T} p_t u_{2t}/T = \dfrac{m_{u_1 u_2} - m_{u_2 u_2}}{(\alpha_1 + \beta_1)}$$

Chapter 11: Simultaneous Equations Model

using the fact that $m_{xu_2} = 0$. Hence,

$$\text{plim}_{T\to\infty} \hat{\beta}_{1,ols} = \beta_1 + \frac{m_{pu_2}}{m_{pp}} = \beta_1 + \frac{(\sigma_{12} - \sigma_{22})(\alpha_1 + \beta_1)}{[\sigma_{11} + \sigma_{22} - 2\sigma_{12} + \alpha_2^2 m_{xx}]}$$

where $m_{u_1 u_2} = \sigma_{12}$, $m_{u_1 u_1} = \sigma_{11}$ and $m_{u_2 u_2} = \sigma_{22}$. The second term gives the simultaneous equation bias of $\hat{\beta}_{1,ols}$.

c. If $\sigma_{12} = 0$, then

$$\text{plim}_{T\to\infty}(\hat{\beta}_{1,ols} - \beta_1) = -\frac{\sigma_{22}(\alpha_1 + \beta_1)}{[\sigma_{11} + \sigma_{22} + \alpha_2^2 m_{xx}]}.$$

The denominator is positive and α_1 and β_1 are positive, hence this bias is negative.

11.21 a. X_2, X_3, X_4 and X_5 are excluded from the first equation. y_2 is the only included right hand side endogenous variable. Therefore, $k_2 = 4$ and $g_1 = 1$, so the first equation is over-identified with degree of over-identification equal to $k_2 - g_1 = 3$. X_1, X_3, X_4 and X_5 are excluded from the second equation, while y_1 and y_3 are the included right hand side endogenous variables. Therefore, $k_2 = 4$ and $g_1 = 2$, so the second equation is over-identified with degree of over-identification equal to $k_2 - g_1 = 2$. X_1 and X_2 are excluded from the third equation and there are no right hand side included endogenous variables. Therefore, $k_2 = 2$ and $g_1 = 0$ and the third equation is over-identified with degree of over-identification equal to $k_2 - g_1 = 2$.

b. Regress y_1 and y_3 on all the X's including the constant, i.e., $[1, X_1, X_2, \ldots, X_5]$. Get $\hat{y}_1$ and $\hat{y}_3$ and regress y_2 on a constant, $\hat{y}_1$, $\hat{y}_3$ and X_2 in the second-step regression.

c. For the first equation, regressing y_2 on X_2 and X_3 yields

$$y_2 = \hat{\pi}_{21} + \hat{\pi}_{22} X_2 + \hat{\pi}_{23} X_3 + \hat{v}_2 = \hat{y}_2 + \hat{v}_2$$

with $\sum_{t=1}^{T} \hat{v}_{2t} = \sum_{t=1}^{T} \hat{v}_{2t} X_{2t} = \sum_{t=1}^{T} \hat{v}_{2t} X_{3t} = 0$ by the property of least squares.

Replacing y_2 by $\hat{y}_2$ in equation (1) yields

$$y_1 = \alpha_1 + \beta_2(\hat{y}_2+\hat{v}_2) + \gamma_1 X_1 + u_1 = \alpha_1 + \beta_2\hat{y}_2 + \gamma_1 X_1 + (\beta_2\hat{v}_2 + u_1).$$

In this case, $\sum_{t=1}^{T} \hat{y}_{2t}\hat{v}_{2t} = 0$ because the predictors and residuals of the same regression are uncorrelated. Also, $\sum_{t=1}^{T} \hat{y}_{2t} u_{1t} = 0$ since $\hat{y}_{2t}$ is a linear combination of X_2 and X_3 and the latter are exogenous. However, $\sum_{t=1}^{T} X_{1t}\hat{v}_{2t}$ is not necessarily zero since X_1 was not included in the first stage regression. Hence, this estimation method does not necessarily lead to consistent estimates.

d. The test for over-identification for equation (1) can be obtained from the F-test given in equation (11.76) with RRSS* obtained from the second stage regression residual sum of squares of 2SLS run on equation (1), i.e., the regression of y_1 on $[1,\hat{y}_2,X_1]$ where $\hat{y}_2$ is obtained from the regression of y_2 on $[1,X_1,X_2,..,X_5]$. The URSS* is obtained from the residual sum of squares of the regression of y_1 on all the set of instruments, i.e., $[1,X_1,..,X_5]$. The URSS is the 2SLS residual sum of squares of equation (1) as reported from any 2SLS package. This is obtained as

$$\text{URSS} = \sum_{t=1}^{T} (y_{1t} - \hat{\alpha}_{1,2SLS} - \hat{\beta}_{2,2SLS} y_{2t} - \hat{\gamma}_{1,2SLS} X_{1t})^2.$$

Note that this differs from RRSS* in that y_{2t} and not $\hat{y}_{2t}$ is used in the computation of the residuals. ℓ, the number of instruments is 6 and $k_1 = 2$ while $g_1 = 1$. Hence, the numerator degrees of freedom of the F-test is $\ell - (g_1+k_1) = 3$ whereas the denominator degrees of freedom of the F-test is $T - \ell = T - 6$. This is equivalent to running 2SLS residuals on the matrix of all predetermined variables $[1,X_1,..,X_5]$ and computing T times the uncentered R^2 of this regression. This is asymptotically distributed as χ^2 with 3 degrees of freedom. Large values of this statistic reject the over-identification restrictions.

11.23 *Identification and Estimation of a Simple Two-Equation Model.* This solution is based upon Singh and Bhat (1988).

a. *Identification with no further information available.* The structural model being studied is $\begin{bmatrix} 1 & -\beta \\ -1 & 1 \end{bmatrix} \begin{bmatrix} y_{t1} \\ y_{t2} \end{bmatrix} = \begin{bmatrix} \alpha \\ \gamma \end{bmatrix} + \begin{bmatrix} u_{t1} \\ u_{t2} \end{bmatrix}$, with

$\Sigma = E \begin{bmatrix} u_{t1}^2 & u_{t1}u_{t2} \\ u_{t2}u_{t1} & u_{t2}^2 \end{bmatrix} = \begin{bmatrix} \sigma_1^2 & \sigma_{12} \\ \sigma_{12} & \sigma_2^2 \end{bmatrix}$. The reduced-form equation is given by

$\begin{bmatrix} y_{t1} \\ y_{t2} \end{bmatrix} = \begin{bmatrix} \pi_{11} \\ \pi_{12} \end{bmatrix} + \begin{bmatrix} v_{t1} \\ v_{t2} \end{bmatrix}$, with $\Omega = E \begin{bmatrix} v_{t1}^2 & v_{t1}v_{t2} \\ v_{t2}v_{t1} & v_{t2}^2 \end{bmatrix} = \begin{bmatrix} \omega_1^2 & \omega_{12} \\ \omega_{12} & \omega_2^2 \end{bmatrix}$.

The structural and reduced form parameters are related as follows:

$\begin{bmatrix} 1 & -\beta \\ -1 & 1 \end{bmatrix} \begin{bmatrix} \pi_{11} \\ \pi_{12} \end{bmatrix} = \begin{bmatrix} \alpha \\ \gamma \end{bmatrix}$

or

$\alpha = \pi_{11} - \beta\pi_{12}$

$\gamma = \pi_{12} - \pi_{11}$

We assume that the reduced-form parameters π_{11}, π_{12}, ω_1^2, ω_{12} and ω_2^2 have been estimated. The identification problem is one of recovering α, γ, β σ_1^2 and σ_{12} and σ_2^2 from the reduced-form parameters.

From the second equation, it is clear that γ is identified. We now examine the relation between Σ and Ω given by $\Sigma = B\Omega B'$ where $B = \begin{bmatrix} 1 & -\beta \\ -1 & 1 \end{bmatrix}$. Thus, we have

$\sigma_1^2 = \omega_1^2 - 2\beta\omega_{12} + \beta^2\omega_2^2$,

$\sigma_{12} = -\omega_1^2 + (1+\beta)\omega_{12} - \beta\omega_2^2$,

$\sigma_2^2 = \omega_1^2 - 2\omega_{12} + \omega_2^2$.

Knowledge of ω_1^2, ω_{12}, and ω_2^2 is enough to recover σ_2^2. Hence, σ_2^2 is identified. Note, however, that σ_1^2 and σ_{12} depend on β. Also, α is dependent on β. Hence, given any $\beta = \beta^*$, we can solve for α, σ_1^2 and σ_{12}. Thus, in principle, we have an infinity of solutions as the choice of β varies. Identification could have been studied using rank and order conditions. However, in a simple framework as this, it is easier to check identification directly.

Identification can be reached by imposing an extra restriction

$f(\beta,\alpha,\sigma_1^2,\sigma_{12}) = 0$, which is dependent of the three equations we have for σ_1^2, σ_{12} and α. For example, $\beta = 0$ gives a recursive structure. We now study the particular case where $f(\beta,\alpha,\sigma_1^2,\sigma_{12}) = \sigma_{12} = 0$.

b. *Identification with $\sigma_{12} = 0$*: When $\sigma_{12} = 0$, one can solve immediately for β from the σ_{12} equation:

$$\beta = \frac{\omega_{12} - \omega_1^2}{\omega_2^2 - \omega_{12}}.$$

Therefore, β can be recovered from ω_1^2, ω_{12}, and ω_2^2. Given β, all the other parameters (σ_1^2, α) can be recovered as discussed above.

c. *OLS estimation of β when $\sigma_{12} = 0$*: The OLS estimator of β is

$$\hat{\beta}_{ols} = \frac{\sum_{t=1}^{T}(y_{t1}-\bar{y}_1)(y_{t2}-\bar{y}_2)}{\sum_{t=1}^{T}(y_{t2}-\bar{y}_2)^2} = \beta + \frac{\sum_{t=1}^{T}(u_{t1}-\bar{u}_1)(y_{t1}-\bar{y}_1)}{\sum_{t=1}^{T}(y_{t2}-\bar{y}_2)^2} + \frac{\sum_{t=1}^{T}(u_{t1}-\bar{u}_1)(u_{t2}-\bar{u}_2)}{\sum_{t=1}^{T}(y_{t2}-\bar{y}_2)^2}.$$

The second term is non-zero in probability limit as y_{t1} and u_{t1} are correlated from the first structural equation. Hence $\hat{\beta}_{ols}$ is not consistent.

d. From the second structural equation, we have $y_{t2} - y_{t1} = \gamma + u_{t2}$, which means that $(y_{t2}-\bar{y}_2) - (y_{t1}-\bar{y}_1) = u_{t2} - \bar{u}_2$, or $z_t = u_{t2} - \bar{u}_2$. When $\sigma_{12} = 0$, this z_t is clearly uncorrelated with u_{t1} and by definition correlated with y_{t2}. Therefore, $z_t = (y_{t2}-\bar{y}_2) - (y_{t1}-\bar{y}_1)$ may be taken as instruments, and thus

$$\tilde{\beta}_{IV} = \frac{\sum_{t=1}^{T} z_t(y_{t1}-\bar{y}_1)}{\sum_{t=1}^{T} z_t(y_{t2}-\bar{y}_2)} = \frac{\sum_{t=1}^{T}(y_{t1}-\bar{y}_1)[(y_{t2}-\bar{y}_2)-(y_{t1}-\bar{y}_1)]}{\sum_{t=1}^{T}(y_{t2}-\bar{y}_2)[(y_{t2}-\bar{y}_2)-(y_{t1}-\bar{y}_1)]}$$

is a consistent estimator of β.

e. From the identification solution of β as a function of reduced-form parameters in part (b), one can obtain an alternate estimator of β by replacing the reduced-form parameters by their sample moments. This yields the indirect least squares estimator of β:

Chapter 11: Simultaneous Equations Model

$$\tilde{\beta}_{ILS} = \frac{\sum_{t=1}^{T}(y_{t1}-\bar{y}_1)(y_{t2}-\bar{y}_2) - \sum_{t=1}^{T}(y_{t1}-\bar{y}_1)^2}{\sum_{t=1}^{T}(y_{t2}-\bar{y}_2)^2 - \sum_{t=1}^{T}(y_{t1}-\bar{y}_1)(y_{t2}-\bar{y}_2)} = \hat{\beta}_{IV}$$

otained in part (d). The OLS estimators of the reduced-form parameters are consistent. Therefore, $\tilde{\beta}_{ILS}$ which is a continuous tranform of those estimators is also consistent. It is however, useful to show this consistency directly since it brings out the role of the identifying restriction.

From the structural equation, we have

$$\begin{bmatrix} 1 & -\beta \\ -1 & 1 \end{bmatrix} \begin{bmatrix} y_{t1}-\bar{y}_1 \\ y_{t2}-\bar{y}_2 \end{bmatrix} = \begin{bmatrix} u_{t1}-\bar{u}_1 \\ u_{t2}-\bar{u}_2 \end{bmatrix}.$$

Using the above relation and with some algebra, it can be shown that for $\beta \neq 1$:

$$\tilde{\beta}_{ILS} = \frac{\sum_{t=1}^{T}(u_{t1}-\bar{u}_1)(u_{t2}-\bar{u}_1) + \beta\sum_{t=1}^{T}(u_{t2}-\bar{u}_2)^2}{\sum_{t=1}^{T}(u_{t1}-\bar{u}_1)(u_{t2}-\bar{u}_2) + \sum_{t=1}^{T}(u_{t2}-\bar{u}_2)^2} \quad \text{since,}$$

$$\text{plim}\frac{1}{T}\sum_{t=1}^{T}(u_{t1}-\bar{u}_1)(u_{t2}-\bar{u}_2) = \sigma_{12}, \quad \text{and} \quad \text{plim}\frac{1}{T}\sum_{t=1}^{T}(u_{t2}-\bar{u}_2)^2 = \sigma_2^2.$$

Therefore, $\text{plim}\,\tilde{\beta}_{ILS} = (\sigma_{12} + b\sigma_2^2)/(\sigma_{12} + \sigma_2^2)$. The restriction $\sigma_{12} = 0$ implies $\text{plim}\,\tilde{\beta}_{ILS} = \beta$ which proves consistency. It is clear that $\tilde{\beta}_{ILS}$ can be interpreted as a method-of-moments estimator.

11.24 *Errors in Measurement and the Wald (1940) Estimator.* This is based on Farebrother (1987).

a. The simple IV estimator of β with instrument z is given by $\hat{\beta}_{IV} = \sum_{i=1}^{n} z_i y_i / \sum_{i=1}^{n} z_i x_i$. But $\sum_{i=1}^{n} z_i y_i$ = sum of the y_i's from the second sample minus the sum of the y_i's from the first sample. If n is even, then $\sum_{i=1}^{n} z_i y_i = \frac{n}{2}(\bar{y}_2 - \bar{y}_1)$. Similarly, $\sum_{i=1}^{n} z_i x_i = \frac{n}{2}(\bar{x}_2 - \bar{x}_1)$, so that $\hat{\beta}_{IV} = (\bar{y}_2 - \bar{y}_1)/(\bar{x}_2 - \bar{x}_1) = \hat{\beta}_W$.

b. Let $\rho^2 = \sigma_u^2/(\sigma_u^2 + \sigma_*^2)$ and $\tau^2 = \sigma_*^2/(\sigma_u^2 + \sigma_*^2)$ then $w_i = \rho^2 x_i^* - \tau^2 u_i$ has mean $E(w_i) = 0$ since $E(x_i^*) = E(u_i) = 0$. Also,

$$\text{var}(w_i) = (\rho^2)^2 \sigma_*^2 + (\tau^2)^2 \sigma_u^2 = \sigma_*^2 \sigma_u^2/(\sigma_*^2 + \sigma_u^2).$$

Since w_i is a linear combination of Normal random variables, it is also Normal.

$$E(x_i w_i) = \rho^2 E(x_i x_i^*) - \tau^2 E(x_i u_i) = \rho^2 \sigma_*^2 - \tau^2 \sigma_u^2 = 0.$$

c. Now $\tau^2 x_i + w_i = \tau^2(x_i^* + u_i) + \rho^2 x_i^* - \tau^2 u_i = (\tau^2 + \rho^2) x_i^* = x_i^*$ since $\tau^2 + \rho^2 = 1$.

Using this result and the fact that $y_i = \beta x_i^* + \epsilon_i$, we get

$$\hat{\beta}_W = \sum_{i=1}^n z_i y_i / \sum_{i=1}^n z_i x_i$$

$$= \beta \tau^2 + \beta \left(\sum_{i=1}^n z_i w_i / \sum_{i=1}^n z_i x_i \right) + \left(\sum_{i=1}^n z_i \epsilon_i / \sum_{i=1}^n z_i x_i \right)$$

so that $E(\hat{\beta}_W/x_1,..,x_n) = \beta \tau^2$. This follows because w_i and ϵ_i are independent of x_i and therefore independent of z_i which is a function of x_i.

Similarly, $\hat{\beta}_{ols} = \sum_{i=1}^n x_i y_i / \sum_{i=1}^n x_i^2 = \beta \tau^2 + \beta \sum_{i=1}^n w_i x_i / \sum_{i=1}^n x_i^2 + \sum_{i=1}^n x_i \epsilon_i / \sum_{i=1}^n x_i^2$

so that $E(\hat{\beta}_{ols}/x_1,..,x_n) = \beta \tau^2$ since w_i and ϵ_i are independent of x_i. Therefore, the bias of $\hat{\beta}_{ols}$ = bias of $\hat{\beta}_W = (\beta \tau^2 - \beta) = -\beta \sigma_u^2/(\sigma_*^2 + \sigma_u^2)$ for all choices of z_i which are a function of x_i.

11.25 *Comparison of t-ratios.* This is based on Farebrother (1991). Let $Z_1 = [y_2, X]$, then by the FWL-Theorem, OLS on the first equation is equivalent to that on $\bar{P}_X y_1 = \bar{P}_X y_2 \alpha + \bar{P}_X u_1$. This yields $\hat{\alpha}_{ols} = (y_2' \bar{P}_X y_2)^{-1} y_2' \bar{P}_X y_1$ and the residual sum of squares are equivalent

$$y_1' \bar{P}_Z y_1 = y_1' \bar{P}_X y_1 - y_1' \bar{P}_X y_2 (y_2' \bar{P}_X y_2)^{-1} y_2' \bar{P}_X y_1$$

so that $\text{var}(\hat{\alpha}_{ols}) = s_1^2 (y_2' \bar{P}_X y_2)^{-1}$ where $s_1^2 = y_1' \bar{P}_Z y_1/(T-K)$. The t-ratio for $H_o^a; \alpha = 0$ is

$$t = \frac{y_2' \bar{P}_X y_1}{y_2' \bar{P}_X y_2} \left(\frac{(T-K) y_2' \bar{P}_X y_2}{y_1' \bar{P}_Z y_1} \right)^{1/2} = (T-K)^{1/2} (y_2' \bar{P}_X y_1) [(y_1' \bar{P}_X y_1)(y_2' \bar{P}_X y_2) - (y_1' \bar{P}_X y_2)^2]^{-1/2}$$

This expresssion is unchanged if the roles of y_1 and y_2 are reversed so that the t-ratio for $H_o^b; \gamma = 0$ in the second equation is the same as that for $H_o^a; \alpha = 0$ in the first equation.

For y_1 and y_2 jointly Normal with correlation coefficient ρ, Farebrother (1991) shows that the above two tests correspond to a test for $\rho = 0$ in the bivariate Normal model so that it makes sense that the two statistics are identical.

REFERENCES

Farebrother, R.W. (1987), "The Exact Bias of Wald's Estimation," *Econometric Theory*, Solution 85.3.1, 3: 162.

Farebrother, R.W. (1991), "Comparison of t-Ratios," *Econometric Theory*, Solution 90.1.4, 7: 145-146.

Singh, N. and A N. Bhat (1988), "Identification and Estimation of a Simple Two-Equation Model," *Econometric Theory*, Solution 87.3.3, 4: 542-545.

CHAPTER 12
Pooling Time-Series of Cross-Section Data

12.1

a. Premultiplying (12.11) by Q one gets

$$Qy = \alpha Q\iota_{NT} + QX\beta + QZ_\mu \mu + Qv$$

But $PZ_\mu = Z_\mu$ and $QZ_\mu = 0$. Also, $P\iota_{NT} = \iota_{NT}$ and $Q\iota_{NT} = 0$. Hence, this transformed equation reduces to (12.12)

$$Qy = QX\beta + Qv$$

Now $E(Qv) = QE(v) = 0$ and $var(Qv) = Q\, var(v)\, Q' = \sigma_v^2 Q$, since $var(v) = \sigma_v^2 I_{NT}$

and Q is symmetric and idempotent.

b. For the general linear model $y = X\beta + u$ with $E(uu') = \Omega$, a necessary and sufficient condition for OLS to be equivalent to GLS is given by $X'\Omega^{-1}\bar{P}_X$ where $\bar{P}_X = I - P_X$ and $P_X = X(X'X)^{-1}X'$, see equation (9.7) of Chapter 9. For equation (12.12), this condition can be written as

$$(X'Q)(Q/\sigma_v^2)\bar{P}_{QX} = 0$$

using the fact that Q is idempotent, the left hand side can be written as

$$(X'Q)\bar{P}_{QX}/\sigma_v^2$$

which is clearly 0, since $\bar{P}_{QX}$ is the orthogonal projection of QX.

One can also use Zyskind's condition $P_X\Omega = \Omega P_X$ given in equation (9.8) of Chapter 9. For equation (12.12), this condition can be written as

$$P_{QX}(\sigma_v^2 Q) = (\sigma_v^2 Q)P_{QX}$$

But, $P_{QX} = QX(X'QX)^{-1}X'Q$. Hence, $P_{QX}Q = P_{QX}$ and $QP_{QX} = P_{QX}$ and the condition is met. Alternatively, we can verify that OLS and GLS yield the same

Chapter 12: Pooling Time-Series of Cross-Section Data   265

estimates. Note that $Q = I_{NT} - P$ where $P = I_N \otimes \bar{J}_T$ is idempotent and is therefore its own generalized inverse. The variance-covariance matrix of the disturbance $\tilde{v} = Qv$ in (12.12) is $E(\tilde{v}\tilde{v}') = E(Qvv'Q) = \sigma_v^2 Q$ with generalized inverse Q/σ_v^2. OLS on (12.12) yields

$$\hat{\beta} = (X'QQX)^{-1}X'QQy = (X'QX)^{-1}X'Qy$$

which is $\tilde{\beta}$ given by (12.13). Also, GLS on (12.12) using generalized inverse yields

$$\hat{\beta} = (X'QQQX)^{-1}X'QQQy = (X'QX)^{-1}X'Qy = \tilde{\beta}.$$

c. The Frisch-Waugh Lovell (FWL) theorem of Davidson and MacKinnon (1993, p. 19) states that for

$$y = X_1\beta_1 + X_2\beta_2 + u \qquad (1)$$

If we premultiply by the orthogonal projection of X_2 given by

$$M_2 = I - X_2(X_2'X_2)^{-1}X_2',$$

then $M_2 X_2 = 0$ and (1) becomes

$$M_2 y = M_2 X_1 \beta_1 + M_2 u \qquad (2)$$

The OLS estimate of β_1 from (2) is the same as that from (1) *and* the residuals from (1) are the same as the residuals from (2). This was proved in section 7.3. Here we will just use this result. For the model in (12.11)

$$y = Z\delta + Z_\mu \mu + v$$

Let $Z = X_1$ and $Z_\mu = X_2$. In this case, $M_2 = I - Z_\mu(Z_\mu' Z_\mu)^{-1}Z_\mu' = I - P = Q$. In this case, premultiplying by M_2 is equivalent to premultiplying by Q and equation (2) above becomes equation (12.12) in the text. By the FWL theorem, OLS on (12.12) which yields (12.13) is the same as OLS on (12.11). Note that $Z = [\iota_{NT}, X]$ and $QZ = [0, QX]$ since $Q\iota_{NT} = 0$.

12.2

a. From (12.17) we get

$$\Omega = \sigma_\mu^2(I_N \otimes J_T) + \sigma_\nu^2(I_N \otimes I_T)$$

Replacing J_T by $T\bar{J}_T$, and I_T by $(E_T + \bar{J}_T)$ where E_T is by definition $(I_T - \bar{J}_T)$, one gets

$$\Omega = T\sigma_\mu^2(I_N \otimes \bar{J}_T) + \sigma_\nu^2(I_N \otimes E_T) + \sigma_\nu^2(I_N \otimes \bar{J}_T)$$

collecting terms with the same matrices, we get

$$\Omega = (T\sigma_\mu^2 + \sigma_\nu^2)(I_N \otimes \bar{J}_T) + \sigma_\nu^2(I_N \otimes E_T) = \sigma_1^2 P + \sigma_\nu^2 Q$$

where $\sigma_1^2 = T\sigma_\mu^2 + \sigma_\nu^2$.

b. $P = Z_\mu(Z'_\mu Z_\mu)^{-1}Z'_\mu = I_N \otimes \bar{J}_T$ is a projection matrix of Z_μ. Hence, it is by definition symmetric and idempotent. Similarly, $Q = I_{NT} - P$ is the orthogonal projection matrix of Z_μ. Hence, Q is also symmetric and idempotent. By definition, $P + Q = I_{NT}$. Also, $PQ = P(I_{NT}-P) = P - P^2 = P - P = 0$.

c. From (12.18) and (12.19) one gets

$$\Omega\Omega^{-1} = (\sigma_1^2 P + \sigma_\nu^2 Q)(\frac{1}{\sigma_1^2}P + \frac{1}{\sigma_\nu^2}Q) = P + Q = I_{NT}$$

since $P^2 = P$, $Q^2 = Q$ and $PQ = 0$ as verified in part (b). Similarly, $\Omega^{-1}\Omega = I_{NT}$.

d. From (12.20) one gets

$$\Omega^{-½}\Omega^{-½} = (\frac{1}{\sigma_1}P + \frac{1}{\sigma_\nu}Q)(\frac{1}{\sigma_1}P + \frac{1}{\sigma_\nu}Q) = \frac{1}{\sigma_1^2}P^2 + \frac{1}{\sigma_\nu^2}Q^2$$

$$= \frac{1}{\sigma_1^2}P + \frac{1}{\sigma_\nu^2}Q = \Omega^{-1}$$

using the fact that $P^2 = P$, $Q^2 = Q$ and $PQ = 0$.

12.3 From (12.20) one gets $\sigma_\nu\Omega^{-½} = Q + (\sigma_\nu/\sigma_1)P$

Therefore,

$$y^* = \sigma_\nu\Omega^{-½}y = Qy + (\sigma_\nu/\sigma_1)Py = y - Py + (\sigma_\nu/\sigma_1)Py = y - (1-(\sigma_\nu/\sigma_1))Py = y - \theta Py$$

where $\theta = 1 - (\sigma_v/\sigma_1)$. Recall that the typical element of Py is $\bar{y}_{i.}$, therefore, the typical element of y^* is $y_{it}^* = y_{it} - \theta \bar{y}_{i.}$.

12.4 $E(u'Pu) = E(tr(uu'P)) = tr(E(uu')P) = tr(\Omega P)$. From (12.21), $\Omega P = \sigma_1^2 P$ since $PQ = 0$. Hence, from (12.18),

$$E(\hat{\sigma}_1^2) = \frac{E(u'Pu)}{tr(P)} = \frac{tr(\sigma_1^2 P)}{tr(P)} = \sigma_1^2.$$

Similarly, $E(u'Qu) = tr(\Omega Q) = tr(\sigma_v^2 Q)$ where the last equality follows from (12.18) and the fact that $\Omega Q = \sigma_v^2 Q$ since $PQ = 0$. Hence, from (12.22),

$$E(\hat{\sigma}_v^2) = \frac{E(u'Qu)}{tr(Q)} = \frac{tr(\sigma_v^2 Q)}{tr(Q)} = \sigma_v^2.$$

12.5

a. $\hat{\sigma}_v^2$ given by (12.23) is the s^2 from the regression given by (12.11). In fact

$$\hat{\sigma}_v^2 = y'Q[I_{NT} - P_{QX}]Qy/[N(T-1)-K]$$

where $P_{QX} = QX(X'QX)^{-1}X'Q$. Substituting Qy from (12.12) into $\hat{\sigma}_v^2$, one gets

$$\hat{\sigma}_v^2 = v'Q[I_{NT} - P_{QX}]Qv/[N(T-1)-K]$$

with $Qv \sim (0, \sigma_v^2 Q)$. Therefore,

$$E[v'Q[I_{NT} - P_{QX}]Qv] = E[tr\{Qvv'Q[I_{NT} - P_{QX}]\}]$$

$$= \sigma_v^2 tr\{Q - QP_{QX}\} = \sigma_v^2 \{N(T-1) - tr(P_{QX})\}$$

where the last equality follows from the fact that $QP_{QX} = P_{QX}$. Also, $tr(P_{QX}) = tr(X'QX)(X'QX)^{-1} = tr(I_K) = K$.

Hence, $E(\hat{\sigma}_v^2) = \sigma_v^2$.

b. Similarly, $\hat{\sigma}_1^2$ given by (12.26) is the s^2 from the regression given by

$$Py = PZ\delta + Pu$$

In fact,

$$\hat{\sigma}_1^2 = y'P[I_{NT} - P_{PZ}]Py/(N-K-1)$$

where $P_{PZ} = PZ(Z'PZ)^{-1}Z'P$. Substituting Py into $\hat{\sigma}_1^2$ one gets

$$\hat{\sigma}_1^2 = u'P[I_{NT} - P_{PZ}]Pu/(N-K-1) = u'P[I_{NT} - P_{PZ}]Pu/(N-K-1)$$

with Pu $\sim (0, \sigma_1^2 P)$ as can be easily verified from (12.18). Therefore,

$$E(u'P[I_{NT} - P_{PZ}]Pu) = E[tr\{Puu'P(I_{NT} - P_{PZ})\}] = \sigma_1^2 tr\{P - PP_{PZ}\} = \sigma_1^2 \{N - tr(P_{PZ})\}$$

where the last equality follows from the fact that $PP_{PZ} = P_{PZ}$. Also,

$tr(P_{PZ}) = tr(Z'PZ)(Z'PZ)^{-1} = tr(I_{K'}) = K'$.

Hence, $E(\hat{\sigma}_1^2) = \sigma_1^2$.

12.6

a. OLS on (12.27) yields

$$\hat{\delta}_{ols} = [(Z'Q, Z'P)\begin{pmatrix} QZ \\ PZ \end{pmatrix}]^{-1}(Z'Q, Z'P)\begin{pmatrix} Qy \\ Py \end{pmatrix}$$
$$= (Z'QZ + Z'PZ)^{-1}(Z'Qy + Z'Py)$$
$$= (Z'(Q+P)Z)^{-1}Z'(Q+P)y = (Z'Z)^{-1}Z'y$$

since $Q + P = I_{NT}$.

b. GLS on (12.27) yields

$$\hat{\delta}_{ols} = \left((Z'Q, Z'P)\begin{bmatrix} \sigma_v^2 Q & 0 \\ 0 & \sigma_1^2 P \end{bmatrix}^{-1}\begin{pmatrix} QZ \\ PZ \end{pmatrix}\right)^{-1}(Z'Q, Z'P)\begin{bmatrix} \sigma_v^2 Q & 0 \\ 0 & \sigma_1^2 P \end{bmatrix}^{-1}\begin{pmatrix} Qy \\ Py \end{pmatrix}$$

Using generalized inverse

$$\begin{bmatrix} \sigma_v^2 Q & 0 \\ 0 & \sigma_1^2 P \end{bmatrix}^{-1} = \begin{bmatrix} Q/\sigma_v^2 & 0 \\ 0 & P/\sigma_1^2 \end{bmatrix}$$

one gets

$$\hat{\delta}_{GLS} = [(Z'QZ)/\sigma_v^2 + (Z'PZ)/\sigma_1^2]^{-1}[(Z'Qy)/\sigma_v^2 + (Z'Py)/\sigma_1^2] = (Z'\Omega^{-1}Z)^{-1}Z'\Omega^{-1}y$$

where Ω^{-1} is given by (12.19).

12.7 From (12.30) we have

$$var(\hat{\beta}_{GLS}) = \sigma_v^2[W_{XX} + \phi^2 B_{XX}]^{-1}$$

where $W_{XX} = X'QX$, $B_{XX} = X'(P-\bar{J}_{NT})X$ and $\phi^2 = \sigma_v^2/\sigma_1^2$. The Within estimator is given by (12.13) with $var(\tilde{\beta}_{Within}) = \sigma_v^2 W_{XX}^{-1}$ Hence,

$$(var(\hat{\beta}_{GLS})^{-1}) - (var(\tilde{\beta}_{Within}))^{-1} = \frac{1}{\sigma_v^2}[W_{XX}+\phi^2 B_{XX}] - \frac{1}{\sigma_v^2}W_{XX} = \phi^2 B_{XX}/\sigma_v^2$$

which is positive semi-definite. Hence, $var(\tilde{\beta}_{Within}) - var(\hat{\beta}_{GLS})$ is positive semi-definite. This last result uses the well known fact that if $A^{-1} - B^{-1}$ is positive semi-definite, then $B - A$ is positive semi-definite.

12.8

a. Differentiating (12.34) with respect to ϕ^2 yields

$$\frac{\partial L_c}{\partial \phi^2} = -\frac{NT}{2} \cdot \frac{d'(P-\bar{J}_{NT})d}{d'[Q+\phi^2(P-\bar{J}_{NT})]d} + \frac{N}{2} \cdot \frac{1}{\phi^2}$$

setting $\partial L_c/\partial \phi^2 = 0$, we get $Td'(P-\bar{J}_{NT})d\phi^2 = d'Qd + \phi^2 d'(P-\bar{J}_{NT})d$

solving for ϕ^2, we get $\hat{\phi}^2(T-1)d'(P-\bar{J}_{NT})d = d'Qd$ which yields (12.35).

b. Differentiating (12.34) with respect to β yields

$$\frac{\partial L_c}{\partial \beta} = -\frac{NT}{2} \cdot \frac{1}{d'[Q+\phi^2(P-\bar{J}_{NT})]d} \cdot \frac{\partial}{\partial \beta}d'[Q+\phi^2(P-\bar{J}_{NT})]d$$

setting $\frac{\partial L_c}{\partial \beta} = 0$ is equivalent to solving $\frac{\partial d'[Q+\phi^2(P-\bar{J}_{NT})]d}{\partial \beta} = 0$

Using the fact that $d = y - X\beta$, this yields

$$-2X'[Q+\phi^2(P-\bar{J}_{NT})](y-X\hat{\beta}) = 0$$

Solving for $\hat{\beta}$, we get $X'[Q+\phi^2(P-\bar{J}_{NT})]X\hat{\beta} = X'[Q+\phi^2(P-\bar{J}_{NT})]y$ which yields (12.36).

12.9

a. From (12.2) and (12.38), $E(u_{i,T+s}u_{jt}) = \sigma_\mu^2$, for $i = j$ and zero otherwise. The only correlation over time occurs because of the presence of the same individual across the panel. The v_{it}'s are not correlated for different time periods. In vector form

$$w = E(u_{i,T+S} u) = \sigma_\mu^2(0,..,0,..,1,..,1,..,0,..,0)'$$

where there are T ones for the i-th individual. This can be rewritten as

$$w = \sigma_\mu^2(\ell_i \otimes \iota_T)$$

where ℓ_i is the i-th column of I_N, i.e., ℓ_i is a vector that has 1 in the i-th position and zero elsewhere. ι_T is a vector of ones of dimension T.

b. $(\ell_i' \otimes \iota_T')P = (\ell_i' \otimes \iota_T')(I_N \otimes \frac{\iota_T \iota_T'}{T}) = (\ell_i' \otimes \iota_T')$. Therefore, in (12.39)

$$w'\Omega^{-1} = \sigma_\mu^2(\ell_i' \otimes \iota_T')[\frac{1}{\sigma_1^2} P + \frac{1}{\sigma_v^2} Q] = \frac{\sigma_\mu^2}{\sigma_1^2} (\ell_i' \otimes \iota_T')$$

since $(\ell_i' \otimes \iota_T')Q = (\ell_i' \otimes \iota_T')(I_{NT}-P) = (\ell_i' \otimes \iota_T') - (\ell_i' \otimes \iota_T') = 0$.

12.10 Using the motor gasoline data on diskette, the following SAS output and program replicates the results in Table 12.1. The program uses the IML procedure (SAS matrix language). This can be easily changed into GAUSS or any other matrix language.

RESULTS OF OLS

LOOK1	PARAMETER	STANDARD ERROR	T-STATISTICS
INTERCEPT	2.39133	0.11693	20.45017
INCOME	0.88996	0.03581	24.85523
PRICE	-0.89180	0.03031	-29.4180
CAR	-0.76337	0.01861	-41.0232

RESULTS OF BETWEEN

LOOK1	PARAMETER	STANDARD ERROR	T-STATISTICS
INTERCEPT	2.54163	0.52678	4.82480
INCOME	0.96758	0.15567	6.21571
PRICE	-0.96355	0.13292	-7.24902
CAR	-0.79530	0.08247	-9.64300

RESULTS OF WITHIN

LOOK1	PARAMETER	STANDARD ERROR	T-STATISTICS
INCOME	0.66225	0.07339	9.02419
PRICE	-0.32170	0.04410	-7.29496
CAR	-0.64048	0.02968	-21.5804

RESULTS OF WALLACE-HUSSAIN

LOOK1	PARAMETER	STANDARD ERROR	T-STATISTICS
INTERCEPT	1.90580	0.19403	9.82195
INCOME	0.54346	0.06353	8.55377

PRICE	-0.47111	0.04550	-10.3546
CAR	-0.60613	0.02840	-21.3425

RESULTS OF AMEMIYA

LOOK1	PARAMETER	STANDARD ERROR	T-STATISTICS
INTERCEPT	2.18445	0.21453	10.18228
INCOME	0.60093	0.06542	9.18559
PRICE	-0.36639	0.04138	-8.85497
CAR	-0.62039	0.02718	-22.8227

RESULTS OF SWAMY-ARORA

LOOK1	PARAMETER	STANDARD ERROR	T-STATISTICS
INTERCEPT	1.99670	0.17824	11.20260
INCOME	0.55499	0.05717	9.70689
PRICE	-0.42039	0.03866	-10.8748
CAR	-0.60684	0.02467	-24.5964

RESULTS OF NERLOVE

LOOK1	PARAMETER	STANDARD ERROR	T-STATISTICS
INTERCEPT	2.40173	1.41302	1.69972
INCOME	0.66198	0.07107	9.31471
PRICE	-0.32188	0.04271	-7.53561
CAR	-0.64039	0.02874	-22.2791

RESULTS OF MAXIMUM LIKELIHOOD

LOOK1	PARAMETER	STANDARD ERROR	T-STATISTICS
INTERCEPT	2.10623	0.20096	10.48088
INCOME	0.58044	0.06286	9.23388
PRICE	-0.38582	0.04072	-9.47515
CAR	-0.61401	0.02647	-23.2001

ONE-WAY ERROR COMPONENT MODEL WITH GASOLINE DATA: BETA, VARIANCES OF BETA, AND THETA

ESTIMATORS	BETA1	BETA2	BETA3	STD_BETA1	STD_BETA2	STD_BETA3	THETA
OLS	0.88996	-0.89180	-0.76337	0.03581	0.03031	0.01861	0.00000
BETWEEN	0.96758	-0.96355	-0.79530	0.15567	0.13292	0.08247	.
WITHIN	0.66225	-0.32170	-0.64048	0.07339	0.04410	0.02968	1.00000
WALLACE & HUSSAIN	0.54346	-0.47111	-0.60613	0.06353	0.04550	0.02840	0.84802
AMEMIYA	0.60093	-0.36639	-0.62039	0.06542	0.04138	0.02718	0.93773
SWAMY & ARORA	0.55499	-0.42039	-0.60684	0.05717	0.03866	0.02467	0.89231
NERLOVE	0.66198	-0.32188	-0.64039	0.07107	0.04271	0.02874	0.99654
IMLE	0.58044	-0.38582	-0.61401	0.06286	0.04072	0.02647	0.92126

NEGATIVE VAR_MHU

NEGA_VAR
OLS ESTIMATOR   .
BETWEEN ESTIMATOR   .
WITHIN ESTIMATOR   .
WALLACE & HUSSAIN ESTIMATOR 0
AMEMIYA ESTIMATOR   0
SWAMY & ARORA ESTIMATOR   0
NERLOVE ESTIMATOR   .
IMLE   .

SAS PROGRAM

```
 Options linesize=162;
  Data Gasoline;
  Infile 'b:\gasoline.dat' firstobs=2;
  Input @1 Country $     @10 Year     @15 Lgaspcar    @29 Lincomep
        @44 Lprpmg    @61 Lcarpcap ;

     Proc IML;
        Use Gasoline; Read all into Temp;
        N=18;  T=19;   NT=N*T;
        One=Repeat(1,NT,1);
        X=Temp[,3:5];  Y=Temp[,2];  Z=One||X;  K=NCOL(X);
        l_t=J(T,1,1);  JT=(l_t*l_t`);  Z_U=I(N)@l_t;
        P=Z_U*INV(Z_U`*Z_U)*Z_U`;    Q=I(NT)-P;
        JNT=Repeat(JT,N,N);          JNT_BAR=JNT/NT;

     *--------- OLS ESTIMATORS ------------*;

        OLS_BETA=INV(Z`*Z)*Z`*Y;
        OLS_RES=Y-Z*OLS_BETA;
        VAR_REG=SSQ(OLS_RES)/(NT-NCOL(Z));
        VAR_COV=VAR_REG*INV(Z`*Z);
        STD_OLS=SQRT(VECDIAG(VAR_COV));
        T_OLS=OLS_BETA/STD_OLS;

        LOOK1=OLS_BETA||STD_OLS||T_OLS;
        CTITLE={'PARAMETER' 'STANDARD ERROR' 'T-STATISTICS'};
        RTITLE={'INTERCEPT' 'INCOME' 'PRICE' 'CAR' };
        PRINT 'RESULTS OF OLS',,
              LOOK1(|COLNAME=CTITLE ROWNAME=RTITLE FORMAT=8.5|);

     *--------- BETWEEN ESTIMATOR -----------------*;

        BW_BETA=INV(Z`*P*Z)*Z`*P*Y;
        BW_RES=P*Y-P*Z*BW_BETA;
        VAR_BW=SSQ(BW_RES)/(N-NCOL(Z));
```

Chapter 12: Pooling Time-Series of Cross-Section Data

```
      V_C_BW=VAR_BW*INV(Z`*P*Z);
      STD_BW=SQRT(VECDIAG(V_C_BW));
      T_BW=BW_BETA/STD_BW;

      LOOK1=BW_BETA||STD_BW||T_BW;
      CTITLE={'PARAMETER' 'STANDARD ERROR' 'T-STATISTICS'};
      RTITLE={'INTERCEPT' 'INCOME' 'PRICE' 'CAR' };
      PRINT 'RESULTS OF BETWEEN',,
               LOOK1(|COLNAME=CTITLE ROWNAME=RTITLE FORMAT=8.5|);

*--------- WITHIN ESTIMATORS ------------*;

     WT_BETA=INV(X`*Q*X)*X`*Q*Y;
     WT_RES=Q*Y-Q*X*WT_BETA;
     VAR_WT=SSQ(WT_RES)/(NT-N-NCOL(X));
     V_C_WT=VAR_WT*INV(X`*Q*X);
     STD_WT=SQRT(VECDIAG(V_C_WT));
     T_WT=WT_BETA/STD_WT;
     LOOK1=WT_BETA||STD_WT||T_WT;
     CTITLE={'PARAMETER' 'STANDARD ERROR' 'T-STATISTICS'};
     RTITLE={'INCOME' 'PRICE' 'CAR' };
     PRINT 'RESULTS OF WITHIN',,
              LOOK1(|COLNAME=CTITLE ROWNAME=RTITLE FORMAT=8.5|);

*-- WALLACE & HUSSAIN ESTIMATOR OF VARIANCE COMPONENTS ---*;

     WH_V_V=(OLS_RES`*Q*OLS_RES)/(NT-N);
     WH_V_1=(OLS_RES`*P*OLS_RES)/N;


     ******* Checking for negative VAR_MHU *******;

     WH_V_MHU=(WH_V_1-WH_V_V)/T;
     IF WH_V_MHU<0 THEN NEGA_WH=1; ELSE NEGA_WH=0;
     WH_V_MHU=WH_V_MHU # (WH_V_MHU>0);
     WH_V_1=(T*WH_V_MHU)+WH_V_V;

     OMEGA_WH=(Q/WH_V_V)+(P/WH_V_1);
          /* Equation 2.19, p.14 */
     WH_BETA=INV(Z`*OMEGA_WH*Z)*Z`*OMEGA_WH*Y;
     THETA_WH=1-(SQRT(WH_V_V)/SQRT(WH_V_1));
     OMEGAWH=(Q/SQRT(WH_V_V))+(P/SQRT(WH_V_1));
     WH_RES=(OMEGAWH*Y)-(OMEGAWH*Z*WH_BETA);
     VAR_WH=SSQ(WH_RES)/(NT-NCOL(Z));
     V_C_WH=INV(Z`*OMEGA_WH*Z);
     STD_WH=SQRT(VECDIAG(V_C_WH));
```

```
    T_WH=WH_BETA/STD_WH;

LOOK1=WH_BETA||STD_WH||T_WH;
CTITLE={'PARAMETER' 'STANDARD ERROR' 'T-STATISTICS'};
RTITLE={'INTERCEPT' 'INCOME' 'PRICE' 'CAR' };
PRINT 'RESULTS OF WALLACE-HUSSAIN',,
        LOOK1(|COLNAME=CTITLE ROWNAME=RTITLE FORMAT=8.5|);
FREE OMEGA_WH OMEGAWH WH_RES;

*-- AMEMIYA ESTIMATOR OF VARIANCE COMPONENTS ---*;

  Y_BAR=Y[:];   X_BAR=X[:,];
  ALPHA_WT=Y_BAR-X_BAR*WT_BETA;
  LSDV_RES=Y-ALPHA_WT*ONE-X*WT_BETA;
  AM_V_V=(LSDV_RES`*Q*LSDV_RES)/(NT-N);
  AM_V_1=(LSDV_RES`*P*LSDV_RES)/N;

  ***** Checking for negative VAR_MHU *********;

  AM_V_MHU=(AM_V_1-AM_V_V)/T;
  IF AM_V_MHU<0 THEN NEGA_AM=1; ELSE NEGA_AM=0;
  AM_V_MHU=AM_V_MHU # (AM_V_MHU>0);
  AM_V_1=(T*AM_V_MHU)+AM_V_V;

  OMEGA_AM=(Q/AM_V_V)+(P/AM_V_1);
      /*Equation 2.19, p.14 */
  AM_BETA=INV(Z`*OMEGA_AM*Z)*Z`*OMEGA_AM*Y;
  THETA_AM=1-(SQRT(AM_V_V)/SQRT(AM_V_1));
  OMEGAAM=(Q/SQRT(AM_V_V))+(P/SQRT(AM_V_1));
  AM_RES=(OMEGAAM*Y)-(OMEGAAM*Z*AM_BETA);
  VAR_AM=SSQ(AM_RES)/(NT-NCOL(Z));
  V_C_AM=INV(Z`*OMEGA_AM*Z);
  STD_AM=SQRT(VECDIAG(V_C_AM));
  T_AM=AM_BETA/STD_AM;

LOOK1=AM_BETA||STD_AM||T_AM;
CTITLE={'PARAMETER' 'STANDARD ERROR' 'T-STATISTICS'};
RTITLE={'INTERCEPT' 'INCOME' 'PRICE' 'CAR' };
PRINT 'RESULTS OF AMEMIYA',,
        LOOK1(|COLNAME=CTITLE ROWNAME=RTITLE FORMAT=8.5|);
FREE OMEGA_AM OMEGAAM AM_RES;


*--- SWAMY & ARORA  ESTIMATOR OF VARIANCE COMPONENTS ----*;

   SA_V_V=(Y`*Q*Y-Y`*Q*X*INV(X`*Q*X)*X`*Q*Y)/(NT-N-K);
```

Chapter 12: Pooling Time-Series of Cross-Section Data 275

```
    SA_V_1=(Y`*P*Y-Y`*P*Z*INV(Z`*P*Z)*Z`*P*Y)/(N-K-1);

****** Checking for negative VAR_MHU ********;

    SA_V_MHU=(SA_V_1-SA_V_V)/T;
    IF SA_V_MHU<0 THEN NEGA_SA=1; ELSE NEGA_SA=0;
    SA_V_MHU=SA_V_MHU # (SA_V_MHU>0);
    SA_V_1=(T*SA_V_MHU)+SA_V_V;

    OMEGA_SA=(Q/SA_V_V)+(P/SA_V_1);
    SA_BETA=INV(Z`*OMEGA_SA*Z)*Z`*OMEGA_SA*Y;
    THETA_SA=1-(SQRT(SA_V_V)/SQRT(SA_V_1));
    OMEGASA=(Q/SQRT(SA_V_V))+(P/SQRT(SA_V_1));
    SA_RES=(OMEGASA*Y)-(OMEGASA*Z*SA_BETA);
    VAR_SA=SSQ(SA_RES)/(NT-NCOL(Z));
    V_C_SA=INV(Z`*OMEGA_SA*Z);
    STD_SA=SQRT(VECDIAG(V_C_SA));
    T_SA=SA_BETA/STD_SA;

  LOOK1=SA_BETA||STD_SA||T_SA;
  CTITLE={'PARAMETER' 'STANDARD ERROR' 'T-STATISTICS'};
  RTITLE={'INTERCEPT' 'INCOME' 'PRICE' 'CAR' };
  PRINT 'RESULTS OF SWAMY-ARORA',,
          LOOK1(|COLNAME=CTITLE ROWNAME=RTITLE FORMAT=8.5|);
  FREE OMEGA_SA OMEGASA SA_RES;

*-- NERLOVE  ESTIMATOR OF VARIANCE COMPONENTS AND BETA --*;

    MHU=P*LSDV_RES;
    MEAN_MHU=MHU[:];
    DEV_MHU=MHU-(ONE*MEAN_MHU);
    VAR_MHU=SSQ(DEV_MHU)/T*(N-1);
    NL_V_V=SSQ(WT_RES)/NT;
    NL_V_1=T*VAR_MHU+NL_V_V;
    OMEGA_NL=(Q/NL_V_V)+(P/NL_V_1);
    NL_BETA=INV(Z`*OMEGA_NL*Z)*Z`*OMEGA_NL*Y;
    THETA_NL=1-(SQRT(NL_V_V)/SQRT(NL_V_1));
    OMEGANL=(Q/SQRT(NL_V_V))+(P/SQRT(NL_V_1));
    NL_RES=(OMEGANL*Y)-(OMEGANL*Z*NL_BETA);
    VAR_NL=SSQ(NL_RES)/(NT-NCOL(Z));
    V_C_NL=INV(Z`*OMEGA_NL*Z);
    STD_NL=SQRT(VECDIAG(V_C_NL));
    T_NL=NL_BETA/STD_NL;

  LOOK1=NL_BETA||STD_NL||T_NL;
  CTITLE={'PARAMETER' 'STANDARD ERROR' 'T-STATISTICS'};
```

```
            RTITLE={'INTERCEPT' 'INCOME' 'PRICE' 'CAR' };
            PRINT 'RESULTS OF NERLOVE',,
                  LOOK1(|COLNAME=CTITLE ROWNAME=RTITLE FORMAT=8.5|);
            FREE OMEGA_NL OMEGANL NL_RES;

            *--- MAXIMUM LIKELIHOOD ESTIMATION ----*;

         /* START WITH WITHIN AND BETWEEN BETA SUGGESTED BY
            BREUSCH(1987) */;

            CRITICAL=1;
            BETA_W=WT_BETA;
            BETA_B=BW_BETA[2:K+1,];
            BETA_MLE=WT_BETA;

            DO WHILE (CRITICAL>0.0001);
               WT_RES=Y-X*BETA_W;
               BW_RES=Y-X*BETA_B;

PHISQ_W=(WT_RES`*Q*WT_RES)/((T-1)*(WT_RES`*(P-JNT_BAR)*WT_RES));
PHISQ_B=(BW_RES`*Q*BW_RES)/((T-1)*(BW_RES`*(P-JNT_BAR)*BW_RES));
CRITICAL=PHISQ_W-PHISQ_B;
BETA_W=INV(X`*(Q+PHISQ_W*(P-JNT_BAR))*X)*X`*(Q+PHISQ_W*(P-JNT_
       BAR))*Y;
BETA_B=INV(X`*(Q+PHISQ_B*(P-JNT_BAR))*X)*X`*(Q+PHISQ_B*(P-JNT_
BAR))*Y;
BETA_MLE=(BETA_W+BETA_B)/2;
END;

D_MLE=Y-X*BETA_MLE;

PHISQ_ML=(D_MLE`*Q*D_MLE)/((T-1)*D_MLE`*(P-JNT_BAR)*D_MLE);
THETA_ML=1-SQRT(PHISQ_ML);
VAR_V_ML=D_MLE`*(Q+PHISQ_ML*(P-JNT_BAR))*D_MLE/NT;
VAR_1_ML=VAR_V_ML/PHISQ_ML;
OMEGA_ML=(Q/VAR_V_ML)+(P/VAR_1_ML);
ML_BETA=INV(Z`*OMEGA_ML*Z)*Z`*OMEGA_ML*Y;
OMEGAML=(Q/SQRT(VAR_V_ML))+(P/SQRT(VAR_1_ML));
ML_RES=(OMEGAML*Y)-(OMEGAML*Z*ML_BETA);
VAR_ML=SSQ(ML_RES)/(NT-NCOL(Z));
V_C_ML=INV(Z`*OMEGA_ML*Z);
STD_ML=SQRT(VECDIAG(V_C_ML));
T_ML=ML_BETA/STD_ML;

LOOK1=ML_BETA||STD_ML||T_ML;
CTITLE={'PARAMETER' 'STANDARD ERROR' 'T-STATISTICS'};
```

```
RTITLE={'INTERCEPT' 'INCOME' 'PRICE' 'CAR' };
PRINT 'RESULTS OF MAXIMUM LIKELIHOOD',,
               LOOK1(|COLNAME=CTITLE ROWNAME=RTITLE FORMAT=8.5|);
FREE OMEGA_ML;

*---------- PRINT AND OUTPUT INFORMATION --------------------*;

BETA=OLS_BETA`[,2:K+1]//BW_BETA`[,2:K+1]//WT_BETA`//WH_BETA`[,
     2:K+1]//
AM_BETA`[,2:K+1]//SA_BETA`[,2:K+1]//NL_BETA`[,2:K+1]//ML_BETA`
     [,2:K+1];
STD_ERR=STD_OLS`[,2:K+1]//STD_BW`[,2:K+1]//STD_WT`//STD_WH`[,2
     :K+1]//
STD_AM`[,2:K+1]//STD_SA`[,2:K+1]//STD_NL`[,2:K+1]//STD_ML`[,2:
     K+1];
THETAS={0,.,1}//THETA_WH//THETA_AM//THETA_SA//THETA_NL//THETA_
     ML;

       NEGA_VAR={.,.,.}//NEGA_WH//NEGA_AM//NEGA_SA//{.,.,.};
       OUTPUT=BETA||STD_ERR||THETAS||NAGA_VAR;
       C2={"BETA1"   "BETA2"   "BETA3"   "STD_BETA1"   "STD_BETA2"
       "STD_BETA3"   "THETA"};
       R={"OLS ESTIMATOR"   "BETWEEN ESTIMATOR"   "WITHIN
         ESTIMATOR"
       "WALLACE & HUSSAIN ESTIMATOR"   "AMEMIYA ESTIMATOR"
       "SWAMY & ARORA ESTIMATOR"   "NERLOVE ESTIMATOR"   "IMLE"};

       PRINT 'ONE-WAY ERROR COMPONENT MODEL WITH GASOLINE DATA:
              BETA, VARIANCES OF BETA, AND THETA'
             ,,OUTPUT (|ROWNAME=R COLNAME=C2   FORMAT=8.5|);

       PRINT 'NEGATIVE VAR_MHU',,NEGA_VAR (|ROWNAME=R|);
```

12.11

a. This solution is based on Baltagi and Krämer (1994). From (12.3), one gets

$\hat{\delta}_{ols} = (Z'Z)^{-1}Z'y$ and $\hat{u}_{ols} = y - Z\hat{\delta}_{ols} = \bar{P}_Z u$ where $\bar{P}_Z = I_{NT} - P_Z$ with $P_Z = Z(Z'Z)^{-1}Z'$. Also,

$E(s^2) = E[\hat{u}'\hat{u}/(NT-K')] = E[u'\bar{P}_Z u/(NT-K')] = tr(\Omega \bar{P}_Z)/(NT-K')$

which from (12.17) reduces to

$E(s^2) = \sigma_v^2 + \sigma_\mu^2 (NT - tr(I_N \otimes J_T)P_Z)/(NT-K')$ since

$\text{tr}(I_{NT}) = \text{tr}(I_N \otimes J_T) = NT$

and $\text{tr}(P_Z) = K'$. By adding and subtracting σ_μ^2, one gets

$$E(s^2) = \sigma^2 + \sigma_\mu^2[K' - \text{tr}(I_N \otimes J_T)P_Z]/(NT - K')$$

where $\sigma^2 = E(u_{it}^2) = \sigma_\mu^2 + \sigma_v^2$ for all i and t.

b. Nerlove (1971) derived the characteristic roots and vectors of Ω given in (12.17). These characteristic roots turn out to be σ_v^2 with multiplicity $N(T-1)$ and $(T\sigma_\mu^2 + \sigma_v^2)$ with multiplicity N. Therefore, the smallest (n-K') characteristic roots are made up of the (n-N) σ_v^2's and (N-K') of the $(T\sigma_\mu^2 + \sigma_v^2)$'s. This implies that the mean of the (n-K') smallest characteristic roots of $\Omega = [(n-N)\sigma_v^2 + (N-K')(T\sigma_\mu^2 + \sigma_v^2)]/(n-K')$. Similarly, the largest (n-K') characteristic roots are made up of the N $(T\sigma_\mu^2 + \sigma_v^2)$'s and (n-N-K') of the σ_v^2's. This implies that the mean of the (n-K') largest characteristic roots of $\Omega = [N(T\sigma_\mu^2 + \sigma_v^2) + (n-N-K')\sigma_v^2]/(n-K')$. Using the Kiviet and Krämer (1992) inequalities, one gets

$$0 \leq \sigma_v^2 + (n-TK')\sigma_\mu^2/(n-K') \leq E(s^2) \leq \sigma_v^2 + n\sigma_\mu^2/(n-K') \leq n\sigma^2/(n-K').$$

As $n \to \infty$, both bounds tend to σ^2, and s^2 is asymptotically unbiased, irrespective of the particular evolution of X.

12.12 $M = I_{NT} - Z(Z'Z)^{-1}Z'$ and $M^* = I_{NT} - Z^*(Z^{*'}Z^*)^{-1}Z^{*'}$ are both symmetric and idempotent. From (12.43), it is clear that $Z = Z^*I^*$ with $I^* = (\iota_N \otimes I_{K'})$, ι_N being a vector of ones of dimension N and $K' = K + 1$.

$$MM^* = I_{NT} - Z(Z'Z)^{-1}Z' - Z^*(Z^{*'}Z^*)^{-1}Z^{*'} + Z(Z'Z)^{-1}Z'Z^*(Z^{*'}Z^*)^{-1}Z^{*'}$$

Substituting $Z = Z^*I^*$, the last term reduces to

$$Z(Z'Z)^{-1}Z'Z^*(Z^{*'}Z^*)^{-1}Z^{*'} = Z(Z'Z)^{-1}I^{*'}Z^{*'}Z^*(Z^{*'}Z^*)^{-1}Z^{*'} = Z(Z'Z)^{-1}Z'$$

Hence, $MM^* = I_{NT} - Z^*(Z^{*'}Z^*)^{-1}Z^{*'} = M^*$.

12.13 This problem differs from problem 12 in that $\dot{Z} = \Sigma^{-1/2}Z$ and $\dot{Z}^* = \Sigma^{-1/2}Z^*$. Since $Z = Z^*I^*$, premultiplying both sides by $\Sigma^{-1/2}$ one gets $\dot{Z} = \dot{Z}^*I^*$. Define

$\dot{M} = I_{NT} - \dot{Z}(\dot{Z}'\dot{Z})^{-1}\dot{Z}'$

and

$\dot{M}^* = I_{NT} - \dot{Z}^*(\dot{Z}^{*\prime}\dot{Z}^*)^{-1}\dot{Z}^{*\prime}$

Both are projection matrices that are symmetric and idempotent. The proof of $\dot{M}\dot{M}^* = \dot{M}^*$ is the same as that of $MM^* = M^*$ given in problem 12 with $\dot{Z}$ replacing Z and $\dot{Z}^*$ replacing Z^*.

12.16

a. $\hat{\beta}_{GLS} = (X'\Omega^{-1}X)^{-1}X'\Omega^{-1}y = \beta + (X'\Omega^{-1}X)^{-1}X'\Omega^{-1}u$ with $E(\hat{\beta}_{GLS}) = \beta$

$\tilde{\beta}_{Within} = (X'QX)^{-1}X'Qy = \beta + (X'QX)^{-1}X'Qu$ with $E(\tilde{\beta}_{Within}) = \beta$

Therefore, $\hat{q} = \hat{\beta}_{GLS} - \tilde{\beta}_{Within}$ has $E(\hat{q}) = 0$.

$\text{cov}(\hat{\beta}_{GLS}, \hat{q}) = E(\hat{\beta}_{GLS} - \beta)(\hat{q}') = E(\hat{\beta}_{GLS} - \beta)[(\hat{\beta}_{GLS} - \beta)' - (\tilde{\beta}_{Within} - \beta)']$

$= \text{var}(\hat{\beta}_{GLS}) - \text{cov}(\hat{\beta}_{GLS}, \tilde{\beta}_{Within})$

$= E[(X'\Omega^{-1}X)^{-1}X'\Omega^{-1}uu'\Omega^{-1}X(X'\Omega^{-1}X)^{-1}]$

$- E[(X'\Omega^{-1}X)^{-1}X'\Omega^{-1}uu'QX(X'QX)^{-1}]$

$= (X'\Omega^{-1}X)^{-1} - (X'\Omega^{-1}X)^{-1}X'\Omega^{-1}\Omega QX(X'QX)^{-1}$

$= (X'\Omega^{-1}X)^{-1} - (X'\Omega^{-1}X)^{-1} = 0$.

b. Using the fact that $\tilde{\beta}_{Within} = -\hat{q} + \hat{\beta}_{GLS}$, one gets

$\text{var}(\tilde{\beta}_{Within}) = \text{var}(-\hat{q} + \hat{\beta}_{GLS}) = \text{var}(\hat{q}) + \text{var}(\hat{\beta}_{GLS})$,

since $\text{cov}(\hat{\beta}_{GLS}, \hat{q}) = 0$. Therefore,

$\text{var}(\hat{q}) = \text{var}(\tilde{\beta}_{Within}) - \text{var}(\hat{\beta}_{GLS}) = \sigma_v^2(X'QX)^{-1} - (X'\Omega^{-1}X)^{-1}$

12.18

a. For the gasoline data given on diskette, OLS was reported in the solution to problem 10 and will not be reproduced here. Using the OLS residuals ($\hat{u}_{it}$) one forms the Prais-Winsten estimate of ρ given by

$$\hat{\rho}_{PW} = \sum_{i=1}^{N} \sum_{t=2}^{T} \hat{u}_{it}\hat{u}_{i,t-1} \bigg/ \sum_{i=1}^{N} \sum_{t=2}^{T-1} \hat{u}_{i,t}^2 = 0.99868$$

With this common ρ estimate one transforms the data using the Prais-Winsten transformation described in section 12.3. The resulting regression yields:

	α	β_1	β_2	β_3
coeff.	1.4429	0.3749	-0.25144	-0.55457
std. error	(0.40401)	(0.07602)	(0.03135)	(0.03471)
t-statistic	3.57	4.93	-8.02	-15.98

Sum of Squared Error = 0.8054504
Degrees of freedom = 338
Mean Square Error = 0.00238299
Std. error of the regression = 0.04881587

Using the residuals ($\hat{u}_{it}^*$) of this regression one computes

$$\hat{\sigma}_{\epsilon i}^2 = \sum_{t=1}^{T} \hat{u}_{it}^{*2}/(T-4) \quad \text{for } i = 1,..,N$$

These $\hat{\sigma}_{\epsilon i}$ are given by

```
0.01995  0.01959  0.05185  0.0548   0.04534  0.04706
0.03953  0.03524  0.11874  0.04832  0.04935  0.04461
0.02562  0.05428  0.03861  0.04482  0.10524  0.03335
```

Transforming the Prais-Winsten regression by dividing by the corresponding standard error for each country's observations, one gets

$$y_{it}^{**} = y_{it}^*/\hat{\sigma}_{\epsilon i} \quad \text{for } i = 1,..,N \text{ and } t = 1,..,T$$

Running this last regression yields

	α	β_1	β_2	β_3
coeff.	1.89466	0.35231	-0.21376	-0.51764
std. error	(0.29655)	(0.06088)	(0.02674)	(0.03134)
t-statistic	6.39	5.79	-7.99	-16.52

Sum of Squared Error = 265.375665

Degrees of freedom = 338
Mean Square Error = 0.785135
Std. error of the regression = 0.886078

The estimates of the β's from the 'common ρ' method are much smaller in absolute value than those obtained from the error component model (0.35 compared to 0.58 for β_1), (-0.21 compared to -0.38 for β_2), and (-0.52 compared to -0.62 for β_3).

b. For the varying ρ method, the OLS residuals obtained in part (a) are used to get N estimates of ρ_i one for each firm as follows:

$$\hat{\rho}_i = \sum_{t=2}^{T} \hat{u}_{it} \hat{u}_{i,t-1} \Big/ \sum_{t=2}^{T-1} \hat{u}_{it}^2 \quad \text{for } i = 1,..,N$$

Some of these $\hat{\rho}_i$ turn out to be larger than one in absolute value and are therefore inadmissible. These are replaced by 0.99999. The N estimates of ρ_i used are given by

0.99999	0.99999	0.99999	0.97853	0.99999	0.95532
0.83592	0.99999	0.99999	0.99999	0.93970	0.97566
0.92108	0.95467	0.99282	0.99999	0.72591	0.65591

With these N estimates of ρ_i, the Prais-Winsten transformation is applied as described in section 12.3. The resulting regression yields

	α	β_1	β_2	β_3
coeff.	0.88293	0.30562	-0.29959	-0.55851
std. error	(0.21336)	(0.06472)	(0.03096)	(0.02508)
t-statistic	4.14	4.72	-9.68	-22.27

Sum of Squared Error = 0.83024367
Degrees of freedom = 338
Mean Square Error = 0.00245634
Std. error of the regression = 0.0495615

Using the residuals $\hat{u}_{it}^*$ of this regression one computes

$$\hat{\sigma}_{\epsilon i}^2 = \sum_{t=1}^{T} \hat{u}_{it}^{*2}/(T-4) \qquad \text{for } i = 1,..,N.$$

These $\hat{\sigma}_{\epsilon i}$ are given by

0.0182	0.0171	0.0477	0.0556	0.0456	0.0497
0.0458	0.0366	0.1212	0.0498	0.0553	0.0435
0.0271	0.0625	0.0381	0.0456	0.0928	0.0497

Transforming the Prais-Winsten regression by dividing by the corresponding standard error for each country's observations, one gets

$$y_{it}^{**} = y_{it}^*/\hat{\sigma}_{\epsilon i} \qquad \text{for } i = 1,..,N \quad t = 1,..,T$$

Running this last regression yields

	α	β_1	β_2	β_3
coeff.	1.14737	0.33727	-0.27635	-0.5512
std. error	(0.18079)	(0.04985)	(0.02776)	(0.02242)
t-statistic	6.35	6.77	-9.96	-24.59

Sum of Squared Error = 267.5907614
Degrees of freedom = 338
Mean Square Error = 267.59076
Std. error of the regression = 0.88976887

Again, these estimates of the β's are much smaller in absolute value than the corresponding estimates from the error component model.

CHAPTER 13
Limited Dependent Variables

13.1 The Linear Probability Model

y_i	u_i	Prob.
1	$1-x_i'\beta$	π_i
0	$-x_i'\beta$	$1-\pi_i$

a. Let $\pi_i = \Pr[y_i=1]$, then $y_i = 1$ when $u_i = 1 - x_i'\beta$ with probability π_i as shown in the table above. Similarly, $y_i = 0$ when $u_i = -x_i'\beta$ with probability $1 - \pi_i$. Hence, $E(u_i) = \pi_i(1-x_i'\beta) + (1-\pi_i)(-x_i'\beta)$.
For this to equal zero, we get, $\pi_i - \pi_i x_i'\beta + \pi_i x_i'\beta - x_i'\beta = 0$ which gives $\pi_i = x_i'\beta$ as required.

b. $\mathrm{var}(u_i) = E(u_i^2) = (1-x_i'\beta)^2 \pi_i + (-x_i'\beta)^2 (1-\pi_i)$
$= [1 - 2x_i'\beta + (x_i'\beta)^2]\pi_i + (x_i'\beta)^2(1-\pi_i)$
$= \pi_i - 2x_i'\beta \pi_i + (x_i'\beta)^2$
$= \pi_i - \pi_i^2 = \pi_i(1-\pi_i) = x_i'\beta(1-x_i'\beta)$

using the fact that $\pi_i = x_i'\beta$.

13.2

a. Since there are no slopes and only a constant, $x_i'\beta = \alpha$ and (13.16) becomes
$\log \ell = \sum_{i=1}^{n} \{y_i \log F(\alpha) + (1-y_i)\log[1-F(\alpha)]\}$ differentiating with respect to α we
get $\dfrac{\partial \log \ell}{\partial \alpha} = \sum_{i=1}^{n} \dfrac{y_i}{F(\alpha)} \cdot f(\alpha) + \sum_{i=1}^{n} \dfrac{(1-y_i)}{1-F(\alpha)}(-f(\alpha))$.

Setting this equal to zero yields $\sum_{i=1}^{n} (y_i - F(\alpha)) f(\alpha) = 0$.

Therefore, $\hat{F}(\alpha) = \sum_{i=1}^{n} y_i/n = \bar{y}$. This is the proportion of the sample with $y_i = 1$.

b. Using $\hat{F}(\alpha) = \bar{y}$, the value of the maximized likelihood, from (13.16), is

$$\log \ell_r = \sum_{i=1}^{n} \{y_i \log \bar{y} + (1-y_i)\log(1-\bar{y})\} = n\bar{y}\log\bar{y} + (n - n\bar{y})\log(1-\bar{y})$$
$$= n[\bar{y}\log\bar{y} + (1-\bar{y})\log(1-\bar{y})] \quad \text{as required.}$$

c. For the empirical example in section 13.8, we know that $\bar{y} = 218/595 = 0.366$. Substituting in (13.33) we get,

$$\log \ell_r = n[0.366 \log 0.366 + (1-0.366)\log(1-0.366)] = -390.918.$$

13.3 Union participation example. See Tables 13.1 and 13.2. These were run using EViews.

a. OLS ESTIMATION

```
LS // Dependent Variable is UNION
Sample: 1 595
Included observations: 595
```

Variable	Coefficient	Std. Error	t-Statistic	Prob.
C	1.195872	0.227010	5.267922	0.0000
EX	-0.001974	0.001726	-1.143270	0.2534
WKS	-0.017809	0.003419	-5.209226	0.0000
OCC	0.318118	0.046425	6.852287	0.0000
IND	0.030048	0.038072	0.789229	0.4303
SOUTH	-0.170130	0.039801	-4.274471	0.0000
SMSA	0.084522	0.038464	2.197419	0.0284
MS	0.098953	0.063781	1.551453	0.1213
FEM	-0.108706	0.079266	-1.371398	0.1708
ED	-0.016187	0.008592	-1.883924	0.0601
BLK	0.050197	0.071130	0.705708	0.4807

R-squared	0.233548	Mean dependent var	0.366387
Adjusted R-squared	0.220424	S.D. dependent var	0.482222
S.E. of regression	0.425771	Akaike info criterion	-1.689391
Sum squared resid	105.8682	Schwarz criterion	-1.608258
Log likelihood	-330.6745	F-statistic	17.79528
Durbin-Watson stat	1.900963	Prob(F-statistic)	0.000000

LOGIT ESTIMATION

```
LOGIT // Dependent Variable is UNION
Sample: 1 595
Included observations: 595
Convergence achieved after 4 iterations
```

Variable	Coefficient	Std. Error	t-Statistic	Prob.
C	4.380828	1.338629	3.272624	0.0011
EX	-0.011143	0.009691	-1.149750	0.2507
WKS	-0.108126	0.021428	-5.046037	0.0000
OCC	1.658222	0.264456	6.270325	0.0000
IND	0.181818	0.205470	0.884888	0.3766
SOUTH	-1.044332	0.241107	-4.331411	0.0000
SMSA	0.448389	0.218289	2.054110	0.0404
MS	0.604999	0.365043	1.657336	0.0980

Chapter 13: Limited Dependent Variables   285

```
FEM     -0.772222       0.489665       -1.577040       0.1153
ED      -0.090799       0.049227       -1.844501       0.0656
BLK      0.355706       0.394794        0.900992       0.3680

Log likelihood   -312.3367
Obs with Dep=1    218
Obs with Dep=0    377
```

Variable	Mean All	Mean D=1	Mean D=0
C	1.000000	1.000000	1.000000
EX	22.85378	23.83028	22.28912
WKS	46.45210	45.27982	47.12997
OCC	0.512605	0.766055	0.366048
IND	0.405042	0.513761	0.342175
SOUTH	0.292437	0.197248	0.347480
SMSA	0.642017	0.646789	0.639257
MS	0.805042	0.866972	0.769231
FEM	0.112605	0.059633	0.143236
ED	12.84538	11.84862	13.42175
BLK	0.072269	0.082569	0.066313

PROBIT ESTIMATION

```
PROBIT // Dependent Variable is UNION
Sample: 1 595
Included observations: 595
Convergence achieved after 3 iterations
```

Variable	Coefficient	Std. Error	t-Statistic	Prob.
C	2.516784	0.762606	3.300242	0.0010
EX	-0.006932	0.005745	-1.206501	0.2281
WKS	-0.060829	0.011785	-5.161707	0.0000
OCC	0.955490	0.152136	6.280522	0.0000
IND	0.092827	0.122773	0.756089	0.4499
SOUTH	-0.592739	0.139100	-4.261243	0.0000
SMSA	0.260701	0.128629	2.026756	0.0431
MS	0.350520	0.216282	1.620664	0.1056
FEM	-0.407026	0.277034	-1.469226	0.1423
ED	-0.057382	0.028842	-1.989533	0.0471
BLK	0.226482	0.228843	0.989683	0.3227

```
Log likelihood   -313.3795
Obs with Dep=1    218
Obs with Dep=0    377
```

d. Dropping the industry variable (IND).

OLS ESTIMATION

```
LS // Dependent Variable is UNION
Sample: 1 595
Included observations: 595
```

Variable	Coefficient	Std. Error	t-Statistic	Prob.
C	1.216753	0.225390	5.398425	0.0000
EX	-0.001848	0.001718	-1.075209	0.2827
WKS	-0.017874	0.003417	-5.231558	0.0000
OCC	0.322215	0.046119	6.986568	0.0000
SOUTH	-0.173339	0.039580	-4.379418	0.0000

SMSA	0.085043	0.038446	2.212014	0.0274
MS	0.100697	0.063722	1.580267	0.1146
FEM	-0.114088	0.078947	-1.445122	0.1490
ED	-0.017021	0.008524	-1.996684	0.0463
BLK	0.048167	0.071061	0.677822	0.4982

R-squared	0.232731	Mean dependent var		0.366387
Adjusted R-squared	0.220927	S.D. dependent var		0.482222
S.E. of regression	0.425634	Akaike info criterion		-1.691687
Sum squared resid	105.9811	Schwarz criterion		-1.617929
Log likelihood	-330.9916	F-statistic		19.71604
Durbin-Watson stat	1.907714	Prob(F-statistic)		0.000000

LOGIT ESTIMATION

LOGIT // Dependent Variable is UNION
Sample: 1 595
Included observations: 595
Convergence achieved after 4 iterations

Variable	Coefficient	Std. Error	t-Statistic	Prob.
C	4.492957	1.333992	3.368053	0.0008
EX	-0.010454	0.009649	-1.083430	0.2791
WKS	-0.107912	0.021380	-5.047345	0.0000
OCC	1.675169	0.263654	6.353652	0.0000
SOUTH	-1.058953	0.240224	-4.408193	0.0000
SMSA	0.449003	0.217955	2.060074	0.0398
MS	0.618511	0.365637	1.691599	0.0913
FEM	-0.795607	0.489820	-1.624285	0.1049
ED	-0.096695	0.048806	-1.981194	0.0480
BLK	0.339984	0.394027	0.862845	0.3886

Log likelihood   -312.7267
Obs with Dep=1   218
Obs with Dep=0   377

Variable	Mean All	Mean D=1	Mean D=0
C	1.000000	1.000000	1.000000
EX	22.85378	23.83028	22.28912
WKS	46.45210	45.27982	47.12997
OCC	0.512605	0.766055	0.366048
SOUTH	0.292437	0.197248	0.347480
SMSA	0.642017	0.646789	0.639257
MS	0.805042	0.866972	0.769231
FEM	0.112605	0.059633	0.143236
ED	12.84538	11.84862	13.42175
BLK	0.072269	0.082569	0.066313

PROBIT ESTIMATION

PROBIT // Dependent Variable is UNION
Sample: 1 595
Included observations: 595
Convergence achieved after 3 iterations

Variable	Coefficient	Std. Error	t-Statistic	Prob.
C	2.570491	0.759181	3.385875	0.0008
EX	-0.006590	0.005723	-1.151333	0.2501
WKS	-0.060795	0.011777	-5.162354	0.0000

Chapter 13: Limited Dependent Variables   287

```
OCC         0.967972      0.151305    6.397481    0.0000
SOUTH      -0.601050      0.138528   -4.338836    0.0000
SMSA        0.261381      0.128465    2.034640    0.0423
MS          0.357808      0.216057    1.656085    0.0982
FEM        -0.417974      0.276501   -1.511657    0.1312
ED         -0.060082      0.028625   -2.098957    0.0362
BLK         0.220695      0.228363    0.966423    0.3342

Log likelihood   -313.6647
Obs with Dep=1    218
Obs with Dep=0    377
```

f. The restricted regressions omitting IND, FEM and BLK are given below:

```
=================================================================
LS // Dependent Variable is UNION
Sample: 1 595
Included observations: 595
=================================================================
   Variable       Coefficien Std. Error t-Statistic    Prob.
=================================================================
      C           1.153900    0.218771    5.274452    0.0000
      EX         -0.001840    0.001717   -1.071655    0.2843
      WKS        -0.017744    0.003412   -5.200421    0.0000
      OCC         0.326411    0.046051    7.088110    0.0000
      SOUTH      -0.171713    0.039295   -4.369868    0.0000
      SMSA        0.086076    0.038013    2.264382    0.0239
      MS          0.158303    0.045433    3.484351    0.0005
      ED         -0.017204    0.008507   -2.022449    0.0436
=================================================================
R-squared              0.229543   Mean dependent var  0.366387
Adjusted R-squared     0.220355   S.D. dependent var  0.482222
S.E. of regression     0.425790   Akaike info criter -1.694263
Sum squared resid    106.4215    Schwarz criterion  -1.635257
Log likelihood      -332.2252    F-statistic         24.98361
Durbin-Watson stat     1.912059   Prob(F-statistic)    0.000000
=================================================================


=================================================================
LOGIT // Dependent Variable is UNION
Sample: 1 595
Included observations: 595
Convergence achieved after 4 iterations
=================================================================
   Variable       Coefficien Std. Error t-Statistic    Prob.
=================================================================
      C           4.152595    1.288390    3.223088    0.0013
      EX         -0.011018    0.009641   -1.142863    0.2536
      WKS        -0.107116    0.021215   -5.049031    0.0000
      OCC         1.684082    0.262193    6.423059    0.0000
      SOUTH      -1.043629    0.237769   -4.389255    0.0000
      SMSA        0.459707    0.215149    2.136659    0.0330
      MS          0.975711    0.272560    3.579800    0.0004
      ED         -0.100033    0.048507   -2.062229    0.0396
=================================================================
Log likelihood   -314.2744
Obs with Dep=1    218
Obs with Dep=0    377
=================================================================
```

Variable	Mean All	Mean D=1	Mean D=0
C	1.000000	1.000000	1.000000
EX	22.85378	23.83028	22.28912
WKS	46.45210	45.27982	47.12997
OCC	0.512605	0.766055	0.366048
SOUTH	0.292437	0.197248	0.347480
SMSA	0.642017	0.646789	0.639257
MS	0.805042	0.866972	0.769231
ED	12.84538	11.84862	13.42175

PROBIT // Dependent Variable is UNION
Sample: 1 595
Included observations: 595
Convergence achieved after 3 iterations

Variable	Coefficien	Std. Error	t-Statistic	Prob.
C	2.411706	0.741327	3.253228	0.0012
EX	-0.006986	0.005715	-1.222444	0.2220
WKS	-0.060491	0.011788	-5.131568	0.0000
OCC	0.971984	0.150538	6.456745	0.0000
SOUTH	-0.580959	0.136344	-4.260988	0.0000
SMSA	0.273201	0.126988	2.151388	0.0319
MS	0.545824	0.155812	3.503105	0.0005
ED	-0.063196	0.028464	-2.220210	0.0268

Log likelihood	-315.1770
Obs with Dep=1	218
Obs with Dep=0	377

13.4 Occupation regression.

  a. OLS Estimation

LS // Dependent Variable is OCC
Sample: 1 595
Included observations: 595

Variable	Coefficient	Std. Error	t-Statistic	Prob.
C	2.111943	0.182340	11.58245	0.0000
ED	-0.111499	0.006108	-18.25569	0.0000
WKS	-0.001510	0.003044	-0.496158	0.6200
EX	-0.002870	0.001533	-1.872517	0.0616
SOUTH	-0.068631	0.035332	-1.942452	0.0526
SMSA	-0.079735	0.034096	-2.338528	0.0197
IND	0.091688	0.033693	2.721240	0.0067
MS	0.006271	0.056801	0.110402	0.9121
FEM	-0.064045	0.070543	-0.907893	0.3643
BLK	0.068514	0.063283	1.082647	0.2794

R-squared	0.434196	Mean dependent var	0.512605
Adjusted R-squared	0.425491	S.D. dependent var	0.500262
S.E. of regression	0.379180	Akaike info criterion	-1.922824
Sum squared resid	84.10987	Schwarz criterion	-1.849067
Log likelihood	-262.2283	F-statistic	49.88075
Durbin-Watson stat	1.876105	Prob(F-statistic)	0.000000

Chapter 13: Limited Dependent Variables   289

LOGIT ESTIMATION

LOGIT // Dependent Variable is OCC
Sample: 1 595
Included observations: 595
Convergence achieved after 5 iterations

Variable	Coefficient	Std. Error	t-Statistic	Prob.
C	11.62962	1.581601	7.353069	0.0000
ED	-0.806320	0.070068	-11.50773	0.0000
WKS	-0.008424	0.023511	-0.358297	0.7203
EX	-0.017610	0.011161	-1.577893	0.1151
SOUTH	-0.349960	0.260761	-1.342073	0.1801
SMSA	-0.601945	0.247206	-2.434995	0.0152
IND	0.689620	0.241028	2.861157	0.0044
MS	-0.178865	0.417192	-0.428735	0.6683
FEM	-0.672117	0.503002	-1.336212	0.1820
BLK	0.333307	0.441064	0.755687	0.4501

Log likelihood   -244.2390
Obs with Dep=1   305
Obs with Dep=0   290

Variable	Mean All	Mean D=1	Mean D=0
C	1.000000	1.000000	1.000000
ED	12.84538	11.10164	14.67931
WKS	46.45210	46.40984	46.49655
EX	22.85378	23.88525	21.76897
SOUTH	0.292437	0.298361	0.286207
SMSA	0.642017	0.554098	0.734483
IND	0.405042	0.524590	0.279310
MS	0.805042	0.816393	0.793103
FEM	0.112605	0.095082	0.131034
BLK	0.072269	0.091803	0.051724

PROBIT ESTIMATION

PROBIT // Dependent Variable is OCC
Sample: 1 595
Included observations: 595
Convergence achieved after 4 iterations

Variable	Coefficient	Std. Error	t-Statistic	Prob.
C	6.416131	0.847427	7.571312	0.0000
ED	-0.446740	0.034458	-12.96473	0.0000
WKS	-0.003574	0.013258	-0.269581	0.7876
EX	-0.010891	0.006336	-1.718878	0.0862
SOUTH	-0.240756	0.147920	-1.627608	0.1041
SMSA	-0.327948	0.139849	-2.345016	0.0194
IND	0.371434	0.135825	2.734658	0.0064
MS	-0.097665	0.245069	-0.398522	0.6904
FEM	-0.358948	0.296971	-1.208697	0.2273
BLK	0.215257	0.252219	0.853453	0.3938

Log likelihood   -246.6581
Obs with Dep=1   305
Obs with Dep=0   290

13.5 Truncated Uniform Density.

$\Pr[x > -\frac{1}{2}] = \int_{-1/2}^{1} \frac{1}{2} dx = \frac{1}{2}[\frac{3}{2}] = \frac{3}{4}$. So that

$f(x/x > -\frac{1}{2}) = \frac{f(x)}{\Pr[x > -\frac{1}{2}]} = \frac{1/2}{3/4} = \frac{2}{3}$ for $-\frac{1}{2} < x < 1$.

b. $E[x/x > -\frac{1}{2}] = \int_{-1/2}^{1} xf(x/x > -\frac{1}{2}) dx = \int_{-1/2}^{1} x \cdot \frac{2}{3} dx = \frac{2}{3} \cdot \frac{1}{2} [x^2]_{-1/2}^{1}$

$= \frac{1}{3}[1 - \frac{1}{4}] = \frac{1}{4}$

$E(x) = \int_{-1}^{1} x \cdot \frac{1}{2} dx = \frac{1}{2} \cdot \frac{1}{2} \cdot [1 - (-1)^2] = 0.$

Therefore, $\frac{1}{4} = E[x/x > -\frac{1}{2}] > E(x) = 0$. Because the density is truncated from below, the new mean shifts to the right.

c. $E(x^2) = \int_{-1}^{1} x^2 \cdot \frac{1}{2} dx = \frac{1}{2} \cdot \frac{1}{3} \cdot [x^3]_{-1}^{1} = \frac{1}{3}$

$\text{var}(x) = E(x^2) - (E(x))^2 = E(x^2) = \frac{1}{3}$

$E(x^2/x > -\frac{1}{2}) = \int_{-1/2}^{1} x^2 \cdot \frac{2}{3} dx = \frac{2}{3} \cdot \frac{1}{3} [x^3]_{-1/2}^{1} = \frac{2}{9}[1 + \frac{1}{8}] = \frac{1}{4}$

$\text{var}(x/x > -\frac{1}{2}) = E(x^2/x > -\frac{1}{2}) - (E[x/x > -\frac{1}{2}])^2 = \frac{1}{4} - (\frac{1}{4})^2$

$= \frac{1}{4} - \frac{1}{16} = \frac{3}{16} < \text{var}(x) = \frac{1}{3}$.

Therefore, as expected, truncation reduces the variance.

13.6 Truncated Normal Density.

a. From the Appendix, equation (A.1), using $c = 1$, $\mu = 1$, $\sigma^2 = 1$ and $\Phi(0) = \frac{1}{2}$, we get, $f(x/x > 1) = \frac{\phi(x-1)}{1 - \Phi(0)} = 2\phi(x-1)$ for $x > 1$

Similarly, using equation (A.2), for $c = 1$, $\mu = 1$ and $\sigma^2 = 1$ with $\Phi(0) = \frac{1}{2}$, we get $f(x/x < 1) = \frac{\phi(x-1)}{\Phi(0)} = 2\phi(x-1)$ for $x < 1$

b. The conditional mean is given in (A.3) and for this example we get

$E(x/x > 1) = 1 + 1 \cdot \frac{\phi(c^*)}{1 - \Phi(c^*)} = 1 + \frac{\phi(0)}{1 - \Phi(0)} = 1 + 2\phi(0)$

Chapter 13: Limited Dependent Variables

$$= 1 + \frac{2}{\sqrt{2\pi}} \approx 1.8 > 1 = E(x)$$

with $c^* = \frac{c-\mu}{\sigma} = \frac{1-1}{1} = 0$. Similarly, using (A.4) we get,

$$E(x/x<1) = 1 - 1 \cdot \frac{\phi(c^*)}{\Phi(c^*)} = 1 - \frac{\phi(0)}{1/2} = 1 - 2\phi(0) = 1 - \frac{2}{\sqrt{2\pi}}$$
$$\approx 1 - 0.8 = 0.2 < 1 = E(x).$$

c. From (A.5) we get, $\mathrm{var}(x/x>1) = 1(1-\delta(c^*)) = 1 - \delta(0)$ where

$$\delta(0) = \frac{\phi(0)}{1-\Phi(0)}\left[\frac{\phi(0)}{1-\Phi(0)} - 0\right] \qquad \text{for} \qquad x > 1$$

$$= 2\phi(0)[2\phi(0)] = 4\phi^2(0) = \frac{4}{2\pi} = \frac{1}{2\pi} = 0.64 \qquad \text{for} \qquad x > 1$$

From (A.6), we get $\mathrm{var}(x/x>1) = 1 - \delta(0)$ where

$$\delta(0) = \frac{-\phi(0)}{\Phi(0)}\left[\frac{-\phi(0)}{\Phi(0)} - 0\right] = 4\phi^2(0) = \frac{4}{2\pi} = 0.64 \qquad \text{for} \qquad x < 1.$$

Both conditional truncated variances are less than the unconditional $\mathrm{var}(x) = 1$.

13.7 *Censored Normal Distribution.*

a. From the Appendix we get,

$$E(y) = \Pr[y=c]\,E(y/y=c) + \Pr[y>c]\,E(y/y>c) = c\Phi(c^*) + (1-\Phi(c^*))\,E(y^*/y^*>c)$$

$$= c\Phi(c^*) + (1-\Phi(c^*))\,[\mu + \sigma\frac{\phi(c^*)}{1-\Phi(c^*)}]$$

where $E(y^*/y^*>c)$ is obtained from the mean of a truncated normal density, see (A.3).

b. Using the result on conditional variance given in Chapter 2 we get,

$\mathrm{var}(y) = E(\text{conditional variance}) + \mathrm{var}(\text{conditional mean})$. But

$E(\text{conditional variance}) = P[y=c]\,\mathrm{var}(y/y=c) + P[y>c]\,\mathrm{var}(y/y>c)$
$$= \Phi(c^*) \cdot 0 + (1-\Phi(c^*))\sigma^2(1-\delta(c^*))$$

where $\mathrm{var}(y/y>c)$ is given by (A.5).

$\mathrm{var}(\text{conditional mean}) = P[y=c] \cdot (c-E(y))^2 + \Pr(y>c)[E(y/y>c) - E(y)]^2$
$$= \Phi(c^*)(c-E(y))^2 + [1-\Phi(c^*)][E(y/y>c) - E(y)]^2$$

where E(y) is given by (A.7) and E(y/y>c) is given by (A.3). This gives

$$\text{var(conditional mean)} = \Phi(c^*)\{c - c\Phi(c^*) - (1-\Phi(c^*))[\mu + \sigma\frac{\phi(c^*)}{1-\Phi(c^*)}]\}^2$$

$$+ [1-\Phi(c^*)]\{\mu + \sigma\frac{\phi(c^*)}{1-\Phi(c^*)} - c\Phi(c^*)\}$$

$$- (1-\Phi(c^*))[\mu + \frac{\sigma\phi(c^*)}{1-\Phi(c^*)}]\}^2$$

$$= \Phi(c^*)\{(1-\Phi(c^*))(c - \mu - \sigma\frac{\phi(c^*)}{1-\Phi(c^*)})\}^2$$

$$+ (1-\Phi(c^*))\{\Phi(c^*)(c - \mu - \sigma\frac{\phi(c^*)}{1-\Phi(c^*)})\}^2$$

$$= \{\Phi(c^*)[1-\Phi(c^*)]\}^2$$

$$+ (1-\Phi(c^*))\Phi^2(c^*)\}\sigma^2[c^* - \frac{\phi(c^*)}{1-\Phi(c^*)}]^2$$

$$= \Phi(c^*)[1-\Phi(c^*)]\sigma^2[c^* - \frac{\phi(c^*)}{1-\Phi(c^*)}]^2.$$

Summing the two terms, we get the expression in (A.8)

$$\text{var}(y) = \sigma^2(1-\Phi(c^*))\{1-\delta(c^*) + [c^* - \frac{\phi(c^*)}{1-\Phi(c^*)}]^2\Phi(c^*)\}.$$

c. For the special case where $c = 0$ and $c^* = -\mu/\sigma$, the mean in (A.7) simplifies to

$$E(y) = 0 \cdot \Phi(c^*) + (1-\Phi(-\frac{\mu}{\sigma}))[\mu + \frac{\sigma\phi(-\mu/\sigma)}{1-\Phi(-\mu/\sigma)}] = \Phi(\frac{\mu}{\sigma})[\mu + \sigma\frac{\phi(\mu/\sigma)}{\Phi(\mu/\sigma)}]$$

as required. Similarly, from part (b), using $c^* = -\mu/\sigma$ and $\Phi(-\mu/\sigma) = \Phi(\mu/\sigma)$,

we get $\text{var}(y) = \sigma^2\Phi(\frac{\mu}{\sigma})[1-\delta(-\frac{\mu}{\sigma}) + (-\frac{\mu}{\sigma} - \frac{\phi(-\mu/\sigma)}{1-\Phi(-\mu/\sigma)})^2\Phi(-\frac{\mu}{\sigma})]$

where $\delta(-\frac{\mu}{\sigma}) = \frac{\phi(-\mu/\sigma)}{1-\Phi(-\mu/\sigma)}[\frac{\phi(-\mu/\sigma)}{1-\Phi(-\mu/\sigma)} + \frac{\mu}{\sigma}] = \frac{\phi(\mu/\sigma)}{\Phi(\mu/\sigma)}[\frac{\phi(-\mu/\sigma)}{\Phi(\mu/\sigma)} + \frac{\mu}{\sigma}].$

13.8 Fixed vs. adjustable mortgage rates. This is based on Dhillon, Shilling and Sirmans (1987).

a. The OLS regression of Y on all variables in the data set is given below. This was done using EViews. The $R^2 = 0.434$ and the F-statistic for the significance

Chapter 13: Limited Dependent Variables

of all slopes is equal to 3.169. This is distributed as $F(15,62)$ under the null hypothesis. This has a p-value of 0.0007. Therefore, we reject H_o and we conclude that this is a significant regression. As explained in section 13.6, using BRMR this also rejects the insignificance of all slopes in the logit specification.

Unrestricted Least Squares

```
=============================================================
LS // Dependent Variable is Y
Sample: 1 78
Included observations: 78
=============================================================
     Variable      Coefficien  Std. Error  t-Statistic   Prob.
=============================================================
        C           1.272832    1.411806    0.901563    0.3708
        BA          0.000398    0.007307    0.054431    0.9568
        BS          0.017084    0.020365    0.838887    0.4048
        NW         -0.036932    0.025320   -1.458609    0.1497
        FI         -0.221726    0.092813   -2.388949    0.0200
        PTS         0.178963    0.091050    1.965544    0.0538
        MAT         0.214264    0.202497    1.058108    0.2941
        MOB         0.020963    0.009194    2.279984    0.0261
        MC          0.189973    0.150816    1.259635    0.2125
        FTB        -0.013857    0.136127   -0.101797    0.9192
        SE          0.188284    0.360196    0.522728    0.6030
        YLD         0.656227    0.366117    1.792399    0.0779
        MARG        0.129127    0.054840    2.354621    0.0217
        CB          0.172202    0.137827    1.249403    0.2162
        STL        -0.001599    0.005994   -0.266823    0.7905
        LA         -0.001761    0.007801   -0.225725    0.8222
=============================================================
R-squared              0.433996  Mean dependent var  0.589744
Adjusted R-squared     0.297059  S.D. dependent var  0.495064
S.E. of regression     0.415069  Akaike info criter -1.577938
Sum squared resid     10.68152   Schwarz criterion  -1.094510
Log likelihood       -33.13764   F-statistic         3.169321
Durbin-Watson stat     0.905968  Prob(F-statistic)   0.000702
=============================================================
```

Plot of Y and YHAT

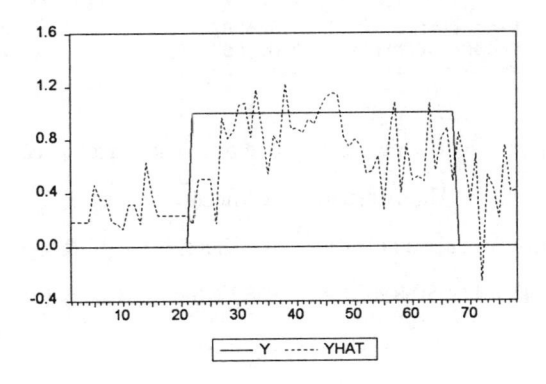

b. The URSS from part (a) is 10.6815 while the RRSS by including only the cost variables is 14.0180 as shown in the enclosed output from EViews. The Chow-F statistic for insignificance of 10 personal characteristics variables is

$$F = \frac{(14.0180 - 10.6815)/10}{10.6815/62} = 1.9366$$

which is distributed as $F(10,62)$ under the null hypothesis. This has a 5 percent critical value of 1.99. Hence, we cannot reject H_o. The principal agent theory suggests that personal characteristics are important in making this mortgage choice. Briefly, this theory suggests that information is asymmetric and the borrower knows things about himself or herself that the lending institution does not. Not rejecting H_o does not provide support for the principal agent theory.

```
TESTING THE EFFICIENT MARKET HYPOTHESIS WITH THE LINEAR PROBABILITY MODEL
Restricted Least Squares
=============================================================
LS // Dependent Variable is Y
Sample: 1 78
Included observations: 78
=============================================================
      Variable    Coefficien Std. Error  t-Statistic  Prob.
=============================================================
         FI       -0.237228   0.078592   -3.018479   0.0035
        MARG       0.127029   0.051496    2.466784   0.0160
         YLD       0.889908   0.332037    2.680151   0.0091
         PTS       0.054879   0.072165    0.760465   0.4495
         MAT       0.069466   0.196727    0.353108   0.7250
          C        1.856435   1.289797    1.439324   0.1544
=============================================================
R-squared            0.257199   Mean dependent var  0.589744
Adjusted R-squared   0.205616   S.D. dependent var  0.495064
S.E. of regression   0.441242   Akaike info criter-1.562522
Sum squared resid   14.01798    Schwarz criterion  -1.381236
Log likelihood     -43.73886    F-statistic         4.986087
Durbin-Watson stat   0.509361   Prob(F-statistic)   0.000562
=============================================================
```

c. The logit specification output using EViews is given below. The unrestricted log-likelihood is equal to -30.8963. The restricted specification output is also given showing a restricted log-likelihood of -41.4729. Therefore, the LR test statistic is given by LR = 2(41.4729-30.8963) = 21.1532 which is distributed as χ^2_{10} under the null hypothesis. This is significant given that the 5 percent critical value of χ^2_{10} is 18.31. This means that the logit specification does not reject the principal agent theory as personal characteristics are not jointly insignificant.

TESTING THE EFFICIENT MARKET HYPOTHESIS WITH THE LOGIT MODEL

Unrestricted Logit Model

```
=================================================================
LOGIT // Dependent Variable is Y
Sample: 1 78
Included observations: 78
Convergence achieved after 5 iterations
=================================================================
     Variable      Coefficien  Std. Error  t-Statistic   Prob.
=================================================================
        C           4.238872    10.47875     0.404521    0.6872
        BA          0.010478     0.075692    0.138425    0.8904
        BS          0.198251     0.172444    1.149658    0.2547
        NW         -0.244064     0.185027   -1.319072    0.1920
        FI         -1.717497     0.727707   -2.360149    0.0214
        PTS         1.499799     0.719917    2.083294    0.0414
        MAT         2.057067     1.631100    1.261153    0.2120
        MOB         0.153078     0.097000    1.578129    0.1196
        MC          1.922943     1.182932    1.625575    0.1091
        FTB        -0.110924     0.983688   -0.112763    0.9106
        SE          2.208505     2.800907    0.788496    0.4334
        YLD         4.626702     2.919634    1.584686    0.1181
        MARG        1.189518     0.485433    2.450426    0.0171
        CB          1.759744     1.242104    1.416744    0.1616
        STL        -0.031563     0.051720   -0.610265    0.5439
        LA         -0.022067     0.061013   -0.361675    0.7188
=================================================================
Log likelihood      -30.89597
Obs with Dep=1             46
Obs with Dep=0             32
=================================================================
     Variable      Mean All    Mean D=1    Mean D=0
=================================================================
        C           1.000000    1.000000    1.000000
        BA         36.03846    35.52174    36.78125
        BS         16.44872    15.58696    17.68750
        NW          3.504013    2.075261    5.557844
        FI         13.24936    13.02348    13.57406
        PTS         1.497949    1.505217    1.487500
        MAT         1.058333    1.027609    1.102500
        MOB         4.205128    4.913043    3.187500
        MC          0.602564    0.695652    0.468750
        FTB         0.615385    0.521739    0.750000
        SE          0.102564    0.043478    0.187500
        YLD         1.606410    1.633261    1.567813
        MARG        2.291923    2.526304    1.955000
        CB          0.358974    0.478261    0.187500
        STL        13.42218    11.72304    15.86469
        LA          5.682692    4.792174    6.962812
=================================================================
```

Restricted Logit Model

```
=================================================================
LOGIT // Dependent Variable is Y
Sample: 1 78
Included observations: 78
Convergence achieved after 4 iterations
=================================================================
     Variable    Coefficien Std. Error  t-Statistic  Prob.
=================================================================
           FI    -1.264608   0.454050   -2.785172   0.0068
         MARG    0.717847   0.313845    2.287265   0.0251
          YLD    4.827537   1.958833    2.464497   0.0161
          PTS    0.359033   0.423378    0.848019   0.3992
          MAT    0.550320   1.036613    0.530883   0.5971
            C    6.731755   7.059485    0.953576   0.3435
=================================================================
Log likelihood   -41.47292
Obs with Dep=1          46
Obs with Dep=0          32
=================================================================
     Variable    Mean All    Mean D=1    Mean D=0
=================================================================
           FI   13.24936    13.02348    13.57406
         MARG    2.291923    2.526304    1.955000
          YLD    1.606410    1.633261    1.567813
          PTS    1.497949    1.505217    1.487500
          MAT    1.058333    1.027609    1.102500
            C    1.000000    1.000000    1.000000
=================================================================
```

d. Similarly, the probit specification output using EViews is given below. The unrestricted log-likelihood is equal to -30.7294. The restricted log-likelihood is -41.7649. Therefore, the LR test statistic is given by LR = 2(41.7649-30.7294) = 22.0710 which is distributed as χ^2_{10} under the null hypothesis. This is significant given that the 5 percent critical value of χ^2_{10} is 18.31. This means that the probit specification does not reject the principal agent theory as personal characteristics are not jointly insignificant.

TESTING THE EFFICIENT MARKET HYPOTHESIS WITH THE PROBIT MODEL

Unrestricted Probit Model

```
=================================================================
PROBIT // Dependent Variable is Y
Sample: 1 78
Included observations: 78
Convergence achieved after 5 iterations
=================================================================
     Variable    Coefficien Std. Error  t-Statistic  Prob.
=================================================================
            C    3.107820   5.954673    0.521913   0.6036
           BA    0.003978   0.044546    0.089293   0.9291
           BS    0.108267   0.099172    1.091704   0.2792
```

```
         NW      -0.128775    0.103438   -1.244943    0.2178
         FI      -1.008080    0.418160   -2.410750    0.0189
        PTS       0.830273    0.379895    2.185533    0.0326
        MAT       1.164384    0.924018    1.260131    0.2123
        MOB       0.093034    0.056047    1.659924    0.1020
         MC       1.058577    0.653234    1.620518    0.1102
        FTB      -0.143447    0.550471   -0.260589    0.7953
         SE       1.127523    1.565488    0.720237    0.4741
        YLD       2.525122    1.590796    1.587332    0.1175
       MARG       0.705238    0.276340    2.552069    0.0132
         CB       1.066589    0.721403    1.478493    0.1443
        STL      -0.016130    0.029303   -0.550446    0.5840
         LA      -0.014615    0.035920   -0.406871    0.6855
==============================================================
Log likelihood    -30.72937
Obs with Dep=1           46
Obs with Dep=0           32
==============================================================


Restricted Probit Model

==============================================================
PROBIT // Dependent Variable is Y
Sample: 1 78
Included observations: 78
Convergence achieved after 3 iterations
==============================================================
     Variable   Coefficien Std. Error t-Statistic  Prob.
==============================================================
         FI      -0.693584    0.244631   -2.835225    0.0059
       MARG       0.419997    0.175012    2.399811    0.0190
        YLD       2.730187    1.099487    2.483146    0.0154
        PTS       0.235534    0.247390    0.952076    0.3442
        MAT       0.221568    0.610572    0.362886    0.7178
          C       3.536657    4.030251    0.877528    0.3831
==============================================================
Log likelihood    -41.76443
Obs with Dep=1           46
Obs with Dep=0           32
==============================================================
```

CHAPTER 14
Time-Series Analysis

14.1 The AR(1) Model. $y_t = \rho y_{t-1} + \epsilon_t$ with $|\rho| < 1$ and $\epsilon_t \sim \text{IIN}(0,\sigma_\epsilon^2)$. Also, $y_0 \sim N(0,\sigma_\epsilon^2/1-\rho^2)$.

a. By successive substitution

$$y_t = \rho y_{t-1} + \epsilon_t = \rho(\rho y_{t-2} + \epsilon_{t-1}) + \epsilon_t = \rho^2 y_{t-2} + \rho\epsilon_{t-1} + \epsilon_t = \rho^2(\rho y_{t-3} + \epsilon_{t-2}) + \rho\epsilon_{t-1}$$
$$+ \epsilon_t = \rho^3 y_{t-3} + \rho^2\epsilon_{t-2} + \rho\epsilon_{t-1} + \epsilon_t = .. = \rho^t y_0 + \rho^{t-1}\epsilon_1 + \rho^{t-2}\epsilon_2 + .. + \epsilon_t$$

Then, $E(y_t) = \rho^t E(y_0) = 0$ for every t, since $E(y_0) = E(\epsilon_t) = 0$.

$\text{var}(y_t) = \rho^{2t}\text{var}(y_0) + \rho^{2(t-1)}\text{var}(\epsilon_1) + \rho^{2(t-2)}\text{var}(\epsilon_2) + .. + \text{var}(\epsilon_t)$

$$= \rho^{2t}(\frac{\sigma_\epsilon^2}{1-\rho^2}) + (\rho^{2(t-1)} + \rho^{2(t-2)} + .. + 1)\sigma_\epsilon^2$$

$$= \frac{\rho^{2t}}{1-\rho^2}\cdot\sigma_\epsilon^2 + \frac{1-\rho^{2t}}{1-\rho^2}\cdot\sigma_\epsilon^2 = \frac{\sigma_\epsilon^2}{1-\rho^2} = \text{var}(y_0) \quad \text{for every t.}$$

If $\rho = 1$, then $\text{var}(y_t) = \sigma_\epsilon^2/0 \to \infty$. Also, if $|\rho| > 1$, then $1 - \rho^2 < 0$ and $\text{var}(y_t) < 0$.

b. The AR(1) series y_t has zero mean and constant variance $\sigma^2 = \text{var}(y_t)$, for $t = 0,1,2,...$ In part (a) we could have stopped the successive substitution at y_{t-s}, this yields $y_t = \rho^s y_{t-s} + \rho^{s-1}\epsilon_{t-s+1} + .. + \epsilon_t$

Therefore, $\text{cov}(y_t,y_{t-s}) = \text{cov}(\rho^s y_{t-s} + \rho^{s-1}\epsilon_{t-s+1} + .. + \epsilon_t, y_{t-s}) = \rho^s \text{var}(y_{t-s}) = \rho^s \sigma^2$
which only depends on s the distance between t and t-s. Therefore, the AR(1) series y_t is said to be covariance-stationary or weakly stationary.

c. First one generates $y_0 = 0.5\, N(0,1)/(1-\rho^2)^{1/2}$ for various values of ρ. Then $y_t = \rho y_{t-1} + \epsilon_t$ where $\epsilon_t \sim \text{IIN}(0, 0.25)$ for $t = 1,2,..,T$.

14.2 The MA(1) Model. $y_t = \epsilon_t + \theta\epsilon_{t-1}$ with $\epsilon_t \sim \text{IIN}(0,\sigma_\epsilon^2)$

a. $E(y_t) = E(\epsilon_t) + \theta E(\epsilon_{t-1}) = 0$ for all t. Also,

$\text{var}(y_t) = \text{var}(\epsilon_t) + \theta^2 \text{var}(\epsilon_{t-1}) = (1+\theta^2)\sigma_\epsilon^2 \quad$ for all t.

Chapter 14: Time-Series Analysis   299

Therefore, the mean and variance are independent of t.

b. $\text{cov}(y_t, y_{t-1}) = \text{cov}(\epsilon_t + \theta\epsilon_{t-1}, \epsilon_{t-1} + \theta\epsilon_{t-2}) = \theta\text{var}(\epsilon_{t-1}) = \theta\sigma_\epsilon^2$

$\text{cov}(y_t, y_{t-s}) = \text{cov}(\epsilon_t + \theta\epsilon_{t-1}, \epsilon_{t-s} + \theta\epsilon_{t-s-1}) = \begin{cases} \theta\sigma_\epsilon^2 & \text{when } s = 1 \\ 0 & \text{when } s > 1 \end{cases}$

Since $E(y_t) = 0$ for all t and $\text{cov}(y_t, y_{t-s}) = \begin{cases} \theta\sigma_\epsilon^2 & \text{when } s = 1 \\ 0 & \text{when } s > 1 \end{cases}$

for all t and s, then the MA(1) process is covariance stationary.

c. First one generates $\epsilon_t \sim \text{IIN}(0, 0.25)$. Then, for various values of θ, one generates

$y_t = \epsilon_t + \theta\epsilon_{t-1}$.

14.3 The consumption - personal disposable income example.

a. For personal disposable income (Y_t), the sample autocorrelation function using EViews is given below:

```
Sample: 1950 1993
Included observations: 44
```

Autocorrelation	Partial Correlation		AC	PAC	Q-Stat	Prob
0.000 . |*******	. |*******	1	0.944	0.944	41.959	
0.000 . |*******	. | .	2	0.885	-0.055	79.740	
0.000 . |******	. | .	3	0.827	-0.032	113.47	
0.000 . |******	.*| .	4	0.764	-0.068	142.99	
0.000 . |*****	.*| .	5	0.698	-0.065	168.27	
0.000 . |*****	. | .	6	0.631	-0.041	189.50	
0.000 . |****	. | .	7	0.567	-0.022	207.08	
0.000 . |****	.*| .	8	0.499	-0.076	221.06	
0.000 . |***	. | .	9	0.429	-0.054	231.71	
0.000 . |***	. | .	10	0.361	-0.042	239.45	
0.000 . |**	. | .	11	0.298	0.007	244.90	
0.000 . |**	. | .	12	0.235	-0.053	248.39	
0.000 . |*.	.*| .	13	0.170	-0.064	250.28	

For the first differenced series (ΔY_t):

```
Sample: 1950 1993
Included observations: 43
```

Autocorrelation	Partial Correlation		AC	PAC	Q-Stat	Prob
0.735 . | .	. | .	1	0.050	0.050	0.1142	
0.920 . | .	. | .	2	0.033	0.031	0.1669	
0.974 . | .	. | .	3	-0.034	-0.037	0.2221	
0.941 .*| .	.*| .	4	-0.106	-0.104	0.7800	
0.961 . |*.	. |*.	5	0.068	0.082	1.0188	
0.881 . |*.	. |*.	6	0.162	0.164	2.3844	
0.872 .*| .	.*| .	7	-0.118	-0.154	3.1374	
0.865 . |*.	. |*.	8	0.118	0.119	3.9092	
0.845 .*| .	.*| .	9	-0.131	-0.114	4.8803	

0.515 **	.	**	.	10	-0.271	-0.269	9.1790
0.474 .	*.	.	*.	11	0.155	0.195	10.634
0.550 .	.	.	.	12	-0.045	-0.049	10.760
0.625 .	.	.*	.	13	-0.034	-0.093	10.835

```
ADF Test Statistic   -2.312770       1%  Critical Value*        -4.1837
                                     5%  Critical Value         -3.5162
                                    10%  Critical Value         -3.1882
```

*MacKinnon critical values for rejection of hypothesis of a unit root.

c. Augmented Dickey-Fuller Test Equation

```
LS // Dependent Variable is D(INCOME)
Sample(adjusted): 1951 1993
Included observations: 43 after adjusting endpoints

Variable    Coefficient    Std. Error    t-Statistic    Prob.

INCOME(-1)    -0.191123     0.082638     -2.312770     0.0260
C              1207.823     452.8069      2.667412     0.0110
@TREND(1950)   41.16729     17.52137      2.349547     0.0238

R-squared              0.121920    Mean dependent var       187.3721
Adjusted R-squared     0.078016    S.D. dependent var       171.1742
S.E. of regression     164.3615    Akaike info criterion    10.27135
Sum squared resid      1080588.    Schwarz criterion        10.39423
Log likelihood        -278.8484    F-statistic               2.776960
Durbin-Watson stat     1.777935    Prob(F-statistic)         0.074248

ADF Test Statistic  -2.383255       1%  critical Value*       -4.1896
                                    5%  Critical Value        -3.5189
                                   10%  Critical Value        -3.1898
```

*MacKinnon critical values for rejection of hypothesis of a unit root.

d. Augmented Dickey-Fuller Test Equation.

```
LS // Dependent Variable is D(INCOME)
Sample(adjusted): 1952 1993
Included observations: 42 after adjusting endpoints

Variable       Coefficient    Std. Error    t-Statistic    Prob.

INCOME(-1)     -0.216469      0.090829     -2.383255      0.0223
D(INCOME(-1))   0.119654      0.155455      0.769707      0.4462
C               1319.300      487.4030      2.706794      0.0101
@TREND(1950)    46.59344      19.40273      2.401386      0.0213

R-squared              0.133954    Mean dependent var       189.3095
Adjusted R-squared     0.065582    S.D. dependent var       172.7713
S.E. of regression     167.0099    Akaike info criterion    10.32650
Sum squared resid      1059908.    Schwarz criterion        10.49199
Log likelihood        -272.4519    F-statistic               1.959191
Durbin-Watson stat     1.996963    Prob(F-statistic)         0.136541
```

Chapter 14: Time-Series Analysis

```
ADF Test Statistic    -2.421416      1%  Critical Value*     -4.1958
                                     5%  Critical Value      -3.5217
                                    10%  Critical Value      -3.1914
```

*MacKinnon critical values for rejection of hypothesis of a unit root.

Augmented Dickey-Fuller Test Equation
LS // Dependent Variable is D(INCOME)
Sample(adjusted): 1953 1993
Included observations: 41 after adjusting endpoints

Variable	Coefficient	Std. Error	t-Statistic	Prob.
INCOME(-1)	-0.246038	0.101609	-2.421416	0.0206
D(INCOME(-1))	0.126208	0.159488	0.791333	0.4339
D(INCOME(-2))	0.125941	0.160424	0.785050	0.4376
C	1450.455	532.3156	2.724803	0.0099
@TREND(1950)	52.94986	21.86536	2.421632	0.0206

R-squared	0.142828	Mean dependent var		191.8293
Adjusted R-squared	0.047587	S.D. dependent var		174.1346
S.E. of regression	169.9408	Akaike info criterion		10.38475
Sum squared resid	1039676.	Schwarz criterion		10.59372
Log likelihood	-266.0639	F-statistic		1.499642
Durbin-Watson stat	1.981901	Prob(F-statistic)		0.222845

```
ADF Test Statistic    -2.511131      1%  Critical Value*     -4.2023
                                     5%  Critical Value      -3.5247
                                    10%  Critical Value      -3.1931
```

*MacKinnon critical values for rejection of hypothesis of a unit root.

Augmented Dickey-Fuller Test Equation
LS // Dependent Variable is D(INCOME)
Sample(adjusted): 1954 1993
Included observations: 40 after adjusting endpoints

Variable	Coefficient	Std. Error	t-Statistic	Prob.
INCOME(-1)	-0.284907	0.113458	-2.511131	0.0170
D(INCOME(-1))	0.151122	0.164977	0.916019	0.3661
D(INCOME(-2))	0.145037	0.165120	0.878372	0.3859
D(INCOME(-3))	0.047550	0.170076	0.279581	0.7815
C	1626.391	580.9674	2.799453	0.0084
@TREND(1950)	61.70187	24.55594	2.512706	0.0169

R-squared	0.160346	Mean dependent var		192.5250
Adjusted R-squared	0.036868	S.D. dependent var		176.2952
S.E. of regression	173.0149	Akaike info criterion		10.44424
Sum squared resid	1017761.	Schwarz criterion		10.69757
Log likelihood	-259.6423	F-statistic		1.298576
Durbin-Watson stat	1.951007	Prob(F-statistic)		0.287607

```
ADF Test Statistic    -5.991433      1%  Critical Value*     -3.5930
                                     5%  Critical Value      -2.9320
                                    10%  Critical Value      -2.6039
```

*MacKinnon critical values for rejection of hypothesis of a unit root.

e. Augmented Dickey-Fuller Test Equation

```
LS // Dependent Variable is D(DY)
Sample(adjusted): 1952 1993
Included observations: 42 after adjusting endpoints
```

Variable	Coefficient	Std. Error	t-Statistic	Prob.
DY(-1)	-0.949762	0.158520	-5.991433	0.0000
C	179.7464	40.46248	4.442299	0.0001

R-squared	0.472972	Mean dependent var	-1.047619
Adjusted R-squared	0.459796	S.D. dependent var	237.6896
S.E. of regression	174.6984	Akaike info criterion	10.37257
Sum squared resid	1220782.	Schwarz criterion	10.45532
Log likelihood	-275.4194	F-statistic	35.89727
Durbin-Watson stat	1.994349	Prob(F-statistic)	0.000000

ADF Test Statistic	-1.923824	1% Critical Value*	-3.5889
		5% Critical Value	-2.9303
		10% Critical Value	-2.6030

*MacKinnon critical values for rejection of hypothesis of a unit root.

f. Augmented Dickey-Fuller Test Equation

```
Augmented Dickey-Fuller Test Equation
LS // Dependent Variable is D(ERROR)
Sample(adjusted): 1951 1993
Included observations: 43 after adjusting endpoints
```

Variable	Coefficient	Std. Error	t-Statistic	Prob.
ERROR(-1)	-0.206801	0.107495	-1.923824	0.0613
C	2.940443	15.41925	0.190699	0.8497

R-squared	0.082797	Mean dependent var	4.507531
Adjusted R-squared	0.060426	S.D. dependent var	104.1658
S.E. of regression	100.9696	Akaike info criterion	9.275034
Sum squared resid	417989.3	Schwarz criterion	9.356951
Log likelihood	-258.4276	F-statistic	3.701099
Durbin-Watson stat	1.932257	Prob(F-statistic)	0.061338

g. ARCH Test

F-statistic	4.210048	Probability	0.022105
Obs*R-squared	7.457683	Probability	0.024021

```
Test Equation:
LS // Dependent Variable is RESID^2
Sample(adjusted): 1952 1993
Included observations: 42 after adjusting endpoints
```

Chapter 14: Time-Series Analysis

Variable	Coefficient	Std. Error	t-Statistic	Prob.
C	13895.91	5753.355	2.415271	0.0205
RESID^2(-1)	0.485770	0.185753	2.615134	0.0126
RESID^2(-2)	-0.039811	0.186061	-0.213968	0.8317

R-squared	0.177564	Mean dependent var	23098.43
Adjusted R-squared	0.135388	S.D. dependent var	29733.26
S.E. of regression	27647.33	Akaike info criterion	20.52332
Sum squared resid	2.98E+10	Schwarz criterion	20.64744
Log likelihood	-487.5851	F-statistic	4.210048
Durbin-Watson stat	1.740367	Prob(F-statistic)	0.022105

h. Using LOG(Y)

Sample: 1950 1993
Included observations: 44

Autocorrelation	Partial Correlation		AC	PAC	Q-Stat	Prob
0.000 . |*******| . |*******|	1	0.943	0.943	41.900
0.000 . |*******| .	.		2	0.885	-0.051	79.603
0.000 . |******| .	.		3	0.825	-0.039	113.17
0.000 . |******| .	.		4	0.762	-0.054	142.58
0.000 . |*****| .*| .		5	0.696	-0.071	167.72
0.000 . |*****| .	.		6	0.631	-0.027	188.93
0.000 . |****| .	.		7	0.568	-0.023	206.55
0.000 . |****| .*| .		8	0.501	-0.073	220.66
0.000 . |***| .*| .		9	0.432	-0.065	231.43
0.000 . |***| .	.		10	0.363	-0.045	239.27
0.000 . |**| .	.		11	0.297	-0.022	244.69
0.000 . |**| .	.		12	0.231	-0.054	248.07
0.000 . |*.| .*| .		13	0.164	-0.057	249.84

Sample: 1950 1993
Included observations: 43

Autocorrelation	Partial Correlation		AC	PAC	Q-Stat	Prob
0.424 . |*.	. |*.		1	0.118	0.118	0.6390
0.627 . |*.	. |*.		2	0.079	0.066	0.9323
0.770 .	.	.	.		3	0.064	0.048	1.1293
0.814 .*| .	.*| .		4	-0.095	-0.114	1.5728
0.885 .	.	.	*.		5	0.056	0.073	1.7305
0.822 . |*.	. |*.		6	0.149	0.152	2.8950
0.824 .*| .	.*| .		7	-0.115	-0.155	3.6061
0.853 . |*.	. |*.		8	0.089	0.086	4.0485
0.857 .	.	.*	.		9	-0.054	-0.062	4.2130
0.610 **| .	**| .		10	-0.261	-0.238	8.1976
0.572 . |*.	. |**		11	0.149	0.201	9.5403
0.652 .	.	.	.		12	-0.029	-0.031	9.5933
0.719 .	.	.	.		13	-0.039	-0.038	9.6932

ADF Test Statistic	-0.605781	1% Critical Value*	-4.1837
		5% Critical Value	-3.5162
		10% Critical Value	-3.1882

*MacKinnon critical values for rejection of hypothesis of a unit root.

Augmented Dickey-Fuller Test Equation
LS // Dependent Variable is D(LNY)
Sample(adjusted): 1951 1993
Included observations: 43 after adjusting endpoints

Variable	Coefficient	Std. Error	t-Statistic	Prob.
LNY(-1)	-0.044020	0.072667	-0.605781	0.5481
C	0.407378	0.632850	0.643720	0.5234
@TREND(1950)	0.000729	0.001579	0.461524	0.6469

R-squared	0.036779	Mean dependent var	0.019189
Adjusted R-squared	-0.011382	S.D. dependent var	0.016504
S.E. of regression	0.016598	Akaike info criterion	-8.129783
Sum squared resid	0.011019	Schwarz criterion	-8.006908
Log likelihood	116.7760	F-statistic	0.763667
Durbin-Watson stat	1.734159	Prob(F-statistic)	0.472627

ADF Test Statistic	-0.814360	1% Critical Value*	-4.1896
		5% Critical Value	-3.5189
		10% Critical Value	-3.1898

*MacKinnon critical values for rejection of hypothesis of a unit root.

Augmented Dickey-Fuller Test Equation
LS // Dependent Variable is D(LNY)
Sample(adjusted): 1952 1993
Included observations: 42 after adjusting endpoints

Variable	Coefficient	Std. Error	t-Statistic	Prob.
LNY(-1)	-0.063811	0.078357	-0.814360	0.4205
D(LNY(-1))	0.143163	0.170168	0.841304	0.4054
C	0.576958	0.680951	0.847282	0.4021
@TREND(1950)	0.001160	0.001716	0.675900	0.5032

R-squared	0.057750	Mean dependent var	0.019247
Adjusted R-squared	-0.016638	S.D. dependent var	0.016699
S.E. of regression	0.016838	Akaike info criterion	-8.077866
Sum squared resid	0.010773	Schwarz criterion	-7.912373
Log likelihood	114.0398	F-statistic	0.776334
Durbin-Watson stat	2.010068	Prob(F-statistic)	0.514499

ADF Test Statistic	-0.921682	1% Critical Value*	-4.1958
		5% Critical Value	-3.5217
		10% Critical Value	-3.1914

*MacKinnon critical values for rejection of hypothesis of a unit root.

Augmented Dickey-Fuller Test Equation
LS // Dependent Variable is D(LNY)
Sample(adjusted): 1953 1993
Included observations: 41 after adjusting endpoints

Variable	Coefficient	Std. Error	t-Statistic	Prob.
LNY(-1)	-0.079805	0.086586	-0.921682	0.3628
D(LNY(-1))	0.138048	0.174027	0.793259	0.4328
D(LNY(-2))	0.106294	0.177783	0.597887	0.5537
C	0.714757	0.750822	0.951966	0.3475
@TREND(1950)	0.001493	0.001913	0.780268	0.4403

Chapter 14: Time-Series Analysis

```
R-squared              0.072240    Mean dependent var       0.019391
Adjusted R-squared    -0.030844    S.D. dependent var       0.016881
S.E. of regression     0.017139    Akaike info criterion   -8.018941
Sum squared resid      0.010575    Schwarz criterion       -7.809968
Log likelihood       111.2118      F-statistic              0.700789
Durbin-Watson stat     1.998974    Prob(F-statistic)        0.596524


ADF Test Statistic    -1.063832     1%  Critical Value*    -4.2023
                                    5%  Critical Value     -3.5247
                                   10%  Critical Value     -3.1931
```

*MacKinnon critical values for rejection of hypothesis of a unit root.

Augmented Dickey-Fuller Test Equation
LS // Dependent Variable is D(LNY)
Sample(adjusted): 1954 1993
Included observations: 40 after adjusting endpoints

```
Variable   Coefficient    Std. Error     t-Statistic     Prob.

LNY(-1)     -0.100545      0.094513      -1.063832       0.2949
D(LNY(-1))   0.152086      0.180253       0.843737       0.4047
D(LNY(-2))   0.112034      0.182683       0.613268       0.5438
D(LNY(-3))   0.112174      0.183646       0.610818       0.5454
C            0.891956      0.818093       1.090287       0.2833
@TREND(1950) 0.001969      0.002099       0.937794       0.3550

R-squared              0.080316    Mean dependent var       0.019250
Adjusted R-squared    -0.054932    S.D. dependent var       0.017071
S.E. of regression     0.017534    Akaike info criterion   -7.949735
Sum squared resid      0.010453    Schwarz criterion       -7.696404
Log likelihood       108.2372      F-statistic              0.593844
Durbin-Watson stat     1.898607    Prob(F-statistic)        0.704719


ADF Test Statistic    -5.710640     1%  Critical Value*    -3.5889
                                    5%  Critical Value     -2.9303
                                   10%  Critical Value     -2.6030
```

*MacKinnon critical values for rejection of hypothesis of a unit root.

Augmented Dickey-Fuller Test Equation
LS // Dependent Variable is D(DLNY)
Sample(adjusted): 1951 1993
Included observations: 43 after adjusting endpoints

```
Variable   Coefficient    Std. Error     t-Statistic     Prob.

DLNY(-1)    -0.879650      0.154037      -5.710640       0.0000
C            0.016891      0.003878       4.355735       0.0001

R-squared              0.443021    Mean dependent var       0.000101
Adjusted R-squared     0.429436    S.D. dependent var       0.021951
S.E. of regression     0.016581    Akaike info criterion   -8.153601
Sum squared resid      0.011272    Schwarz criterion       -8.071685
Log likelihood       116.2881      F-statistic             32.61141
Durbin-Watson stat     1.998889    Prob(F-statistic)        0.000001
```

```
ADF Test Statistic    -2.253192    1%   Critical Value*         -3.5889
                                   5%   Critical Value          -2.9303
                                   10%  Critical Value          -2.6030
```

*MacKinnon critical values for rejection of hypothesis of a unit root.

Augmented Dickey-Fuller Test Equation
LS // Dependent Variable is D(LN_ERROR)
Sample(adjusted): 1951 1993
Included observations: 43 after adjusting endpoints

Variable	Coefficient	Std. Error	t-Statistic	Prob.
LN_ERROR(-1)	-0.251905	0.111799	-2.253192	0.0297
C	0.000119	0.001513	0.078500	0.9378

R-squared	0.110183	Mean dependent var		0.000286
Adjusted R-squared	0.088480	S.D. dependent var		0.010379
S.E. of regression	0.009909	Akaike info criterion		-9.183243
Sum squared resid	0.004026	Schwarz criterion		-9.101327
Log likelihood	138.4254	F-statistic		5.076875
Durbin-Watson stat	1.918035	Prob(F-statistic)		0.029660

ARCH Test:

F-statistic	4.628023	Probability	0.015722
Obs*R-squared	8.056067	Probability	0.017809

Test Equation:
LS // Dependent Variable is RESID^2
Sample(adjusted): 1952 1993
Included observations: 42 after adjusting endpoints

Variable	Coefficient	Std. Error	t-Statistic	Prob.
C	0.000124	4.77E-05	2.605322	0.0129
RESID^2(-1)	0.512581	0.177962	2.880280	0.0064
RESID^2(-2)	-0.090366	0.177743	-0.508406	0.6140

R-squared	0.191811	Mean dependent var		0.000200
Adjusted R-squared	0.150366	S.D. dependent var		0.000240
S.E. of regression	0.000221	Akaike info criterion		-16.76529
Sum squared resid	1.91E-06	Schwarz criterion		-16.64117
Log likelihood	295.4756	F-statistic		4.628023
Durbin-Watson stat	1.783145	Prob(F-statistic)		0.015722

USING LOG(C)

Correlogram of LNC

Sample: 1950 1993
Included observations: 44

```
Autocorrelation  Partial Correlation       AC      PAC    Q-Stat   Prob

0.000 . |*******|       . |*******|   1   0.942   0.942   41.796
0.000 . |*******|       .  |.     |   2   0.882  -0.054   79.277
0.000 . |******ˍ|       .  |.     |   3   0.821  -0.040   112.51
0.000 . |******ˍ|       .*|.     |   4   0.756  -0.064   141.41
```

Chapter 14: Time-Series Analysis   307

```
0.000 .  |*****       |    .*|  .       5    0.688   -0.065   165.96
0.000 .  |*****       |      .  .       6    0.622   -0.017   186.57
0.000 .  |****        |      .  .       7    0.557   -0.037   203.53
0.000 .  |****        |    .*|  .       8    0.490   -0.058   217.00
0.000 .  |***         |      .  .       9    0.421   -0.057   227.26
0.000 .  |***         |      .  .      10    0.356   -0.013   234.82
0.000 .  |**          |      .  .      11    0.293   -0.037   240.09
0.000 .  |**          |      .  .      12    0.231   -0.039   243.46
0.000 .  |*.          |      .  .      13    0.168   -0.054   245.31
```

Correlogram of D(LNC)
Sample: 1950 1993
Included observations: 43

Autocorrelation Partial Correlation   AC   PAC   Q-Stat   Prob

```
0.057 .  |**    |          .  |**  |    1     0.281    0.281   3.6326
0.146 .*|  .    |         .*|  .   |    2    -0.067   -0.159   3.8474
0.218 .*|  .    |          .  |  . |    3    -0.110   -0.049   4.4356
0.333 .  |  .   |          .  |  . |    4    -0.054   -0.017   4.5819
0.401 .*|  .    |         .*|  .   |    5    -0.103   -0.112   5.1229
0.500 .*|  .    |          .  |  . |    6    -0.065   -0.017   5.3455
0.610 .  |  .   |          .  |  . |    7     0.035    0.041   5.4118
0.646 .*|  .    |         .*|  .   |    8    -0.104   -0.175   6.0152
0.730 .  |  .   |          .  |*.  |    9     0.038    0.137   6.0975
0.793 .  |  .   |          .  |  . |   10     0.052   -0.034   6.2552
0.838 .  |  .   |          .  |  . |   11     0.064    0.040   6.5064
0.884 .  |  .   |          .  |  . |   12    -0.036   -0.047   6.5853
0.922 .  |  .   |          .  |  . |   13     0.005    0.029   6.5869
```

ADF Test Statistic   -1.493977   1%   Critical Value*   -4.1837
                                     5%   Critical Value   -3.5162
                                    10%   Critical Value   -3.1882

*MacKinnon critical values for rejection of hypothesis of a unit root.

Augmented Dickey-Fuller Test Equation
LS // Dependent Variable is D(LNC)
Sample(adjusted): 1951 1993
Included observations: 43 after adjusting endpoints

Variable	Coefficient	Std. Error	t-Statistic	Prob.
LNC(-1)	-0.126815	0.084884	-1.493977	0.1430
C	1.113426	0.731302	1.522525	0.1357
@TREND(1950)	0.002649	0.001830	1.447199	0.1556

R-squared	0.055544	Mean dependent var	0.019379
Adjusted R-squared	0.008322	S.D. dependent var	0.016317
S.E. of regression	0.016249	Akaike info criterion	-8.172211
Sum squared resid	0.010561	Schwarz criterion	-8.049337
Log likelihood	117.6882	F-statistic	1.176219
Durbin-Watson stat	1.323619	Prob(F-statistic)	0.318883

ADF Test Statistic   -2.172601   1%   Critical Value*   -4.1896
                                     5%   Critical Value   -3.5189
                                    10%   Critical Value   -3.1898

*MacKinnon critical values for rejection of hypothesis of a unit root.

Augmented Dickey-Fuller Test Equation
LS // Dependent Variable is D(LNC)
Sample(adjusted): 1952 1993
Included observations: 42 after adjusting endpoints

Variable	Coefficient	Std. Error	t-Statistic	Prob.
LNC(-1)	-0.187287	0.086204	-2.172601	0.0361
D(LNC(-1))	0.388757	0.153854	2.526787	0.0158
C	1.627277	0.741296	2.195179	0.0343
@TREND(1950)	0.003939	0.001872	2.104237	0.0420

R-squared	0.187844	Mean dependent var		0.019746
Adjusted R-squared	0.123727	S.D. dependent var		0.016334
S.E. of regression	0.015290	Akaike info criterion		-8.270708
Sum squared resid	0.008884	Schwarz criterion		-8.105216
Log likelihood	118.0895	F-statistic		2.929687
Durbin-Watson stat	1.976745	Prob(F-statistic)		0.045857

ADF Test Statistic	-1.816750	1%	Critical Value*	-4.1958
		5%	Critical Value	-3.5217
		10%	Critical Value	-3.1914

*MacKinnon critical values for rejection of hypothesis of a unit root.

Augmented Dickey-Fuller Test Equation
LS // Dependent Variable is D(LNC)
Sample(adjusted): 1953 1993
Included observations: 41 after adjusting endpoints

Variable	Coefficient	Std. Error	t-Statistic	Prob.
LNC(-1)	-0.181181	0.099728	-1.816750	0.0776
D(LNC(-1))	0.392399	0.163234	2.403898	0.0215
D(LNC(-2))	-0.025274	0.172098	-0.146855	0.8841
C	1.575129	0.856083	1.839926	0.0740
@TREND(1950)	0.003807	0.002178	1.748121	0.0890

R-squared	0.184420	Mean dependent var		0.019921
Adjusted R-squared	0.093800	S.D. dependent var		0.016497
S.E. of regression	0.015704	Akaike info criterion		-8.193790
Sum squared resid	0.008879	Schwarz criterion		-7.984818
Log likelihood	114.7962	F-statistic		2.035091
Durbin-Watson stat	1.990159	Prob(F-statistic)		0.110113

ADF Test Statistic	-1.794862	1%	Critical Value*	-4.2023
		5%	Critical Value	-3.5247
		10%	Critical Value	-3.1931

*MacKinnon critical values for rejection of hypothesis of a unit root.

Augmented Dickey-Fuller Test Equation
LS // Dependent Variable is D(LNC)
Sample(adjusted): 1954 1993
Included observations: 40 after adjusting endpoints

Variable	Coefficient	Std. Error	t-Statistic	Prob.
LNC(-1)	-0.198099	0.110370	-1.794862	0.0816
D(LNC(-1))	0.410648	0.173964	2.360534	0.0241
D(LNC(-2))	-0.019901	0.183248	-0.108604	0.9142
D(LNC(-3))	0.050011	0.181281	0.275873	0.7843
C	1.718661	0.946020	1.816727	0.0781
@TREND(1950)	0.004194	0.002418	1.734686	0.0919

R-squared	0.188456	Mean dependent var		0.019847
Adjusted R-squared	0.069111	S.D. dependent var		0.016700
S.E. of regression	0.016113	Akaike info criterion		-8.118792
Sum squared resid	0.008827	Schwarz criterion		-7.865460
Log likelihood	111.6183	F-statistic		1.579091
Durbin-Watson stat	1.968921	Prob(F-statistic)		0.192391

ADF Test Statistic	-4.794885	1%	Critical Value*	-3.5930
		5%	Critical Value	-2.9320
		10%	Critical Value	-2.6039

*MacKinnon critical values for rejection of hypothesis of a unit root.

Augmented Dickey-Fuller Test Equation
LS // Dependent Variable is D(LNC,2)
Sample(adjusted): 1952 1993
Included observations: 42 after adjusting endpoints

Variable	Coefficient	Std. Error	t-Statistic	Prob.
D(LNC(-1))	-0.719057	0.149963	-4.794885	0.0000
C	0.014314	0.003794	3.772965	0.0005

R-squared	0.364988	Mean dependent var		0.000411
Adjusted R-squared	0.349113	S.D. dependent var		0.019653
S.E. of regression	0.015856	Akaike info criterion		-8.241987
Sum squared resid	0.010056	Schwarz criterion		-8.159240
Log likelihood	115.4863	F-statistic		22.99092
Durbin-Watson stat	1.916015	Prob(F-statistic)		0.000023

14.5 Data Description: This data is obtained from the Citibank data base.

M1: is the seasonally adjusted monetary base. This a monthly average series. We get the quarterly average of M1 by using $(M1_t + M1_{t+1} + M1_{t+2})/3$.

TBILL3: is the T-bill - 3 month - rate. This is a monthly series. We calculate a quarterly average of TBILL3 by using $(TBILL3_t + TBILL3_{t+1} + TBILL3_{t+2})/3$. Note that TBILL3 is an annualized rate (per annum).

GNP: This is Quarterly GNP. All series are transformed by taking their natural logarithm.

a. VAR with two lags on each variable

Sample(adjusted): 1959:3 1995:2
Included observations: 144 after adjusting endpoints
Standard errors & t-statistics in parentheses

	LNGNP	LNM1	LNTBILL3
LNGNP(-1)	1.135719	-0.005500	1.437376
	(0.08677)	(0.07370)	(1.10780)
	(13.0886)	(-0.07463)	(1.29751)
LNGNP(-2)	-0.130393	0.037241	-1.131462
	(0.08750)	(0.07431)	(1.11705)
	(-1.49028)	(0.50115)	(-1.01290)
LNM1(-1)	0.160798	1.508925	1.767750
	(0.07628)	(0.06478)	(0.97383)
	(2.10804)	(23.2913)	(1.81525)
LNM1(-2)	-0.163492	-0.520561	-1.892962
	(0.07516)	(0.06383)	(0.95951)
	(-2.17535)	(-8.15515)	(-1.97284)
LNTBILL3(-1)	0.001615	-0.036446	1.250074
	(0.00645)	(0.00547)	(0.08230)
	(0.25047)	(-6.65703)	(15.1901)
LNTBILL3(-2)	-0.008933	0.034629	-0.328626
	(0.00646)	(0.00549)	(0.08248)
	(-1.38286)	(6.31145)	(-3.98453)
C	-0.011276	-0.179754	-1.656048
	(0.07574)	(0.06433)	(0.96696)
	(-0.14888)	(-2.79436)	(-1.71264)
R-squared	0.999256	0.999899	0.946550
Adj. R-squared	0.999223	0.999895	0.944209
Sum sq. resids	0.009049	0.006527	1.474870
S.E. equation	0.008127	0.006902	0.103757
Log likelihood	492.2698	515.7871	125.5229
Akaike AIC	-9.577728	-9.904358	-4.484021
Schwarz SC	-9.433362	-9.759992	-4.339656
Mean dependent	8.144045	5.860579	1.715690
S.D. dependent	0.291582	0.672211	0.439273

Determinant Residual Covariance   2.67E-11
Log Likelihood   1355.989
Akaike Information Criteria   -24.24959
Schwarz Criteria   -24.10523

b. VAR with three lags on each variable

Sample(adjusted): 1959:4 1995:2
Included observations: 143 after adjusting endpoints
Standard errors & t-statistics in parentheses

	LNGNP	LNM1	LNTBILL3
LNGNP(-1)	1.133814	-0.028308	1.660761
	(0.08830)	(0.07328)	(1.11241)
	(12.8398)	(-0.38629)	(1.49295)

Chapter 14: Time-Series Analysis

```
LNGNP(-2)      -0.031988       0.103428       0.252378
               (0.13102)      (0.10873)      (1.65053)
              (-0.24414)      (0.95122)      (0.15291)

LNGNP(-3)      -0.105146      -0.045414      -1.527252
               (0.08774)      (0.07281)      (1.10526)
              (-1.19842)     (-0.62372)     (-1.38180)

LNM1(-1)        0.098732       1.375936       1.635398
               (0.10276)      (0.08528)      (1.29449)
               (0.96081)     (16.1349)       (1.26335)

LNM1(-2)       -0.012617      -0.134075      -3.555324
               (0.17109)      (0.14198)      (2.15524)
              (-0.07375)     (-0.94432)     (-1.64962)

LNM1(-3)       -0.085778      -0.253402       1.770995
               (0.09254)      (0.07680)      (1.16577)
              (-0.92693)     (-3.29962)      (1.51917)

LNTBILL3(-1)    0.001412      -0.041461       1.306043
               (0.00679)      (0.00564)      (0.08555)
               (0.20788)     (-7.35638)     (15.2657)

LNTBILL3(-2)   -0.013695       0.039858      -0.579077
               (0.01094)      (0.00908)      (0.13782)
              (-1.25180)      (4.38997)     (-4.20158)

LNTBILL3(-3)    0.006468       0.000144       0.207577
               (0.00761)      (0.00632)      (0.09588)
               (0.84990)      (0.02281)      (2.16504)

C               0.037812      -0.166320      -2.175434
               (0.07842)      (0.06508)      (0.98789)
               (0.48217)     (-2.55566)     (-2.20210)

R-squared          0.999271       0.999907       0.950041
Adj. R-squared     0.999222       0.999901       0.946661
Sum sq. resids     0.008622       0.005938       1.368186
S.E. equation      0.008051       0.006682       0.101425
Log likelihood   491.8106       518.4767       129.5215
Akaike AIC        -9.576480      -9.949432      -4.509499
Schwarz SC        -9.369288      -9.742240      -4.302307
Mean dependent     8.148050       5.866929       1.718861
S.D. dependent     0.288606       0.670225       0.439160

Determinant Residual Covariance      2.18E-11
Log Likelihood                    1360.953
Akaike Information Criteria        -24.40808
Schwarz Criteria                   -24.20088
```

d. Pairwise Granger Causality Tests

```
Sample: 1959:1 1995:2
Lags: 3
```

Null Hypothesis:	Obs	F-Statistic	Probability
LNTBILL3 does not Granger Cause LNM1	143	20.0752	7.8E-11
LNM1 does not Granger Cause LNTBILL3		1.54595	0.20551

e. Pairwise Granger Causality Tests

```
Pairwise Granger Causality Tests
Sample: 1959:1 1995:2
Lags: 2
```

Null Hypothesis:	Obs	F-Statistic	Probability
LNTBILL3 does not Granger Cause LNM1	144	23.0844	2.2E-09
LNM1 does not Granger Cause LNTBILL3		3.99777	0.02051

14.6 *The Simple Deterministic Time Trend Model.* This is based on Hamilton (1994).

$$y_t = \alpha + \beta t + u_t \qquad t = 1,..,T \text{ where } u_t \sim IIN(0,\sigma^2).$$

In vector form, this can be written as

$$y = X\phi + u, \quad X = [1,t], \quad \phi = \begin{bmatrix}\alpha\\\beta\end{bmatrix}.$$

a. $\hat{\phi}_{ols} = (X'X)^{-1}X'y$ and $\hat{\phi}_{ols} - \phi = (X'X)^{-1}X'u$

Therefore,

$$\hat{\phi}_{ols} - \phi = \begin{bmatrix}\hat{\alpha}_{ols}-\alpha\\\hat{\beta}_{ols}-\beta\end{bmatrix} = [(\begin{smallmatrix}1'\\t'\end{smallmatrix})(1,t)]^{-1}(\begin{smallmatrix}1'\\t'\end{smallmatrix})u$$

$$= \begin{bmatrix}1'1 & 1't\\t'1 & t't\end{bmatrix}^{-1}\begin{bmatrix}1'u\\t'u\end{bmatrix}$$

$$= \begin{bmatrix}\sum_{t=1}^{T}1 & \sum_{t=1}^{T}t\\ \sum_{t=1}^{T}t & \sum_{t=1}^{T}t^2\end{bmatrix}^{-1}\begin{bmatrix}\sum_{t=1}^{T}u_t\\ \sum_{t=1}^{T}tu_t\end{bmatrix} = \begin{bmatrix}T & \sum_{t=1}^{T}t\\ \sum_{t=1}^{T}t & \sum_{t=1}^{T}t^2\end{bmatrix}^{-1}\begin{bmatrix}\sum_{t=1}^{T}u_t\\ \sum_{t=1}^{T}tu_t\end{bmatrix}$$

b. $\dfrac{X'X}{T} = T^{-1}\begin{bmatrix}T & \sum_{t=1}^{T}t\\ \sum_{t=1}^{T}t & \sum_{t=1}^{T}t^2\end{bmatrix} = T^{-1}\begin{bmatrix}T & T(T+1)/2\\ T(T+1)/2 & T(T+1)(2T+1)/6\end{bmatrix}$

$$= \begin{bmatrix}1 & (T+1)/2\\ (T+1)/2 & (T+1)(2T+1)/6\end{bmatrix}$$

Therefore, $\text{plim}(\dfrac{X'X}{T})$ diverges as $T \to \infty$ and is not a positive definite matrix.

c. Note that

Chapter 14: Time-Series Analysis 313

$$A^{-1}(X'X)A^{-1} = \begin{bmatrix} T^{-\frac{1}{2}} & 0 \\ 0 & T^{-\frac{3}{2}} \end{bmatrix} \begin{bmatrix} \sum_{t=1}^{T} 1 & \sum_{t=1}^{T} t \\ \sum_{t=1}^{T} t & \sum_{t=1}^{T} t^2 \end{bmatrix} \begin{bmatrix} T^{-\frac{1}{2}} & 0 \\ 0 & T^{-\frac{3}{2}} \end{bmatrix} = \begin{bmatrix} \frac{1}{T}\sum_{t=1}^{T} 1 & \frac{1}{T^2}\sum_{t=1}^{T} t \\ \frac{1}{T^2}\sum_{t=1}^{T} t & \frac{1}{T^3}\sum_{t=1}^{T} t^2 \end{bmatrix}$$

$$= \begin{bmatrix} 1 & \frac{T(T+1)}{2T^2} \\ \frac{T(T+1)}{2T^2} & \frac{T(T+1)(2T+1)}{6T^3} \end{bmatrix} = \begin{bmatrix} 1 & \frac{1}{2} + \frac{1}{2T} \\ \frac{1}{2} + \frac{1}{2T} & \frac{1}{3} + \frac{1}{2T} + \frac{1}{6T^2} \end{bmatrix}$$

Therefore plim $A^{-1}(X'X)A^{-1} = \begin{bmatrix} 1 & \frac{1}{2} \\ \frac{1}{2} & \frac{1}{3} \end{bmatrix} = Q$ as $T \to \infty$ which is a finite positive definite matrix. Also

$$A^{-1}(X'u) = \begin{bmatrix} \frac{1}{\sqrt{T}} & 0 \\ 0 & \frac{1}{T\sqrt{T}} \end{bmatrix} \begin{bmatrix} 1'u \\ t'u \end{bmatrix} = \begin{bmatrix} \frac{1}{\sqrt{T}} & 0 \\ 0 & \frac{1}{T\sqrt{T}} \end{bmatrix} \begin{bmatrix} \sum_{t=1}^{T} u_t \\ \sum_{t=1}^{T} tu_t \end{bmatrix} = \begin{bmatrix} \frac{1}{\sqrt{T}}\sum_{t=1}^{T} u_t \\ \frac{1}{T\sqrt{T}}\sum_{t=1}^{T} tu_t \end{bmatrix}$$

d. Show that $z_1 = \frac{1}{\sqrt{T}} \sum_{t=1}^{T} u_t \sim N(0,\sigma^2)$. $u_t \sim N(0,\sigma^2)$, so that $\sum_{t=1}^{T} u_t \sim N(0, T\sigma^2)$.

Therefore, $\frac{1}{\sqrt{T}} \sum_{t=1}^{T} u_t \sim N(0, \frac{1}{\sqrt{T}} \cdot T\sigma^2 \cdot \frac{1}{\sqrt{T}}) = N(0,\sigma^2)$.

Also, show that $z_2 = \frac{1}{T\sqrt{T}} \sum_{t=1}^{T} tu_t \sim N[0, \frac{\sigma^2}{6T^2} \cdot (T+1)(2T+1)]$.

Let $\sigma_t^2 = \text{var}(\frac{t}{T}u_t) = E(\frac{t}{T}u_t)^2 = \frac{t^2}{T^2}\sigma^2$. Then,

$$\frac{1}{T}\sum_{t=1}^{T} \sigma_t^2 = \frac{\sigma^2}{T^3}\sum_{t=1}^{T} t^2 = \frac{\sigma^2}{T^3} \cdot \frac{T(T+1)(2T+1)}{6} = \frac{\sigma^2(T+1)(2T+1)}{6T^2}$$

Since $\text{var}(\frac{1}{T\sqrt{T}}\sum_{t=1}^{T} tu_t) = \frac{1}{T}\text{var}(\sum_{t=1}^{T}\frac{t}{T}u_t) = \frac{1}{T}\sum_{t=1}^{T}\sigma_t^2 = \frac{\sigma^2(T+1)(2T+1)}{6T^2}$

Therefore $\frac{1}{T\sqrt{T}}\sum_{t=1}^{T} tu_t \sim N(0, \frac{\sigma^2(T+1)(2T+1)}{6T^2})$

Now $z_1 = \frac{1}{\sqrt{T}}\sum_{t=1}^{T} u_t \sim N(0,\sigma^2)$ and

$$z_2 = \frac{1}{T\sqrt{T}} \sum_{t=1}^{T} tu_t \sim N(0, \frac{(T+1)(2T+1)}{6T^2}\sigma^2)$$

with

$$\text{cov}(z_1, z_2) = E(z_1 z_2) = E(\sum_{t=1}^{T} \frac{t}{T^2} u_t^2) = \sigma^2 \sum_{t=1}^{T} t/T^2$$

$$= T(T+1)\sigma^2/2T^2 = (T+1)\sigma^2/2T.$$

Hence,

$$\begin{pmatrix} z_1 \\ z_2 \end{pmatrix} \sim N\left(0, \sigma^2 \begin{bmatrix} 1 & \frac{T+1}{2T} \\ \frac{T+1}{2T} & \frac{(T+1)(2T+1)}{6T^2} \end{bmatrix}\right)$$

Therefore as $T \to \infty$, $\begin{pmatrix} z_1 \\ z_2 \end{pmatrix}$ has an asymptotic distribution that is

$$N\left(0, \sigma^2 \begin{bmatrix} 1 & \frac{1}{2} \\ \frac{1}{2} & \frac{1}{3} \end{bmatrix}\right) \text{ or } N(0, \sigma^2 Q).$$

e. $\begin{bmatrix} \sqrt{T}(\hat{\alpha}_{ols} - \alpha) \\ T\sqrt{T}(\hat{\beta}_{ols} - \beta) \end{bmatrix} = [A^{-1}(X'X)A^{-1}]^{-1}[A^{-1}(X'u)]$

But from part (c), we have plim $A^{-1}(X'X)A^{-1}$ is Q which is finite and positive definite. Therefore plim$[A^{-1}(X'X)A^{-1}]^{-1}$ is Q^{-1}. Also, from part (d)

$$A^{-1}(X'u) = \begin{bmatrix} \frac{1}{\sqrt{T}} \sum_{t=1}^{T} u_t \\ \frac{1}{T\sqrt{T}} \sum_{t=1}^{T} tu_t \end{bmatrix} = \begin{bmatrix} z_1 \\ z_2 \end{bmatrix}$$

has an asymptotic distribution $N(0, \sigma^2 Q)$. Hence, $[A^{-1}(X'X)A^{-1}]^{-1}[A^{-1}(X'u)]$ has an asymptotic distribution $N(0, Q^{-1}\sigma^2 QQ^{-1})$ or $N(0, \sigma^2 Q^{-1})$. Thus,

$$\begin{bmatrix} \sqrt{T}(\hat{\alpha}_{ols} - \alpha) \\ T\sqrt{T}(\hat{\beta}_{ols} - \beta) \end{bmatrix}$$

has an asymptotic distribution $N(0, \sigma^2 Q^{-1})$. Since $\hat{\beta}_{ols}$ has the factor $T\sqrt{T}$ rather than the usual $\sqrt{T}$, it is said to be *superconsistent*. This means that not only does $(\hat{\beta}_{ols} - \beta)$ converge to zero in probability limits, but so does $T(\hat{\beta}_{ols} - \beta)$. Note that the normality assumption is not needed for this result. Using the

Chapter 14: Time-Series Analysis 315

central limit theorem, all that is needed is that u_t is White noise with finite fourth moments, see Sims, Stock and Watson (1990) or Hamilton (1994).

14.7 *Test of Hypothesis with a Deterministic Time Trend Model.* This is based on Hamilton (1994).

a. Show that plim $s^2 = \frac{1}{T-2} \sum_{t=1}^{T} (y_t - \hat{\alpha}_{ols} - \hat{\beta}_{ols} t)^2 = \sigma^2$. By the law of large numbers $\frac{1}{T-2} \sum_{t=1}^{T} (y_t - \hat{\alpha}_{ols} - \hat{\beta}_{ols} t)^2 = \frac{1}{T-2} \sum_{t=1}^{T} u_t^2$.

Hence, plim $s^2 = $ plim $\frac{1}{T-2} \sum_{t=1}^{T} u_t^2 = \text{var}(u_t) = \sigma^2$

b. Show that $t_\alpha = \dfrac{\hat{\alpha}_{ols} - \alpha_o}{[s^2(1,0)(X'X)^{-1}\begin{bmatrix}1\\0\end{bmatrix}]^{\frac{1}{2}}}$

has the same asymptotic $N(0,1)$ distribution as $t_\alpha^* = \dfrac{\sqrt{T}(\hat{\alpha}_{ols} - \alpha_o)}{\sigma\sqrt{q^{11}}}$

Multiplying both the numerator and denominator of t_α by $\sqrt{T}$ gives

$$t_\alpha = \dfrac{\sqrt{T}(\hat{\alpha}_{ols} - \alpha_o)}{[s^2(\sqrt{T},0)(X'X)^{-1}\begin{bmatrix}\sqrt{T}\\0\end{bmatrix}]^{\frac{1}{2}}}$$

From part (c) in problem 6, we already showed that for $A = \begin{bmatrix} \sqrt{T} & 0 \\ 0 & T\sqrt{T} \end{bmatrix}$ the plim$[A^{-1}(X'X)A^{-1}]^{-1} = Q^{-1}$ where

$$Q = \begin{bmatrix} 1 & \frac{1}{2} \\ \frac{1}{2} & \frac{1}{3} \end{bmatrix}.$$

Now, $[\sqrt{T},0]$ in t_α can be rewritten as $[1,0]A$, because

$[1,0]A = [1,0]\begin{bmatrix} \sqrt{T} & 0 \\ 0 & T\sqrt{T} \end{bmatrix} = [\sqrt{T},0]$.

Therefore, $t_\alpha = \dfrac{\sqrt{T}(\hat{\alpha}_{ols} - \alpha_o)}{[s^2[1,0]A(X'X)^{-1}A\begin{pmatrix}1\\0\end{pmatrix}]^{\frac{1}{2}}}$

Using plim $s^2 = \sigma^2$ and plim $A(X'X)^{-1}A = Q^{-1}$, we get

$$\text{plim}[s^2[1,0]A(X'X)^{-1}A\begin{bmatrix}1\\0\end{bmatrix}]^{1/2} = [\sigma^2[1,0]Q^{-1}\begin{bmatrix}1\\0\end{bmatrix}]^{1/2} = \sigma\sqrt{q^{11}}$$

where q^{11} is the (1,1) element of Q^{-1}. Therefore, t_α has the same asymptotic distribution as

$$\frac{\sqrt{T}(\hat{\alpha}_{ols}-\alpha_o)}{\sigma\sqrt{q^{11}}} = t_\alpha^*$$

In part (e) of problem 6, we showed that $\begin{bmatrix}\sqrt{T}(\hat{\alpha}_{ols}-\alpha_o)\\T\sqrt{T}(\hat{\beta}_{ols}-\beta_o)\end{bmatrix}$ has an asymptotic distribution $N(0,\sigma^2 Q^{-1})$, so that $\sqrt{T}(\hat{\alpha}_{ols}-\alpha_o)$ is asymptotically distributed $N(0,\sigma^2 q^{11})$. Thus,

$$t_\alpha^* = \frac{\sqrt{T}(\hat{\alpha}_{ols}-\alpha_o)}{\sigma\sqrt{q^{11}}}$$

is asymptotically distributed as $N(0,1)$ under the null hypothesis of $\alpha = \alpha_o$. Therefore, both t_α and t_α^* have the same asymptotic $N(0,1)$ distribution.

c. Similarly, for testing H_o: $\beta = \beta_o$, show that

$$t_\beta = (\hat{\beta}_{ols}-\beta_o) / [s^2(0,1)(X'X)^{-1}(0,1)']^{1/2}$$

has the same asymptotic $N(0,1)$ distribution as $t_\beta^* = T\sqrt{T}(\hat{\beta}_{ols}-\beta_o)/\sigma\sqrt{q^{22}}$. Multiplying both numerator and denominator of t_β by $T\sqrt{T}$ we get,

$$t_\beta = T\sqrt{T}(\hat{\beta}_{ols}-\beta_o) / [s^2(0,T\sqrt{T})(X'X)^{-1}(0,T\sqrt{T})']^{1/2}$$
$$= T\sqrt{T}(\hat{\beta}_{ols}-\beta_o) / [s^2(0,1)A(X'X)^{-1}A(0,1)']^{1/2}$$

Now $[0,T\sqrt{T}]$ in t_β can be rewritten as $[0,1]A$ because

$$[0,1]A = [0,1]\begin{bmatrix}\sqrt{T} & 0\\ 0 & T\sqrt{T}\end{bmatrix} = [0,T\sqrt{T}].$$

Therefore,

$$\text{plim } t_\beta = T\sqrt{T}(\hat{\beta}_{ols}-\beta_o)/[\sigma^2(0,1)Q^{-1}(0,1)']^{1/2} = T\sqrt{T}(\hat{\beta}_{ols}-\beta_o)/\sigma\sqrt{q^{22}}.$$

Using $\text{plim } s^2 = \sigma^2$ and $\text{plim } A(X'X)^{-1}A = Q^{-1}$, we get that

$$\text{plim } [s^2(0,1)A(X'X)^{-1}A(0,1)']^{1/2} = [\sigma^2(0,1)Q^{-1}(0,1)']^{1/2} = \sigma\sqrt{q^{22}}$$

where q^{22} is the (2,2) element of Q^{-1}. Therefore, t_β has the same asymptotic distribution as

$$T\sqrt{T}(\hat{\beta}_{ols}-\beta_o)/\sigma\sqrt{q^{22}} = t_\beta^*$$

From part (e) of problem 6, $T\sqrt{T}(\hat{\beta}_{ols}-\beta_o)$ has an asymptotic distribution $N(0,\sigma^2 q^{22})$. Therefore, both t_β and t_β^* have the same asymptotic $N(0,1)$ distribution. Also, the usual OLS t-tests for $\alpha = \alpha_o$ and $\beta = \beta_o$ will give asymptotically valid inference.

14.8 *A Random Walk Model.* This is based on Fuller (1976) and Hamilton (1994).
$y_t = y_{t-1} + u_t$, $t = 0,1,..,T$ where $u_t \sim IIN(0,\sigma^2)$ and $y_o = 0$.

a. Show that $y_t = u_1 +..+ u_t$ with $E(y_t) = 0$ and $var(y_t) = t\sigma^2$. By successive substitution, we get

$$y_t = y_{t-1} + u_t = y_{t-2} + u_{t-1} + u_t = .. = y_o + u_1 + u_2 +..+ u_t$$

substituting $y_o = 0$ we get $y_t = u_1 +..+ u_t$.

Hence, $E(y_t) = E(u_1) +..+ E(u_t) = 0$

$var(y_t) = var(u_1) +..+ var(u_t) = t\sigma^2$

and $y_t \sim N(0,t\sigma^2)$.

b. Squaring the random walk equation, we get

$$y_t^2 = (y_{t-1}+u_t)^2 = y_{t-1}^2 + 2y_{t-1}u_t + u_t^2.$$

Solving for $y_{t-1}u_t$ yields $y_{t-1}u_t = \frac{1}{2}(y_t^2 - y_{t-1}^2 - u_t^2)$

Summing over $t = 1,2..,T$, we get

$$\sum_{t=1}^{T}(y_{t-1}u_t) = \frac{1}{2}\sum_{t=1}^{T}(y_t^2-y_{t-1}^2) - \frac{1}{2}\sum_{t=1}^{T}u_t^2.$$

But

$$\sum_{t=1}^{T} y_t^2 = y_1^2 +..+ y_T^2$$

$$\sum_{t=1}^{T} y_{t-1}^2 = y_0^2 +..+ y_{T-1}^2$$

Hence, by subtraction $\sum_{t=1}^{T}(y_t^2-y_{t-1}^2) = y_T^2 - y_0^2$ Substituting this result above,

we get $\sum_{t=1}^{T} y_{t-1}u_t = \frac{1}{2}(y_T^2-y_0^2) - \frac{1}{2}\sum_{t=1}^{T} u_t^2 = \frac{1}{2}y_T^2 - \frac{1}{2}\sum_{t=1}^{T} u_t^2$

Dividing by $T\sigma^2$ we get $\dfrac{1}{T\sigma^2}\sum_{t=1}^{T} y_{t-1}u_t = \dfrac{1}{2}\left(\dfrac{y_T}{\sqrt{T}\sigma}\right)^2 - \dfrac{1}{2T\sigma^2}\sum_{t=1}^{T} u_t^2$

But, from part (a), $y_T \sim N(0, T\sigma^2)$ and $y_T/\sqrt{T}\sigma \sim N(0,1)$. Therefore, $(y_T/\sqrt{T}\sigma)^2 \sim \chi_1^2$. Also, by the law of large numbers,

$$\text{plim}\,\dfrac{\sum_{t=1}^{T} u_t^2}{T} = \text{var}(u_t) = \sigma^2. \quad \text{Hence, plim}\,\dfrac{1}{2\sigma^2}\cdot\dfrac{\sum_{t=1}^{T} u_t^2}{T} = \dfrac{1}{2}. \quad \text{Therefore,}$$

$$\dfrac{1}{T\sigma^2}\sum_{t=1}^{T} y_{t-1}u_t = \dfrac{1}{2}\left(\dfrac{y_T}{\sqrt{T}\sigma}\right)^2 - \dfrac{1}{2\sigma^2}\cdot\dfrac{1}{T}\sum_{t=1}^{T} u_t^2$$

is asymptotically distributed as $\dfrac{1}{2}(\chi_1^2 - 1)$.

c. Show that $E(\sum_{t=1}^{T} y_{t-1}^2) = \dfrac{T(T-1)}{2}\sigma^2$. Using the results in part (a), we get

$$y_{t-1} = y_0 + u_1 + u_2 + \ldots + u_{t-1}$$

Substituting $y_0 = 0$, squaring both sides and taking expected values, we get $E(y_{t-1}^2) = E(u_1^2) + \ldots + E(u_{t-1}^2) = (t-1)\sigma^2$ since the u_t's are independent. Therefore,

$$E(\sum_{t=1}^{T} y_{t-1}^2) = \sum_{t=1}^{T} E(y_{t-1}^2) = \sum_{t=1}^{T}(t-1)\sigma^2 = \dfrac{T(T-1)}{2}\sigma^2$$

where we used the fact that $\sum_{t=1}^{T} t = T(T+1)/2$ from problem 6.

d. For the AR(1) model, $y_t = \rho y_{t-1} + u_t$, show that OLS estimate of ρ satisfies

$$\text{plim}\,T(\hat{\rho} - \rho) = \text{plim}\,\dfrac{\sum_{t=1}^{T} y_{t-1}u_t/T\sigma^2}{\sum_{t=1}^{T} y_{t-1}^2/T^2\sigma^2} = 0 \quad \text{where}$$

$$\hat{\rho} = \dfrac{\sum_{t=1}^{T} y_{t-1}y_t}{\sum_{t=1}^{T} y_{t-1}^2} = \rho + \dfrac{\sum_{t=1}^{T} y_{t-1}u_t}{\sum_{t=1}^{T} y_{t-1}^2}.$$

From part (b), $\dfrac{1}{T\sigma^2}\sum_{t=1}^{T} y_{t-1}u_t$ has an asymptotic distribution $\dfrac{1}{2}(\chi_1^2 - 1)$. This implies that $\sum_{t=1}^{T} y_{t-1}u_t/\sigma^2$ converges to an asymptotic distribution of $\dfrac{1}{2}(\chi_1^2 - 1)$

Chapter 14: Time-Series Analysis   319

at the rate of T. Also, from part (c), $E(\sum_{t=1}^{T} y_{t-1}^2) = \frac{\sigma^2 T(T-1)}{2}$ implies that $\text{plim} \frac{1}{T^2} \sum_{t=1}^{T} y_{t-1}^2$ is $\sigma^2/2$. This means that $\sum_{t=1}^{T} y_{t-1}^2/\sigma^2$ converges to $\frac{1}{2}$ at the rate of T^2. One can see that the asymptotic distribution of $\hat{\rho}$ when $\rho = 1$ is a ratio of a $\frac{1}{2}(\chi_1^2 - 1)$ random variable to a non-standard distribution in the denominator which is beyond the scope of this book, see Hamilton (1994) or Fuller (1976) for further details. The object of this exercise is to show that if $\rho = 1$, $\sqrt{T}(\hat{\rho} - \rho)$ is no longer Normal as in the standard stationary least squares regression with $|\rho| < 1$. Also, to show that for the non-stationary (random walk) model, $\hat{\rho}$ converges at a faster rate (T) than for the stationary case ($\sqrt{T}$). From part (c) it is clear that one has to divide the denominator of $\hat{\rho}$ by T^2 rather than T to get a convergent distribution.

14.9 Cointegration Example

a. Solving for the *reduced form* from (14.14) and (14.15) we get

$$Y_t = \frac{u_t - v_t}{(\alpha - \beta)} = \frac{1}{(\alpha - \beta)} u_t - \frac{1}{(\alpha - \beta)} v_t$$

and

$$C_t = \frac{\beta(u_t - v_t)}{(\alpha - \beta)} + u_t = \frac{\alpha}{(\alpha - \beta)} u_t - \frac{\beta}{(\alpha - \beta)} v_t$$

In this case, u_t is I(0) and v_t is I(1). Therefore both Y_t and C_t are I(1). Note that there are no excluded exogenous variables in (14.14) and (14.15) and only one right hand side endogenous variable in each equation. Hence both equations are unidentified by the order condition of identification. However, a linear combination of the two structural equations will have a mongrel disturbance term that is neither AR(1) nor random walk. Hence, both equations are identified. If $\rho = 1$, then both u_t and v_t are random walks and the mongrel disturbance is also a random walk. Therefore, the system is unidentified. In such a case, there is no cointegrating relationship between C_t and Y_t. Let $(C_t - \gamma Y_t)$ be another cointegrating relationship, then subtracting it from the first cointegrating relationship, one gets $(\gamma - \beta) Y_t$ which should be I(0).

Since Y_t is I(1), this can only happen if $\gamma = \beta$. Differencing both equations in (14.14) and (14.15) we get

$$\Delta C_t - \beta \Delta Y_t = \Delta u_t = (\rho-1)u_{t-1} + \epsilon_t = \epsilon_t + (\rho-1)(C_{t-1} - \beta Y_{t-1})$$

$$= \epsilon_t + (\rho-1)C_{t-1} - \beta(\rho-1)Y_{t-1}$$

and $\Delta C_t - \alpha \Delta Y_t = \Delta v_t = \eta_t$. Writing them as a VAR, we get (14.18)

$$\begin{bmatrix} 1 & -\beta \\ 1 & -\alpha \end{bmatrix} \begin{bmatrix} \Delta C_t \\ \Delta Y_t \end{bmatrix} = \begin{bmatrix} \epsilon_t + (\rho-1)C_{t-1} - \beta(\rho-1)Y_{t-1} \\ \eta_t \end{bmatrix}$$

Post-multiplying by the inverse of the first matrix, we get

$$\begin{bmatrix} \Delta C_t \\ \Delta Y_t \end{bmatrix} = \left(\frac{1}{\beta-\alpha}\right) \begin{bmatrix} -\alpha & \beta \\ -1 & 1 \end{bmatrix} \begin{bmatrix} \epsilon_t + (\rho-1)C_{t-1} - \beta(\rho-1)Y_{t-1} \\ \eta_t \end{bmatrix}$$

$$= \frac{1}{\beta-\alpha} \begin{bmatrix} -\alpha\epsilon_t - \alpha(\rho-1)C_{t-1} + \alpha\beta(\rho-1)Y_{t-1} + \beta\eta_t \\ -\epsilon_t - (\rho-1)C_{t-1} + \beta(\rho-1)Y_{t-1} + \eta_t \end{bmatrix}$$

$$= \frac{1}{(\beta-\alpha)} \begin{bmatrix} -\alpha(\rho-1) & \alpha\beta(\rho-1) \\ -(\rho-1) & \beta(\rho-1) \end{bmatrix} \begin{pmatrix} C_{t-1} \\ Y_{t-1} \end{pmatrix} + \begin{pmatrix} h_t \\ g_t \end{pmatrix}$$

where h_t and g_t are linear combinations of ϵ_t and η_t. This is equation (14.19). This can be rewritten as

$$\Delta C_t = \frac{-\alpha(\rho-1)}{(\beta-\alpha)}(C_{t-1} - \beta Y_{t-1}) + h_t = -\alpha\delta Z_{t-1} + h_t$$

$$\Delta Y_t = \frac{-(\rho-1)}{(\beta-\alpha)}(C_{t-1} - \beta Y_{t-1}) + g_t = -\delta Z_{t-1} + g_t$$

where

$\delta = (\rho-1)/(\beta-\alpha)$ and $Z_t = C_t - \beta Y_t$.

These are equations (14.20) and (14.21). This is the *Error-Correction Model* (ECM) representation of the original model. Z_{t-1} is the error correction term. It represents a disequilibrium term showing the departure from long-run equilibrium. Note that if $\rho = 1$, then $\delta = 0$ and Z_{t-1} drops from both ECM equations.

b. $\hat{\beta}_{ols} = \dfrac{\sum_{t=1}^{T} C_t Y_t}{\sum_{t=1}^{T} Y_t^2} = \beta + \dfrac{\sum_{t=1}^{T} Y_t u_t}{\sum_{t=1}^{T} Y_t^2}$

Since u_t is I(0) if $\rho \ne 1$ and Y_t is I(1), we have plim $\sum_{t=1}^{T} Y_t^2 / T^2$ is O(1), while plim $\sum_{t=1}^{T} Y_t u_t / T$ is O(1). Hence $T(\hat{\beta}_{ols} - \beta)$ is O(1) or $(\hat{\beta}_{ols} - \beta)$ is O(T).

Printing and binding: Druckerei Triltsch, Würzburg